Advancing Global Bioethics

Volume 18

The book series Global Bioethics provides a forum for normative analysis of a vast range of important new issues in bioethics from a truly global perspective and with a cross-cultural approach. The issues covered by the series include among other things sponsorship of research and education, scientific misconduct and research integrity, exploitation of research participants in resource-poor settings, brain drain and migration of healthcare workers, organ trafficking and transplant tourism, indigenous medicine, biodiversity, commodification of human tissue, benefit sharing, bio-industry and food, malnutrition and hunger, human rights, and climate change.

More information about this series at https://link.springer.com/bookseries/10420

Henk ten Have

The Covid-19 Pandemic and Global Bioethics

Springer

Henk ten Have
Department of Bioethics
Anahuac University
Mexico City, Mexico

ISSN 2212-652X ISSN 2212-6538 (electronic)
Advancing Global Bioethics
ISBN 978-3-030-91493-6 ISBN 978-3-030-91491-2 (eBook)
https://doi.org/10.1007/978-3-030-91491-2

This Springer imprint is published by the registered company Springer Nature Switzerland AG
The registered company address is: Gewerbestrasse 11, 6330 Cham, Switzerland

For Felix Jansen, and all those born in the era of corona

Preface

The emergence of Covid-19 is a perfect example of a global phenomenon that affects everybody across the world. In 2020, it demonstrated its powerful effects on human well-being and healthcare, but also on all dimensions of human existence. It does not take long to notice that the pandemic is associated with challenging ethical issues. Exploring the Covid-19 pandemic from the perspective of global bioethics is therefore not an unusual decision. As many others have observed, writing a book during subsequent lockdowns, and even a curfew, is a bizarre experience. On the one hand, you are working in almost monastic circumstances, which is beneficial for reflection and study, on the other hand, opportunities to test and discuss ideas with students and colleagues are limited. At the same time, scientific research, at least for an expert in ethics and philosophy, has never been so untroubled since all relevant information is easily and freely accessible online, so that the latest scientific information about Covid-19 can be directly read without consulting libraries or providers.

It is obvious that this book is work in progress. Covid-19 is a moving target, and the virus continues to surprise us. New study data and research results are published every day. Yesterday's certainties become questionable tomorrow. Information in this book has been updated up to November 2021, and I am sure that numerous findings will change in the near future. However, the bioethical challenges will not alter so rapidly because they often relate to longer-term trends and moral principles with a long history of concern which have now found new relevancy in exceptional circumstances.

Although I have published extensively in the field of global bioethics, and particularly explored the ethical dimensions of vulnerability as well as the bioethical concerns in regard to biodiversity loss, Covid-19 is obviously a novel topic, and therefore an interesting subject to test and trace the theoretical and practical apparatus of global bioethics. I am grateful to Springer Nature for permission to use parts of earlier publications, notably "Sheltering at Our Common Home" (*Journal of Bioethical Inquiry* 2020; 17 (4): 525–529) in Chap. 6 of this book, and "Vulnerability in the Light of the Covid-19 Crisis" (*Medicine, Health Care and Philosophy* 2021; 24 (2): 153–154) in Chap. 9.

This book is dedicated to our grandson Felix who was born just at the beginning of the Covid-19 pandemic. He started to move around and talk when the country was in lockdown and social contacts were minimal. As grandparents, we created a “bubble” with his parents so that we could cuddle and watch him grow and explore the world while Amsterdam was exceptionally quiet for a long time. Following Millennials and Generation Z, he and others born during the Covid-19 pandemic will undoubtedly be a demographic cohort, a new generation that will face unprecedented challenges which are already present but will be more articulate. His early experiences may hopefully foster his sensitivity to solidarity and relationality.

Amsterdam, The Netherlands
1 November 2021

Henk ten Have

We stand and embrace at the window, they watch us from
the street:
it is time, for this to be known!
It is time that the stone took the trouble to bloom,
that unrest's heart started to beat.
It's time for it to be time.

Paul Celan (1920–1970), *Corona*

Contents

Chapter 1
Introduction: The Perspective of Global Bioethics

Abstract The Covid-19 pandemic is a global threat that requires collective and global action. The global nature of the disease calls for international cooperation and solidarity. It illustrates that a global bioethics perspective is needed to address the challenges of the pandemic that go beyond an individually focused ethics. The chapter shows how the book will provide a broader ethics perspective in order to highlight the significant dimensions of this global disease phenomenon.

Keywords Global bioethics · Globalization · Global South · Infectious diseases · Social ethics

1.1 Introduction

Covid-19 is an illustration *par excellence* of globalization. The virus does not respect national borders or political affiliations. Even when it started in a particular country, it cannot be contained. That highlights the responsibility of countries to be as transparent as possible because the threat is global. This is particularly urgent to prevent future pandemics, and not to blame or settle scores. In fact, many countries have enacted inadequate policies so that the whole of humanity has to learn lessons. In almost all countries, policy makers started with denial, disbelief and defensive strategies, underestimating the need for preparedness and prevention. Concerns about economic activity and productivity, as reasonable and understandable as they could be, seemed to come as first for too long. What is not recognized is that globalization has amplified human vulnerability. It has also exacerbated inequalities and injustices which is now reflected in the extent to which Covid-19 especially harms poor and marginalized people, particularly in the Global South.

H. ten Have, *The Covid-19 Pandemic and Global Bioethics*, Advancing Global Bioethics 18, https://doi.org/10.1007/978-3-030-91491-2_1

1.2 Global Threats

Early in 2019, the World Health Organization published a list of ten threats to global health requiring attention for the next decade [1]. Air pollution and climate is on the top of the list. It also includes infectious diseases such as global influenza, Ebola and other high-threat pathogens, dengue and HIV. The list has been the basis for the new 5-year strategic plan of the Organization (the 13th General Programme of Work), allocating three billion US dollars to transform the future of public health to ensure more access to health care, better protection from health emergencies, and make more people experience improved health and well-being. The 2019 list differed from the one published one year earlier. The number one on this 2018 list was pandemic influenza. In fact, the majority of threats on this list were infectious diseases, including cholera, diphtheria, malaria, meningitis, and yellow fever. The 2019 list was revised in January 2020. The most urgent health challenge in 2020 is now elevating health in the climate debate. Number 5 is stopping infectious disease, and number 6 preparing for epidemics [2]. It is obvious that the Covid-19 pandemic not only was not expected but also disturbed priorities and planning of global organizations such as the WHO. However, the WHO Director-General issued an auspicious warning just before the pandemic disturbed the world: "We need to realize that health is an investment in the future. Countries invest heavily in protecting their people from terrorist attacks, but not against the attack of a virus, which could be far more deadly, and far more damaging economically and socially. A pandemic could bring economies and nations to their knees. Which is why health security cannot be a matter for ministries of health alone" [3].

1.3 International Cooperation

Health care challenges as identified in the past are characterized by their global nature. Many challenges are interconnected and will require a coordinated international effort. As a global threat, Covid-19 requires collective and global action. Without international cooperation the infection can perhaps be mitigated in some countries but it can never be controlled as long as it is not carefully managed in other countries. It is therefore important to coordinate activities, to engage in international research, to compare and improve policies, and bring together scholars and researchers. The global nature of the pandemic also calls for international solidarity, focussing not merely on domestic interests but on the well-being of the human community.

Responsibility, vulnerability, solidarity and cooperation are fundamental values. Their importance is now articulated and rediscovered in the experiences with Covid-19. The usual ethical emphasis on individual autonomy and personal responsibility is no longer sufficient in the face of this global threat. What is needed is a broader perspective of bioethics that emphasizes that the planet is our common home, and that human beings are interconnected in a world-wide community.

1.4 Ethics and Health Care

Health care practitioners are not objective technicians intervening in the patient's body, attempting to repair physical defects in the corporeal mechanism. They are not like engineers or architects who work with materials or develop new constructions. Their work is focused on patients as persons who suffer, are impaired, and sometimes mortally ill. Of course, health professionals need to be experts, have the latest scientific knowledge, and base their interventions on evidence and experience. But, first of all, they need to care, be engaged with what bothers their patients, and show empathy and concern with their predicament. That means that they have to establish relationships with patients, and be primarily driven by the interests of their patients. This explains why medicine, and health care in general, has been associated with ethics from its very beginning, not only in Western cultures but also in Islamic, Chinese, Indian and traditional cultures. While it has more recently developed into a science, it has always remained a moral profession.

The concern with ethics in health care was significantly transformed in the 1960s and 1970s. Until that time, ethics was considered as professional ethics. At least some health care practitioners wanted to distinguish themselves from others by a commitment to high moral standards so that patients could trust them. In the nineteenth century, medical professional organizations adopted codes of conduct, emphasizing professional responsibilities and duties rather than personal commitment and virtues. Professional medical ethics came under pressure after the Second World War. Health care professionals were increasingly criticized since they were often concerned with self-interest and protection of their profession. Advances in medical research and technology enormously increased the power of medicine. At the same time, the social and cultural environment changed, with concerns for the rights and values of patients, contrary to what has more and more been regarded as medical paternalism. These developments led to a transition from medical ethics into a broader ethical discourse. This transition became evident in a new vocabulary. In the 1970s, 'bioethics' was the new catchword. Coined in 1970 by Van Rensselaer Potter, the new term rapidly expanded in scholarly and public debate [4]. It indicated that ethical concerns were not only the business of health care professionals but of everyone. Many ethical challenges were beyond the professional orientation on good conduct, duties and virtues. They were often associated with questions of life and death, reproduction, limits to treatment and intervention, and allocation of scarce resources. These questions were the concern of all citizens. Ethical debate therefore increasingly took place in the media and policy fora. This broadening of the ethics debate encouraged the growth of bioethics as a new discipline with specific institutions, journals, committees, and experts [5].

The expansion of bioethics during the second part of the last century unquestionably influenced health care. Guidelines, regulations and legislation proliferated in most countries, while medical practice was also shaped by ethics committees and moral case deliberation. Ethics teaching was generally introduced in medical education. However, bioethics was increasingly criticized, particularly since the 1990s. Its dominant approach applied a limited set of ethical principles, often prioritizing the principle of respect for individual autonomy. The social, economic, and political context of bioethical problems was not frequently addressed. Environmental ethics was considered as a separate area of applied ethics. Simultaneously, under the influence of processes of globalization, medical research and healthcare internationalized, and new problems emerged such as biopiracy, brain drain, pandemics and organ trafficking. This global dimension of health care and its challenges put into question the feasibility of the approach of mainstream bioethics as originating in developed countries. This criticism of bioethics was anticipated by Potter. In his view, bioethics may signify a new approach but it was in fact medical ethics under a new name, and not innovative enough. First of all, it is restricted to the medical, especially clinical setting, focused on individual survival and short-term solutions. Second, and in connection to this point, it emphasizes individual autonomy, rather than public health or the common good. Third, it does not address environmental and social issues. And fourth, it is concerned with problems that are specific for developed countries, lacking a global perspective. In Potter's view, the most important problems of our time are population growth, war and violence, pollution and environmental degradation, poverty, short-term politics and the uncritical belief in progress. These problems are vital since they jeopardize the survival of humanity. To guarantee that there is a viable future for our children and grandchildren, a new type of wisdom is required, based on combining biological knowledge and human values. This is, according to Potter, the basic mission of bioethics. In order to clarify this, and to show that more is needed than a reformulated version of traditional medical ethics (under the now fashionable label of bioethics), he launched the new term of 'global bioethics' [6].

1.5 Bioethics as Social Ethics

Global bioethics is an emerging discipline with a broader scope than mainstream bioethics. In ethics in general, and bioethics in particular, the primary moral question is: what should *I* do? In global bioethics on the other hand, the basic question is: what should *we* do? Global bioethics, in other words, is social ethics. This social-ethical nature of the global ethics discourse is first due to the fact that global bioethical problems have a specific character. If a poor or uninsured Covid-19 patient does not receive appropriate care, there is a moral question: why is that patient not receiving care or treatment that can benefit him or her, and how can this be justified? But the moral problem is beyond the level of individual cases. Global bioethics assesses the more general issue: why do people not have access to care and medication when they are poor or marginalized? The fundamental moral challenge is generic and global. The problem manifests itself at the level of individual patients, but it cannot be restricted in ethical discourse to individual cases. Normative analysis requires a broader perspective. Second, global bioethical problems are not addressable by individuals. Even if individuals could address them, it would be unfair to leave the problem to individual management since individuals have not caused these problems. This is clear in the case of climate change. Global warming affects many countries, particularly in the Global South, when the changing climate is primarily associated with the life styles of populations in the Global North. Global threats cannot be addressed by individuals but require cooperation and solidarity. Third, while the global bioethical principle of respect for diversity demands that people take into account diverging moral views, international cooperation and global governance aimed at addressing global challenges such as pandemics require agreement on basic ethical principles. The need for global governance, therefore, implies the search and development of common perspectives that provide a basis for practical actions. Global bioethics is not merely theoretical reflection. Due to the confrontation with serious challenges to health, global bioethics is forced to apply its ethical framework in practices and policies, even if this framework is shaky, uncertain, and provisionary. The need for governance will require continuous reflection on the basis of everyday experiences. Fourth, the social nature of global bioethics is articulated in its basic ethical principles. Being ‘global’ means that a bioethical problem is affecting in principle all human beings wherever they live. Of course, problems challenge individuals, who have to determine what they ought to do in response. It is evident that global problems can be translated into individual ethical problems. But as typically global phenomena they do not first of all present ethical challenges at an individual level. One conclusion from the global nature of contemporary bioethics problems is that they force us to go beyond an individually focused ethics. The implication is that global bioethics needs to articulate ethical principles that transcend the point of view of individual moral agents. This is also why global bioethics differs from mainstream bioethics. The debate cannot merely focus on potential treatments and new vaccines that will benefit individuals but also has to take into account the social, political, and economic contexts in which for example pandemics emerge and expand, and in which some groups of people are more affected than others. The final reference should be humanity, not the autonomous individual. The fifth and last reason why global bioethics has a social-ethical nature has to do with the sources and roots of the global problems with which it is confronted. Contemporary global bioethical problems are not simply the result of globalization as such but of specific processes of globalization that are primarily driven by economic motivations. In contemporary economics and politics ‘market’ has become the dominant metaphor for the organization of social life, encompassing everything, from transportation and research to healthcare and education. Competition rather than cooperation is the core value. It is clear now that neoliberal policies have had a negative impact on global health. Social inequalities have increased. The damaging effects of these policies are particularly clear in the breakdown of healthcare systems in many developing countries. Reduced expenditures for health and social services, privatization of care, lower salaries for healthcare workers, and introduction of user fees for patients have reduced access to health services for the majority of populations. In Latin America for example, health reforms have increased inequity and inefficiency. Entire populations are deprived from necessary treatment and medication, simply because the prices are unaffordable [7]. Globally, each year, two to three million people die of tuberculosis. Effective treatment exists but 79% of tuberculosis patients do not have access to appropriate medication [8]. Against this backdrop, global bioethics is redefined as a *social* ethics. If global problems are produced within the context of broader political choices and patterns for human relationships, it will be inevitable to critically address this context. If global bioethics wants to understand and resolve the global problems that human health is facing, it, therefore, needs a specific normativity. This book will explore this normativity and will argue that a broader perspective on moral challenges is needed on the basis of philosophical and anthropological reflections.

Global bioethics is not a ready-made product but a process. It is the aspiration to realize the universal in the local. But it is, first of all, a social ethics that goes beyond the view that ethics is primarily a matter of personal commitment and individual lifestyle. Global bioethics presents a horizon of reflection, analysis, and action that brings ethical principles associated with cooperation, future generations, justice, protection of the environment, solidarity, social responsibility, and vulnerability (back) into the debate of globalization.

1.6 Images of Globalization

The notion of global bioethics has two meanings. One is 'encompassing,' i.e. comprehensive and bringing together medical, social and environmental concerns. The other is 'worldwide' or 'planetary' [9]. Frequently, the planet is visualized as a lonely globe in outer space, articulating the experience that it is the fragile and finite common home of human beings within the universe. This image of the globe, powerful as it is, posits the Earth as an external object. It does not provoke the sense that it is in fact the habitat of human beings so that our relationship to 'environing' conditions is internal rather to external; we cannot disengage ourselves from our habitat; our lifeworld cannot be disconnected from the planet. The image of globe risks therefore to separate humans from the context within which they dwell. A more appropriate metaphor to emphasize this internal relation is 'sphere' [10]. Using this metaphor evokes interconnectedness, relatedness, and interdependency. This is also expressed in notions such as 'atmosphere,' 'biosphere,' 'ecosphere,' and 'virosphere.' The planet is not just the dwelling location but the world within which humans live, in which they feel at home. For human beings, embedded in spheres, the environment is part of their lifeworld. The notion of sphere presents the world as lived experience, perceived and understood from within. The human world begins in the local rather than the global because the spherical view accentuates embeddedness, and thus locality. Globalization therefore is not an external process that impacts our common globe; it concerns the human world, the *mundus*, expanding the life world through global interaction and cultural diffusion. This is why *mundalization* is sometimes proposed as a better term for global processes [11]. Mundialization (as expressed in the French 'mondialisation' and 'bioéthique mondiale') underlines the interchange and integration of ideas and values of people and cultures around the globe, rather than the spatial and geographic dimensions of the world as our home.

1.7 Overview of the Book

The first part of the book will discuss the experiences with Covid-19, showing why a move towards a perspective of global bioethics is inescapable. The pandemic has not only emerged as global phenomenon but is also associated with how human beings are interacting with the environing world. It is not the first time that humanity is confronted with pandemic diseases. Human life has always been marked by infections, since humans, animals and microbes cohabitate in the same world. But the advances of medical science have promoted the belief that these diseases can be managed and controlled, and sometimes eradicated through vaccinations and medications (especially early in life). Infectious diseases as lethal threats have become less frightening for many people. However, this is a cultural prejudice since populations in less developed countries are continuously threatened by infectious diseases. In fact, especially in developing countries, many more people are infected by malaria, dengue, and tuberculosis than by coronavirus until now. In 2019, just before the Covid-19 outbreak, 409,000 people have died from malaria, and 1.4 million from tuberculosis [12]. The second chapter will discuss lessons from history. Previous pandemics such as cholera in the nineteenth century have had a major impact on society and culture. The third chapter will examine the emergence of infectious diseases during the past decades. Diseases such as Avian flu, Ebola, and Zika have been an early warning for the current pandemic but the lessons have not been taken seriously in most countries. This is clarified in the fourth chapter, analyzing the policy responses across the globe. These responses are diverging and uncoordinated as if the viral threat has come as a surprise. Another characteristic of the pandemic is the gap between facts and values, elaborated in the fifth chapter. While public health measures need to be taken, scientific knowledge of the new disease is inadequate, and only slowly increasing. Facts are changing and evolving while often not taken seriously. In this predicament of uncertainty and ambiguity, policy responses are based on values such as individual autonomy, privacy, economic growth, and public security. What is most striking about the pandemic is its influence on human experiences. Human interactions are interrupted and intimate connections such as touching and meeting are disrupted and regarded as threatening. Especially older, sick and disabled people feel isolated and deprived of human communication. These effects of the pandemic on human experiences are discussed in the sixth chapter. It highlights the need to reflect on these experiences, and interpret them from the perspective of philosophy and anthropology so that it can be clarified what they mean not only as a medical event but as a fundamental human reality.

The second part of the book (Chaps. 7 and 8) will examine the ethical challenges of the pandemic. It is clear that widespread infectious disease has confronted all societies, more or less developed, with serious trials of the healthcare system. First of all, in the domain of treatment. The overwhelming number of seriously ill patients initially focused the ethical debate on life-saving interventions. Hospitals and especially intensive care units saw a rising number of dying people as well as patients who needed long-term care. Less attention was given in the early stages to people in care and nursing homes. Basic protective materials such as face masks, gloves and protective clothing were not available. Most care homes were closed for visitors, isolating patients and reinforcing psychological problems. Thirdly, what is striking is the lack of preventive efforts, not merely prior to but also during the pandemic. The lack of testing equipment is illustrative for the impact of globalization; it shows that basic ingredients are now mostly produced in a limited number of countries. Similar challenges have confronted countries in the development, production, distribution and application of Covid-19 vaccines.

The third part of the book (Chap. 9) analyzes the implications of the Covid-19 experiences for bioethics as well as globalization. The Covid pandemic highlights fundamental experiences which have not received sufficient attention in bioethical discourse since this is primarily focused on individual autonomy, and the balancing of harms and benefits. Anthropological experiences such as vulnerability, connectedness and community, solidarity and cooperation have been articulated in the perspective of global bioethics, and should be better elaborated in post-pandemic bioethical debate. Pandemic experiences furthermore direct attention to the positive and negative dimensions of globalization. Especially neoliberal policies of globalization are nowadays scrutinized [13]. These policies have increased inequalities, so that marginalized populations are more severely hit by disease. They have also made health care systems less equipped to deal with public health challenges. The experiences with Covid-19 are raising questions about how the world will look like after the pandemic. Economies of all countries are depressed, and many businesses are severely affected. The global system of travel and tourism has been upset. The challenge is how these systems will be rebuilt when the pandemic is over. This question is particularly important in view of the other major global threat: climate change. Answers will depend on ethical articulation: what kind of values should be guiding the global community, and what kind of community should humankind pursue?

References

1. Scheres, J., and K. Kuszewski. 2019. The ten threats to global health in 2018 and 2019. A welcome and informative communication of WHO to everybody. *Zdrowie Publiczne i Zarządzanie* 17 (1): 2–8.

2. World Health Organization. 2020. *Urgent health challenges for the next decade*. Geneva: WHO.
3. World Health Organization, *Urgent health challenges for the next decade.*
4. Potter, V.R. 1971. *Bioethics: Bridge to the future*. Englewood Cliffs: Prentice Hall.
5. Ten Have, H. 2016. *Global bioethics. An introduction*. London/New York: Routledge.
6. Potter, V.R. 1988. *Global bioethics: Building on the Leopold legacy*. East Lansing: Michigan State University Press.
7. Ten Have, H. *Global bioethics*, 67–71.
8. Ten Have, H. 2020. La pandemia de COVID-19 vista por los expertos de bioética. *Bioetica Complutense* 39: 18–21.
9. This last meaning is illustrated by the image of Earth on the cover of Potter's first book on bioethics. See Potter, *Bioethics: Bridge to the future*, front cover.
10. Ingold, T. 2000. Globes and spheres. The topology of environmentalism. In *The perception of the environment. Essays on livelihood, dwelling and skill*, ed. T. Ingold, 209–218. London and New York: Routledge.
11. Cha, I. 2012. *The mundialization of home in the age of globalization. Towards a transcultural ethics*. Műnchen: LIT Verlag.
12. It is estimated that in 2019 there were worldwide 10 million cases of tuberculosis, and 229 million cases of malaria. World Health Organization. 2020. *Tuberculosis. Key facts*; World Health Organization. 2021. Malaria. *Key facts*.
13. See, for example, Van den Brink, G. 2020. *Ruw ontwaken uit een neoliberale droom en de eigenheid van het Europese continent*. Amsterdam: Prometheus.

Chapter 2
Pandemic Pasts. Experiences from History

Abstract The Covid-19 pandemic is not the first time that humanity is confronted with a sudden and lethal global disease threat. This chapter discusses previous lethal pandemics in human history. Examples of the Black Death in the fourteenth century, the cholera pandemics in the nineteenth century, and the Spanish flu in the twentieth century show that not only millions of people have died but that these scourges have also led to significant changes in society and culture. From these examples, patterns in the manifestations of epidemic diseases and in the responses to them are identified and examined.

Keywords Black Death · Cholera · History · Contagionism · Miasmatism · Pandemics · Plague · Public health · Spanish flu · Thucydides

2.1 Introduction

"In the first days of summer the Lacedaemonians and their allies … invaded Attica… Not many days after their arrival in Attica the plague first began to show itself among the Athenians. It was said that it has broken out in many places previously … but a pestilence of such extent and mortality was nowhere remembered. Neither were the physicians at first of any service, ignorant as they were of the proper way to treat it, but they died themselves the most thickly, as they visited the sick most often; not did any human art succeed any better. Supplications in the temples, divinations, and so forth were found equally futile, till the overwhelming nature of the disaster at last put a stop to them altogether" [1]. With this description, the Greek historian Thucydides begins his story of the plague of Athens, in the second year of the Peloponnesian War (430 BCE). He observed the sudden outbreak and rapid dissemination of a contagious disease, killing many people (possible one-third of the population) while no remedies were effective, and everybody, strong or weak, young or old, was affected. In modern times, Thucydides is acclaimed for his eyewitness account and his cool, accurate and detailed reporting as he carefully

H. ten Have, *The Covid-19 Pandemic and Global Bioethics*, Advancing Global Bioethics 18, https://doi.org/10.1007/978-3-030-91491-2_2

described the symptoms, their evolution and complications. In the Greek language, the word ‘plague’ or ‘pest’ is used for any epidemic disease in general, not a specific infection [2]. The precise symptomatology has led to many attempts to identify the nature of the disease. Now it seems most likely that the plague of Athens has been measles or perhaps smallpox [3]. Thucydides is most famous for his description of the moral and political consequences of the epidemic. Since the disease makes no differences among people, and no remedies are available, despair and a sense of doom are prevailing. Some people are abandoned, others receive care but all die. If they are not near to others, they expire in isolation. Courageous persons who care for one another face the same destiny. People lose all hope as soon as they realize that they are ill. Management of the disease is absent. Thucydides does not mention any efforts to control or mitigate the infection. Instead, scapegoating is usual. The disease is regarded as an outside enemy, arriving at the port of Athens, coming from ‘Ethiopia.’ It is also believed that the wells are poisoned by the Spartans as a kind of biological warfare. Leading statesman Pericles becomes the subject of growing animosity, and is put on trial but reinstalled a year later when he unfortunately died of the plague. His death led to the eventual loss of the war. Thucydides outlines how the disease was the beginning of the decline of the Athenian democracy. People became concerned with immediate pleasure and profit rather than sacrifices for the common good. Social disintegration produced increasing lawlessness and individualism as well as loss of moral standards. For example, funeral rituals which have been important in Greek civilization were abandoned. So many corpses accumulated that bodies were simply disposed off as quickly as possible without proper ceremonies.

In the Covid-19 pandemic Thucydides’ book has become an icon for the interpretation of disaster experience, much like *The Plague* of Albert Camus and *The Decameron* of Giovanni Boccaccio [4]. *History of the Peloponnesian War* is read not simply as a candid narrative of what happened long ago but as a warning about what might result from general calamity. Thucydides vividly portrays the demoralization of the population and the collapse of the social, moral and political order. The way of life that characterized Athenian civilization was corrupted and traditional values undermined. The balance between private and public interests was compromised. Unregulated passions, short-term interests, and indifference to law and religion came to dominate social life [5]. The approach used by Thucydides strengthens this cautionary dimension. His reporting is considered as objective, although he has been a victim himself. He does not want to speculate about the origin and the causes of the plague: “I shall simply set down its nature, and explain the symptoms by which perhaps it may be recognized by the student, if it should ever break out again.” The purpose of history in his view is instructive: it shows the world as it is or has been so that future generations might learn from the past.

This chapter discusses previous lethal pandemics in human history. Examples of the Black Death in the fourteenth century, the cholera pandemics in the nineteenth century, and the Spanish flu in the twentieth century show that not only millions of people have died but that these scourges have also led to significant changes in society and culture. While its causes were unknown for a long time, cholera for example has been a major impetus to the development of public health, safe water provision and sewerage as well as hygienic practices.

2.2 The Black Death

The deadliest pandemic in the past has been the plague in the fourteenth century [6]. The disease originated in Central Asia (especially around the Tibetan Plateau), possibly disseminated through the Mongol conquest and spreading over Western Asia, the Middle East, Europe and North Africa between 1346 and 1353. Bubonic plague was known to exist in North Africa and the Middle East (first described in the Bible as well as in the Hippocratic writings), but it was not mentioned in Chinese and Indian history. The point of departure of the fourteenth century pandemic was the Italian trading city of Kaffa, a port at the Black Sea. From there it was transported by ships to commercial seaports such as Constantinople, Alexandria, Messina, Genoa and Marseille (in 1347). After arrival, the pestilence followed the main routes of trade and travel by land and sea, reaching Florence (March 1348), Paris (August 1348), London (September 1348) and Vienna (February 1349). Some cities such as Milan that took drastic measures (closing the city gates and isolating patients and their families in their houses) avoided a major outbreak [7].

The fatality rate of the plague is 80%. It is estimated that 60% of the European population died (50 million people) [8]. Especially the countryside was severely ravaged, more than the towns since there were comparatively more rats per human in rural environments. At this time, almost 90% of the population in Europe lived in the countryside. Between 1347 and 1497, the population declined with 60–65%. The disease was also dreadful because of its fast course. After the onset of symptoms, patients on average die within 14.5 h [9].

The horror of the pandemic was amplified because its cause and dissemination mechanism were unknown. Nonetheless, there was no lack of explanations: divine punishment, astrological events, and miasma (foul air). Interpreted as punishment by God, the disease necessitated penance and prayer, religious rituals and ceremonies [10]. The medical framework of miasma led to emphasis on sanitary measures such as purifying contaminated air. Physicians used special protective gear (with a beak filled with smelling herbs). It also led to the invention of quarantine in Venice, requiring arriving ships to wait 40 days before disembarking. The disease, however, was unstoppable, undermining religious and medical authority. God-fearing people died as much as other persons. Physicians could only advice to avoid the sick. The sense of helplessness was expressed by Petrarch who wrote in 1348: "When has any such thing been ever heard or seen; in what annals has it ever been read that houses were left vacant, cities deserted, the country neglected, the fields too small for the dead, and a fearful and universal solitude over the whole earth? Consult your historians, they are silent; question your doctors, they are dumb; seek an answer from your philosophers, they shrug their shoulders and frown, and with their fingers to their lips bid you be silent" [11].

Fear and the general sense of doom provoked by the pest fomented scapegoating. Especially foreigners, lepers and Jews were blamed and pogroms were common [12]. There also was the story of biological warfare. When the Mongol army besieged Kaffa, plague broke out. The Mongols then catapulted the bodies of dead victims into the city. This story of biological warfare cannot be true since bodies of plague victims are not contagious [13].

The effect of the plague was that everything stopped. People locked themselves in, and towns were closed. It was evident that the disease moved by contact and travel. But since the cause was unknown, countermeasures were not effective. Rats and their fleas could easily penetrate locked-down communities. People who fled, carry rat fleas in their clothing and luggage, helping to disseminate the disease. Many studies have described the long-term consequences of what was called the 'great mortality.' The normal way of life, human behaviors and attitudes changed. In the city of Florence (where 60% of the population died), behavior changed in two directions. Some people closed themselves in while others engaged in partying and drinking and the rich escaped to the countryside [14]. The social and economic impact of the plague was significant. The massive depopulation created shortages of labor to work on the land. This caused lack of food and starvation. It also increased wages, destabilized the existing serfdom system thus hastening the end of the medieval feudal system. Because the plague weakened the authority of institutions such as the Church, it is argued that it encouraged the growth of individualism, personal mysticism, and privatization of faith, resulting in the Renaissance and the Reformation [15]. Finally, the pandemic generated cultural changes. The dread of deadly infection intensified concerns with human mortality and transformed attitudes towards death. Pessimism and surge in sense of sin as well as obsession with death and decay of the body found artistic expression in the Danse Macabre and in literature on the art of dying [16]. In the perspective that worldly goods are frail and futile, veneration of saints and their relics became more popular (with specific plague saints such as St. Roch and St. Sebastian).

With historical hindsight, the plague is regarded as a transformative experience: "A great catastrophe … breaks many links with the past…" [17]. Whatever lessons may be presented, a pestilence on this scale is a so-called liminal event. It signifies a transition to a new phase, leaving behind the traditional ways of living and thinking. The experience of lethal threat is a time in between the old and the new, triggering people to redefine experiences and customs, and thus engendering a new era. Historians usually indicate that the plague demarcate the last phase of the Middle Ages (1350–1520) [18]. The fourteenth century was the age of calamities: famines, wars, brigandage, rebellions and uprisings. The plague returning five times before the end of the century certainly magnified the feeling that the end of the world was near [19]. Only in 1894 it was discovered that plague was caused by the bacterium *Yersinia pestis*. The pathogen is transmitted by rat fleas with the black rat as the normal host. If rats die, the fleas attack human beings and infect them. It takes on average 15 days before an infected rat population is so decimated that fleas attack human beings. It then takes 3–5 days before infected humans get symptoms. It is now known that the bacterium can be hosted by many species of rodents as well as other animals (e.g. marmots and camels) [20]. Dissemination of the pathogen is comparatively slow compared to viral spread through droplets. As long as the mechanism of transmission is unknown, effective countermeasures cannot be taken. Although plague is nowadays almost forgotten and regarded as a historical event, it is still reported today (with 243 cases in 2018, mostly in Africa). In 2020, cases were reported in China, Congo, and the United States [21]. It can effectively be treated with antibiotics.

2.3 The Blue Death

Memories of the plague returned five centuries later. A new disease entered European countries in 1831, creating an atmosphere of crisis as would a foreign invasion. The unpredictable nature and sudden eruption of cholera, the so-called 'Blue Death,' revived the horrors of the medieval Black Death, without empirical justification since the number of deaths was small compared to earlier plague death rates, and nearly equal to the contemporary death rates of diseases such as typhus and common diarrhoea [22]. The pestilence was also called 'Asiatic cholera' because it emerged from India. It has been endemic in the Asian Subcontinent for thousands of years. The first pandemic started in 1817, originating in the Ganges Delta and affecting Asia and the Middle East. The second pandemic also originating in Bengal in 1826 infected Europe and the United States in 1831–1832.

Although the plague has reemerged in Europe as a pandemic in the seventeenth century, the sudden and massive strike of the cholera overwhelmed societies that have assumed that the time for deadly pandemics in Europe was over [23]. After infection, 10–20% of people show symptoms and fall ill. Without adequate treatment, 40–60% of the victims die mainly because of dehydration. The lethality of cholera, however, is less than that of the plague (killing approximately 1% of the population). It was also clear that cities were more severely affected than rural areas, and that within cities poor people were more afflicted [24].

The invisible threat that suddenly makes many people ill and rapidly spreads among populations caused a flood of publications with advices, recommendations and remedies. Medical doctors engaged in disputes and quarrels about the nature of the disease and its best treatments. Many remedies were advertised as cures (e.g. brandy in Britain) [25]. Especially, those arguing that 'Asiatic cholera' did not really exist, or was not contagious, found a large audience [26]. Conspiracy theories flourished, for example the belief that foreign enemies or immigrants were responsible, or that the rich disseminated the disease in order to reduce the population. Distrust of medical doctors was increasing and sometimes resulted in intimidations and attacks. In Berlin in 1831 physicians could only visit patients in some neighborhoods with a military escort.

Because cholera moved relatively slow, there was time for preparation. When the disease broke out in Germany in 1831, the Dutch King sent three experts to Prussia and Hamburg to study the most effective way to treat the disease. They published their report in April 1832, just 2 months before the disease reached the Netherlands. However, the experts concluded that cholera was not contagious and they did not provide helpful recommendations [27]. Nonetheless, the government applied quarantine measures. It also instructed provincial authorities to set up cholera commissions to coordinate policy responses such as special hospitals and cleaning activities in cities. In many cases, responses were different and sometimes absent. Quarantine and isolation were ineffective because authorities could not really enforce them and because citizens did not implement them. Many patients resisted admission to cholera hospitals which was not mandatory. Local authorities, dominated by commercial interests, frequently delayed stringent measures such as quarantines. Authorities were also reluctant to intervene because of ideological reasons. The political philosophy of liberalism prevented stringent government intervention while poverty was often moralized and regarded as the consequence of immoral conduct. Furthermore, as soon as the threat was over, policy measures were reversed until the next pandemic appeared [28]. Slowly, there was a learning effect with a shift towards prevention and the elimination of unhealthy conditions, moving from a medical to a social approach of the disease. After several pandemics, more strict health legislation was adopted and more sanitary environments created with new infrastructure (sewerage and drinking water systems) and sanitation of urban living conditions but in many European countries these were realized only after the last pandemic was over.

It is now known that cholera is caused by a bacterium (*Vibrio Cholerae* identified by Koch in 1883) and transmitted through contaminated water (or food) and spread by ingestion. Although the cause was unknown, there was no lack of explanatory frameworks. Like in the era of the plague, the disease was attributed to sinful behavior and failings in morality. To many, cholera was the manifestation of a moral defect: intemperance, filth, wickedness and impiety [29]. The dominant medical framework was the miasma theory. The idea that epidemic disease was caused by contaminated air was based on the observation that especially cities and poor neighborhoods were affected. In 1830, most cities in Western Europe were more densely populated than hundred years earlier or later. Overpopulation produced enormous pollution and degradation of the living environment. Stench was everywhere. Without sewer systems, systematic waste removal and clean drinking water, symptoms of diarrhoea were common, and usually diagnosed as 'cholera' or 'tyfus.' In the early stages of the pandemic, symptoms were therefore not recognized, but the intensity and rapid course made clear that this was not the 'common cholera' but a new disease.

The theory that disease is produced by odors and emanations assumes that unhealthy miasmas are generated by several sources in the environment: dirt, bad drainage, crowded housing, lack of ventilation and polluted water. The poisonous miasmas are propagated through the air. There is no patient-to-patient transmission; individuals, and groups of people are affected by widely diffused, atmospheric agents. This theory has ancient roots in the writings of Hippocrates and Galen. Malaria was a paradigm case, showing the disordered relation between humans and the environment. The usual response was to purify the air from obnoxious emanations. In the eighteenth century, miasmatism was associated with active applications that attempted to modify the environmental conditions themselves and removing the causes of noxious air. An alternative explanatory framework was contagionism. This theory attributed epidemic disease to 'contagia', minute living organisms that were transmitted between individuals. It was influential in Greek and Roman medicine, and became more important after the sixth century with the spread of leprosy. As a coherent theory it was first formulated by Fracastoro in the sixteenth century, postulating the existence of small imperceptible particles as causes of disease. On the basis of this theory the only effective measure against an epidemic disease was stringent isolation of infected persons. The theory was corroborated by the successful inoculation against smallpox in the early nineteenth century.

The arrival of an unknown, deadly disease such as the cholera challenged these explanatory frameworks. At first, while most medical experts believed that the disease was due to atmospheric conditions and not contagious, contagionism was the common-sense view [30]. The general population regarded the epidemic as contagious. It provided some measure of control: contamination could be avoided by abstaining from any personal contact. However, policy measures such as isolation and quarantine were not stopping the spread of cholera. These experiences discredited contagionism and made the theory of a miasmatic origin more plausible. During the third pandemic (1848–1849), miasmatic thinking dominated medical research and professional opinion. It was also attractive for policy-makers since measures were necessary that not impeded trade and commerce, were not socially disruptive, and that not restricted individual freedom. Miasmatism was also the expression of belief in progress and social activism. It produced a sanitary movement, inspired by the utilitarian philosophy of Jeremy Bentham, that emphasized the social origin of disease, the need for scientific study of environmental conditions (using the new science of statistics), and the priority of prevention. The idea was that sanitation of the physical environment would benefit the greatest number of the population. It would have economic benefits since it reduced expenses for curative health care and poor relief. Furthermore, it produced moral benefits, encouraging changes in lifestyle and morals, promoting hygienic habits [31].

The subsequent cholera pandemics signify a fundamental transition in medicine. The controversy between competing frameworks to explain the etiology, transmission and prevention of epidemic disease reflect different styles of thinking, each producing their own facts and policies [32]. The choice between thought-styles is not merely a matter of argumentation, empirical evidence and rationality. In some instances, physicians vote to determine whether cholera is contagious or not [33]. Opinions oscillate between the two theories. In the second part of the nineteenth century, the majority of European physicians adhere to a contagionist origin of the cholera, even before the major bacteriological discoveries in the 1870s and 1880s are made. In the confrontation with contagionism, miasma theory becomes more specific and sophisticated. Instead of uncontrollable conditions and ill-defined emanations as sources of disease, it specifies the nature and origin of miasma, compiling disease histories, meteorological journals, mortality registers and topographical data, gathering specific evidence of the association between environmental factors and diseases. Moreover, miasmatism transforms into a program of social reform. It becomes more than a medical theory, as practical alliance of scientists, technicians and politicians, much in line with the liberalism espoused in France, England and Germany [34]. Most European countries face significant social and economic changes due to the Industrial Revolution, especially mechanization of production and transportation [35]. Railways and steamships increase mobility while steam power intensifies industrial production. The Industrial Revolution is associated with demographic changes: exponential growth of the population and rapid urbanization. At the same time, poverty is growing. Conditions of labor and housing are appalling, and child labor is common. Nutrition for the majority of the population is deficient. In 1859 in Amsterdam, 26% of children die before their first year of life [36]. It is clear that cholera is primarily a disease of the poor. Although authorities often blame the poor for their condition (attributed to alcohol use, laziness and immoral behavior), they also realize that some measure of governmental intervention is necessary [37]. In the mid-nineteenth century context, autocratic policies (based on contagionism and the traditional idea of medical policing) are no longer acceptable. At the same time, assistance and healthcare cannot be left to charity and philanthropy. Rational approaches of disease and indigence shall focus on sanitizing social circumstances, and governments at various levels must have a central role in this. Simultaneously, it is important to civilize the habits and lifestyles of the masses. Not repression or coercion can transmit norms of behavior, such as cleanliness, soberness and moderation but the subtle means of advice, persuasion and education will pass the norms of a healthy, regular and disciplined conduct into domestic life. Medical doctor and local politician Samuel Sarphati can serve as an example. He initiates many practical projects to improve the living conditions of the lower classes in Amsterdam, and to create a healthy city without medieval walls and with green parks, walking space, and sanitation [38]. Sarphati personifies the mission of the sanitary movement and of hygienism; the first to produce clean and healthy social circumstances, the second to reform private morality and promote personal hygiene. Instead of the coercive external control of contagionism, a less visible system of internal control emerges, based on internalized norms. This emphasis on moralization and normalization is at the same time responsible for the decline of hygienism (and the miasmatic thought-style) in the last decades of the nineteenth century. The emergence of the germ theory of disease and the results of laboratory science articulate the idea that medicine is a value-free, objective science.

While miasmatism in retrospect was an incorrect theory, its practical implications were significant since it had a positive effect on public health. Better sanitation, housing, waste collection, sewage systems, clean drinking water, removal of graveyards outside inner cities all contributed to improvement of the health of populations. It can be argued that the dynamic interaction of thought-styles has promoted the growth of medical knowledge [39]. Both styles make intensive use of modern sciences such as chemistry, physiology, pathology and mathematics. The controversial interaction between different styles of thought demonstrates that the progress of medical science is not a linear process. An example is the work of John Snow in London during the outbreak of 1848–1849. With meticulous empirical studies (mortality registers, interviews of patients, identifying sources of drinking water, making maps of cases) Snow shows that cholera is not inhaled but disseminated by ingestion of contaminated water. He applies scientific methods to public health challenges, founding the science of epidemiology, but his waterborne theory is not accepted by many contemporary colleagues [40]. During the subsequent waves of cholera pandemics, the philosophical foundations of contemporary medicine and health care become firmly established. In the final part of the nineteenth century, medicine conceives itself as a natural science with a powerful paradigm to explain and manage infectious diseases, assuming a specific etiology (each disease has a specific microbial cause) as well as a specific treatment for particular diseases.

The interaction of thought-styles furthermore illustrates that medical and social perspectives on disease cannot be clearly demarcated. An epidemic is not merely a biological phenomenon but also a social event. Medical disputes about causality had enormous economic and ideological implications. Isolation and quarantine seriously damaged commerce, and this gave representatives of trade a preference for the miasmatic theory. The way societies interpret and understand infectious diseases is only partially based on medical observation and evidence.

Finally, an important consequence of the cholera pandemics is visible in the arena of international cooperation. Cholera comes to be regarded as a threat common to all countries. The pathogen is an invader that is using international transport and colonial enterprises as vehicles of attack. The velocity of transport has increased with railways and steamships, so that countries feel more vulnerable. The heterogeneity of national measures provides cholera too many opportunities to advance as long as better coordination and transborder regulation is not effectuated. Since 1851, eight international sanitary conferences took place. Although the success of these conferences was limited, they demonstrated the growing awareness of global interconnectedness. However, international efforts were very Eurocentric. Convinced that the origin of cholera was outside Europe, policies were aimed at defending the Continent against the threat from Asia and the Orient. From a European perspective, the main challenge was to allow trade and colonial expansion and at the same time to block contagion from the East. This resulted in double standards. For example, military transports (to defend the colonial systems) were not subjected to controls, while much emphasis was put on regulation of pilgrim traffic (especially in Muslim countries). A distinction was made between good and bad mobility in order not to close borders but make them semipermeable, i.e. blocking contagion and permitting commerce [41].

After the last wave of cholera (1881–1896), pandemics have not occurred. But cholera still is a global threat. The World Health Organization reports in 2019 that worldwide each year there are approximately 1.3–4 million cases with 21,000–143,000 death [42]. Cholera is endemic in many countries with local transmission and seasonal outbreaks. From time to time, epidemic outbreaks happen (for example in 1999 in Madagascar and in 2010 in Haiti). A large cholera outbreak is affecting Yemen since 2017 with more than 2,5 million cases. The disease can now be treated with intravenous fluids and antibiotics, and oral vaccines are available. While it is possible to reduce cholera, it is impossible to eradicate it. Due to climate change, there is always a risk for major outbreaks. The best cure for cholera still is providing safe drinking water and improved sanitation [43].

2.4 The Spanish Flu

In absolute numbers, the pandemic of influenza in 1918–1919 was the deadliest. Estimations are that between 50 and 100 million people died [44]. Relative to the total population, the mortality rate of the Black Death however was much more severe. The fatality rate of the influenza pandemic was estimated at 2,5% (compared to 0.1% of influenza today). However, in cases where reliable statistics were available (for example in army camps in the U.S.) mortality was often more than 5%, and sometimes higher than 10% [45]. Influenza could strike suddenly and massively since it is a very contagious disease with a short incubation period (24–72 h).

The origin of the worldwide flu pandemic is not known. Presumably, it started in Kansas in the United States, rapidly infecting soldiers in army camps, and from there it reached France. Another option is that the pandemic begun in France. It is not likely that the origin was in China [46]. Certain is that the movements of the pathogen were accelerated by the First World War, bringing together enormous numbers of young men in training camps and on frontlines. The pandemic came in three waves. The first wave begun in Spring 1918 and mainly infected recruits in training camps, beginning in February in the United States. The disease arrived in France in early April, affecting the French and German armies, and started to spread across countries. Since Spain was neutral in the war and the press not censored, reports of the pandemic first appeared in Spanish newspapers, and the disease was therefore known as the 'Spanish flu' [47]. The illness was usually mild, lasting 2 or 3 days, and almost everyone recovered. By July the authorities and physicians assumed that the threat was over. But the virus had not disappeared. The second wave started in September 1918 and subsided in late Spring 2019. This time the infection was more lethal. Perhaps the virus had adapted to human beings and had mutated in the meantime so that the human immune system was incapacitated. Many victims died of secondary bacterial pneumonia (occurring in 10–20% of all cases). The third wave lasted from January to April 1919 and was lethal again.

While the disease could affect everyone, it was quickly observed that morbidity and mortality rates were higher among poor populations who lived in crowded conditions. Also, populations that had not been exposed to influenza viruses before were severely hit (for instance, Eskimos, Native Americans and people in Africa). A curious phenomenon was furthermore that young adults were most severely hit. While the seasonal influenza usually attacks the very old and very young, this pandemic especially killed healthy people between 20 and 40 years old. There still is no explanation for this excess mortality [48].

The cause of the disease was unknown but it was generally assumed to be a germ. The dominant view considered influenza as a bacterial infection, caused by a bacillus discovered by Pfeiffer in 1892. The bacillus, however, could not be found in many cases so that doubts about its causative role increased. Only at the end of the nineteenth century a distinction was made between bacteria and viruses. Using laboratory filters that did not allow bacteria to pass, scientists discovered infectious agents that were filterable. Martinus Beijerinck called this new class of agents *contagium vivum fluidum* or 'virus' [49]. The first human virus was identified in 1901 (yellow fever virus). The human influenza virus was described in 1933 (following the discovery of the swine influenza virus in 1931).

More clarity existed about the mode of transmission of the pathogen. Influenza is airborne and thus spreads through inhalation. Three to six days after a person is infected, other people are infected through droplets from the throat, nose and mouth. The virus can survive for a limited time on surfaces so that it can also be spread through contact. Effective treatment was not available, except aspirin to lower fever. However, many doctors engaged in heroic interventions such as traditional bloodletting. Numerous vaccines were prepared and administered but without significant effect. Folk remedies and disinfectants were widely advertised. In general, especially after the first wave of the pandemic, a feeling of futility prevailed [50]. In fact, the best approach was prevention. Influenza was considered 'a crowd disease' so that interfering with person-to-person transmission was the best strategy [51]. Policy-makers responded with quarantine, closing schools, theaters, churches and stores, and building temporary hospitals for flu victims. They recommended avoiding crowds and using face masks to prevent infection. Nonetheless, quarantine and closing borders, although commonly used, were ineffective and delaying the disease for a few days. Only places that completely isolated themselves escaped. The reason that policies were not stringently implemented is connected to the war effort of many countries. Authorities and newspapers provided inadequate information to the public and minimized the danger of the pandemic in order to avoid fear-mongering and panic which could undermine social mobilization for the war. In the United States that had entered the war in April 1917, the Wilson Administration never publicly acknowledged the disease, and the federal government did not show any leadership. The military, aware of the disease, did not take any action either. The transports of soldiers to France continued, with devastating effects. All countries involved in the war tried to control public perception. In the first stage of the flu outbreak, they were silent (only the Spanish press was reporting it). Then, as the disease was evident and could no longer be denied, people were told not to worry, arguing that influenza was nothing more than the 'old-fashioned grippe' [52]. And when the number of victims was surging, authorities assured the population that the worst was over. Misinformation and the clear gap between reality and political reassurances made that people lost trust in authorities. That undermined the credibility of policy-makers and the adherence to their recommendations.

The story of the Spanish flu pandemic is at the same time the story of scientific progress. In the decades prior to the outbreak the germ theory of disease had been confirmed in a series of discoveries (e.g. the tuberculosis bacillus in 1882 and the cholera bacillus in 1883, both by Robert Koch). The criteria for establishing microbial causality have been formulated in 'Koch's postulates.' Laboratory investigations and experiments became common in medicine. Medical research expanded with the creation of special institutes such as the Pasteur Institute in 1887 and the Rockefeller Institute in 1901. Virology as a separate scientific discipline developed somewhat later. The virus as a new type of pathogen was largely a theoretical construct until the invention of the electron microscope (in 1931) made it possible to visualize viruses. Since the notion of virus was not compatible with the prevailing dogma of Koch's criteria, it took two decades before it was accepted by the scientific community [53]. After the flu virus was identified in 1933, three different influenza viruses have been described. Only influenza A viruses produce epidemics or pandemics in humans. These viruses do not originate in human beings but in wild aquatic birds. Avian virus can directly infect humans but it only is transmitted from person-to-person when it is changed and adapted to human beings. The virus can also be hosted in mammals, particularly swine, and then jump to humans. As an RNA virus (like HIV and coronavirus) it can mutate rapidly. The Spanish flu was caused by influenza A virus of the H_1N_1 subtype. The genetic make-up of this virus has been determined in the last decade of the twentieth century [54].

2.5 Pandemic Experiences

It is questionable whether any lessons can be learned from history. As the Black Death, the Blue Death and the Spanish flu illustrate, pandemics emerge in specific conditions, and they are addressed from a constellation of ideological, economic and cultural factors that are typical for a particular time and place. There is moreover the general doubt whether humans will ever learn from the past, following Hegel's adage that the only thing we learn from history is that we learn nothing from history [55]. Nonetheless, it is possible to discern several patterns in the manifestations of epidemic diseases and in the responses to them.

- Pandemics are sudden, explosive and widespread events [56]. They usually overpower and surprise populations and governments. Their infectiousness and severity provoke fear and panic. With the Spanish flu healthy persons could die within 12 h. Ordinary life was suspended; work and meeting places closed. People try to escape, for example fleeing to rural areas or other countries which they assume are not infected. These movements usually further disseminate the disease.
- Pandemics elicit similar patterns of response. At first is denial, then the argument that it is not so bad or that the disease is not different from known and earlier ones. The story of an epidemic unfolds as a dramaturgic event with a specific sequence, as argued by Charles Rosenberg [57]. The first stage is progressive revelation. The reality of the disease is slowly accepted. There is hesitancy to sound the alarm. Authorities in the initial phase do not recognize that the disease is present, usually in order to avoid the economic consequences. During the second cholera pandemic the lag between the time of contamination and recognition was usually 6 weeks [58]. Action was only taken when the disease presence has

become obvious and mortality undeniable. In the cholera outbreak in Hamburg in 1892 (the only city in Western Europe hit by cholera) the town administrators first suppressed the news. They did not want to create anxiety and panic. The balance between providing information and avoiding public anxiety was even more disturbed during the Spanish flu because many countries were involved in conducting war. Furthermore, in the early stages the symptoms were often not distinguished from those of known diseases. The cholera was regarded as common diarrhoea, and the influenza as common flu. During the plague and cholera, the contagiousness of the disease was questioned by physicians while the authorities reassured the population that there was no danger. The second stage, according to Rosenberg, is managing randomness. With accumulating cases and deaths, the reality of disease is gradually recognized. In order to understand the disease, explanatory frameworks are necessary that promise some measure of control. The threat is often differentiated so that vulnerabilities can be determined. It is assumed that for example particular groups are more susceptible. Identifying risk factors allows to see death not as a random accident but as related to constitution, behavior, lifestyle and environment. During the plague, it was common to regard the pestilence as punishment for sin. During the cholera, the poor were especially vulnerable because of unhygienic habits but the focus could also be on environmental conditions so that social criticism was more appropriate. The third stage in the response to epidemic disease is negotiating public response. Because pandemics create a sense of crisis, there is a need for a visible response as an expression of community solidarity. Collective rituals such as quarantine, purifying the air, disinfection of public spaces or joint prayer demonstrate the belief that control is possible. Whether or not they are effective, is less important than the visible demonstration of control. The use of face masks during the Spanish flu was a case in point; there was serious doubt about their effect but they symbolized the fight against contagion. The fourth and last stage is subsidence and retrospection. When the disease gradually declines, the question is considered how the community has dealt with the pestilence. Lessons learned are identified and retrospective moral judgments of the policies are formulated.

- Pandemics raise the need for explanation. The above examples show that competing frameworks have been used to explain the pestilence. Religious interpretations have been important during the plague. Medical explanations have been rather powerless and unconvincing. During the cholera competing medical theories were at work. Medical explanations have become more powerful during the Spanish flu, with the growing influence of the germ theory of disease. Whatever the explanatory theories and interpretive frameworks, popular belief was attached to contagion. Fear of contagion motivated isolation and avoidance of contacts, restriction of mobility, and flight to 'safer' areas. In fact, two types of disease explanation can be distinguished: the contamination view emphasizing person-to-person transmission [59], and the configuration view, regarding disease as a specific configuration of circumstances.

  At the same time, in all pandemics moral frameworks were used. During plague and cholera, victims were morally blamed because immoral behavior made them vulnerable to the pestilence. However, during subsequent cholera pandemics, the disease was increasingly regarded as a social rather than moral phenomenon, especially as the product of environment [60]. But moral views persisted since due to the common use of the war metaphor, others were blamed for the contagion and conspiracy theories flourished. The infection is regarded as invasion, the pathogen as enemy. The armies of the Black Death marched slowly, and irresistibly conquered the regions of Europe [61]. During the Spanish flu courageous warriors attacked the mysteries of the deadly micro-organism [62]. The naming of pandemics reflects this perspective. Especially Asia was regarded as a source of infectious disease, as expressed in the label 'Asiatic cholera' while on the other hand the Spanish flu, emerging in the United States was not blamed as an American pestilence.

- Another feature common to the above pandemics is the type of spread: through human communication (via sea, roads and rivers). The plague was spreading slowly. There was no person-to-person transmission in the majority of cases (only in cases of pneumonic plague). Benedictow calculated that the Black Death moved with 0.5–2 km per day by land (and with 40 km per day by ship) [63]. The cholera moved faster because of the railways and steamships. The Spanish flu disseminated even more rapidly because it was airborne. This type of spread could be controlled by preventing person-to-person transmission. This was called at that time 'crowding control' rather than 'social distancing.' However, identifying contacts and isolation is almost impossible as long as the infecting organism could not be identified. The type of spread of pandemics finally allows the possibility of preparing for the disease. Since the pandemics come in waves, there is some time for preparatory measures. However, as the examples demonstrate, this opportunity was hardly used. Policies have mostly been reactive rather than proactive.
- Pandemics affect special groups in the population more heavily than others. In general, the poor were seriously afflicted like all crowded populations. During the plague, religious communities had extreme high mortality rates. In monasteries, often the majority of inhabitants died [64]. Women were also more susceptible because they spend more time in flea-infected houses. Research of plague cemeteries showed that the disease especially targeted people of all ages who were frail because of malnutrition, prior disease or poor immune systems [65]. The cholera particularly infected poor and unhygienic parts of crowded cities. The Spanish flu ravaged through army camps and frontlines crowded with young soldiers.
- In the plague, cholera and flu there was an upsurge of what we nowadays would call 'fake news.' False cures, remedies and prophylactics were promoted. In desperation people used all kinds of medical and chemical substances, like brandy during the cholera or mercury during the Spanish flu.
- All pandemics had social and economic effects. Although it is questionable to regard pandemics as the cause of such effects, they undoubtedly accelerated and

modified trends which already existed. The Black Death for example was a catalyst for the growth of education in the vernacular and the disappearance of Latin as the international medium of communication [66]. Plague as well as Spanish flu led to labor shortage and demands for higher wages. The plague precipitated the collapse of the medieval feudal system while the cholera promoted 'hygienization' of human behavior and the social environment. The flu outbreak in the last phase of World War I probably accelerated the end of fighting. All pandemics produced economic devastation because of the closure of many commercial establishments and the interruption of trade. At the same time, the massive morbidity and mortality overwhelmed the healthcare system with shortages of hospital beds, face masks and protective equipment.

- Pandemics demand major government intervention. The Black Death initiated hectic activities of local governments, for example instituting quarantine. Since individuals are powerless in the face of an epidemic disease, and many persons are affected at the same time, only the community as represented by secular or religious authorities can take action in regard to purifying the air, closing town gates and providing decent burials. During the Blue Death there was continuous tension between local and central governments. A patchwork of local authorities with different and incoherent policies approached the disease in multiple ways, and did not consistently enforced isolation and quarantine measures in order to protect local commercial interests. This put increasing pressure on central governments to develop legal frameworks to implement and make local interventions more effective. When the civilian population was much affected by the Spanish flu, many countries involved in the war did not engage in containing policies and prevention. In the United States, the federal government did not take any initiative to counter the pandemic. States and cities were on their own to cope with the disease. Because of the insufficient and haphazard responses of authorities, pandemics may undermine the legitimacy and credibility of governments and administrations because the population no longer trust authorities.

Although common patterns can be found in the three examined pandemics, we cannot simply apply these patterns on a new pandemic in the twenty-first century. There are major differences between the time and context in which the past pandemics emerged and our present time. One is that the current situation is characterized by sophisticated science and technology. While cholera was a mystery in the 1830s, in the 1860s the medical profession had adequate empirical knowledge about its mode of transmission. The germ theory of diseases was increasingly accepted in the 1880s. History demonstrates the impotence and even harmfulness of medicine, but that is no longer the case today. Medical sciences provide better knowledge of causation, dissemination, treatment and prevention of epidemic diseases. The coronavirus as causative agent of the Covid-19 pandemic has quickly been identified. This allowed the rapid development of diagnostic tests and the early start of research on vaccines and treatments. Another difference is that the science and practice of epidemiology has advanced. Research and surveillance of existing and new microorganisms have expanded. Most countries have virology laboratories, and many cities and regions public health services that can perform tests and contact tracing. Public health legislation has been adopted long ago while pandemic preparedness plans were completed at the instigation of the World Health Organization. Much has improved since the Spanish flu when data collection was inadequate since reporting cases was not required so that estimations of dead, infected and recovered patients were unreliable. Another difference with the past is that current societies are medicalized. Due to the growth of medical knowledge since the nineteenth century, diseases are now understood as specific entities with discrete causes. This has reshaped the role of physicians and scientists because they have the tools to understand, explain and treat diseases. This focus on specific diseases and their treatment explains the priority of hospital interventions (with current focus on ICUs in Covid care). It has also expanded medical authority, increasing the gap with lay knowledge. Nowadays, healthcare is regarded as a human right, and many countries have healthcare systems that are accessible for the majority, and sometimes all citizens. Medicalization is also a cultural process: modern people often define challenges in life as medical problems that need to be addressed by the healthcare system. For many, health is the most appreciated value in human life. People expect that states intervene and take action to control diseases. A final difference is that in past pandemics the focus was on the European perspective. Most histories of the plague, cholera and the flu studied their occurrence and effects in European countries. This traditional Eurocentric view is nowadays replaced by a global view, articulating that pandemics are events that affect all countries across the world. The idea that Europe can be attacked from the outside, and need to protect itself from pathogenic invaders by isolation from the global context, is no longer tenable.

Regardless of these differences between the past and the present, there are two basic realities that are still the same. One is the reality of microorganisms. They remind us that human beings are embedded in nature. Human beings cannot survive without viruses. They constitute a virosphere that not merely surrounds them but that is within them [67]. The other reality are human beings themselves. It is not clear how much their nature and behavior have changed over time. It seems that in view of a lethal challenge, humans continue to show the same behavior as in previous times. Even if we have now more medical knowledge as ever before, it needs to be used by human beings. Policy recommendations only work if they are followed and implemented. Healthcare information is never completely certain so that there are always doubts and uncertainties. It also is applied within a social and cultural context which can be authoritarian or liberal, so determining limits and constraints on how stringent measures such as quarantine, isolation or testing can be applied or enforced. As a form of drama, a pandemic is not just a medical event but a social phenomenon with a particular evolvement in time. Knowledge of the pathogen and the etiology of the disease is not sufficient to control an epidemic disease [68].

2.6 Conclusion

In this chapter three examples have been discussed of past pandemics. They are reminders that the current Covid-19 pandemic is not the first time that humanity is confronted with a sudden and lethal global disease threat. They also caution that pandemics can be expected but not predicted. As will be examined in the following chapter, many experts have warned that pandemic diseases can be expected sooner or later so that the international community has to be prepared. They also pointed out that the idea that infectious diseases are no longer threatening is false. Pestilences from the past such as plague and cholera are now re-emerging while many strains of bacilli have become resistant to antibiotics. New threats are emerging with novel viruses. In any case, infectious diseases have never disappeared: malaria, tuberculosis and dengue are devastating experiences in many countries around the globe. The AIDS pandemic in the 1980s had already shattered the belief that infectious disease can be controlled.

These warmings refer to the importance of an historical perspective. First of all, past pandemics can teach us about failures. What has gone wrong in the past? Why were some policy measures working and others not? Pandemics, as argued above, present recognizable patterns such as particular responses like denial, blaming others, and conspiracy theories that might be expected and therefore can be avoided. Past experiences also highlight that pandemic threats are not only a medical and scientific challenge but simultaneously a moral and social confrontation. They provoke, and sometime reawaken cultural narratives of crisis. Furthermore, questions of meaning are raised. A pandemic makes people aware of a new reality. They cannot continue to do what they used to do. As a crisis, a pandemic not only involves healthcare but all dimensions of human existence: economy, education, culture, sports, leisure, safety, and well-being. As they are often produced by social life and human communication, in their turn they seriously impact both. As the Covid-19 pandemic demonstrates, these experiences activate reflection on our way of living, re-thinking the ethical foundations of society. Historic examples of catastrophic disease indicate that they have been a point of transition between the past and the future, instilling new awareness of the fragility of human existence.

References

1. Thucydides, *The history of the Peloponnesian War*. Second Book, Chapter VI.
2. Gervais, A. 1972. À propos de la 'Peste' d'Athènes: Thucydide et la littérature de l'épidémie. *Bulletin de l'Association Guillaume Budé: Lettres d'humanité* 31: 395–429.
3. Cunha, B.A. 2004. The cause of the plague of Athens: Plague, typhoid, typhus, smallpox, or measles? *Infectious Disease Clinics of North America* 18: 29–43.
4. See for example: Fins, J.J. 2020. Pandemics, protocols, and the plague of Athens: Insights from Thucydides. *Hastings Center Report* 50 (3): 50–53.

5. Longrigg, J. 1980. The great plague of Athens. *History of Science* 18: 209–225; Nielsen, D.A. 1996. Pericles and the plague: Civil religion, anomie, and injustice in Thucydides. *Sociology of religion* 57 (4): 397–407; Soupios, M.A. 2004. Impact of the plague in Ancient Greece. *Infectious Disease Clinics of North America* 18: 45–51.
6. The disease was later called the Black Death, after the Latin *atra mors* (meaning terrible but also black death).
7. Ziegler, P. 2009. *The Black Death*, 54. New York: HarperCollins Publishers.
8. Benedictow, O.J. 2004. The Black Death 1346–1353. In *The complete history*, 380–384. Woodbridge: The Boydell Press.
9. Benedictow, *The Black Death 1346–1353*, 25–26.
10. Bedyński, W. 2020. Liminality: Black Death 700 years later. What lessons are for us from the medieval pandemic? *Society Register* 4 (3): 129–144.
11. Cockerell, T.D.A. 1916. The Black Death, and its lessons for to-day. *The Scientific Monthly* 3 (1): 82.
12. Cantor, N.F. 2001. *In the wake of the plague. The Black Death and the world it made*. New York: Simon & Schuster.
13. Benedictow, *The Black Death 1346–1353*, 52–53.
14. Boccaccio, G. 1353. *The Decameron.*
15. Tuchman, B. 1978. *A distant mirror. The calamitous 14th century*. London: Penguin Books. See also: McNeill, W.H. 1998. *Plagues and people*. New York: Anchor Books, 194 ff.
16. Huizinga, J. 1924. *The waning of the Middle Ages*. New York: St. Martin's Press, 124 ff.
17. Bedyński, Liminality: Black Death 700 years later, 85.
18. Benedictow, *The Black Death 1346–1353*, 62.
19. Tuchman, *A distant mirror.*
20. Ziegler, M. 2014. The Black Death and the future of the plague. *The Medieval Globe* 1: 259–283; Green, M.H. 2004. Taking 'pandemic' seriously: Making the Black Death global. *The Medieval Globe* 1: 27–61.
21. Bertherat, E. 2019. Plague around the world in 2019. *Weekly Epidemiological Record* 25: 289–292.
22. Due to severe dehydration, the skin of cholera victims often turns blue, hence the name 'Blue Death'.
23. Ten Have, H. 1983. Geneeskunde en filosofie. *De invloed van Jeremy Bentham op het medisch denken en handelen*. Lochem-Poperinge: Uitgeversmaatschappij De Tijdstroom, 29.
24. Thomas, A. J. 2020. *Cholera. The Victorian plague*. Barnsley: Pen and Sword Books, 60 ff.
25. Thomas, *Cholera*, 31 ff.
26. Ten Have, *Geneeskunde en filosofie*, 32.
27. Boshart, M. 2016. *De blauwe dood. Cholera in Nederland*. Soesterberg: Uitgeverij Aspekt.
28. Boshart, *De blauwe dood*, 177 ff.
29. Rosenberg, C. E. 1987. *The cholera years. The United States in 1832, 1849, and 1866*. Chicago/London: The University of Chicago Press, 40 ff.
30. ———. 1992. *Explaining epidemics and other studies in the history of medicine*. Cambridge: Cambridge University Press, 295. See also: Rosenberg, *The cholera years*, 75 ff.
31. Ten Have, H.A.M.J. 1990. Knowledge and practice in European medicine: The case of infectious diseases. In *The growth of medical knowledge*, ed. H.A.M.J. ten Have, G.K. Kimsma, and S. Spicker, 15–40. Dordrecht/Boston/London: Kluwer Academic Publishers.
32. Fleck, L. 1935. *Entstehung und Entwicklung einer wissenschaftliche Tatsache: Einführung in die Lehre vom Denkstil und Denkkollektiv*. Suhrkamp Verlag: Frankfurt am Main.
33. Ten Have, Knowledge and practice in European medicine, 27.
34. Ten Have, Knowledge and practice in European medicine, 32 ff.
35. "... the specific menace of cholera was a product of the Industrial Age and its global shipping networks." Johnson, S. 2006. *The ghost map. The story of London's most terrifying epidemic – and how it changed science, cities, and the modern world*. New York: Riverhead Books, 33.
36. Ten Have, Knowledge and practice in European medicine, 24.

37. Boshart, *De blauwe dood*, 130 ff.
38. Sarphati initiated the first bread factory in the Netherlands, and organized a system of waste collection in Amsterdam, among many other things. Slingeland, A. 2020. Dr. Samuel Sarphati. *Hektoen International. A Journal of Medical Humanities* 12(3); De Bruin, W. 2010. Samuel Sarphati (1813–1866). Schepper van een nieuwe stad, *Historisch Nieuwsblad* 8.
39. See, Ten Have, Knowledge and practice in European medicine.
40. See, Johnson, *The Ghost map*.
41. Huber, V. 2006. The unification of the globe by disease? The international sanitary conferences on cholera, 1851–1894. *The Historical Journal* 49 (2): 453–476.
42. WHO, *Cholera fact sheet*, January 2019.
43. Waldman, R.J., E.D. Mintz, and H.E. Papowitz. 2013. The cure for cholera – Improving access to safe water and sanitation. *New England Journal Medicine* 368 (7): 592–594.
44. Honigsbaum, M. 2020. *The pandemic century. One hundred years of panic, hysteria, and hubris*. New York: W. W. Norton & Company, 26; Brown, J. 2018. Influenza. *The hundred-year hunt to cure the deadliest disease in history*. New York: Simon & Schuster, 5.
45. Barry, J.M. 2005. 1918 revisited: Lessons and suggestions for further inquiry. In *The threat of pandemic influenza: Are we ready? Workshop summary*, edited by S. L. Knobler, A. Mack, A. Mahmoud et al, 58-68. Washington: Institute of Medicine (US) Forum on microbial threats, 61.
46. Barry, 1918 revisited, 67; Barry, J.M. 2018 (original 2004). *The great influenza. The story of the deadliest pandemic in history*. New York: Random House, 156 ff.
47. Barry, *The great influenza*, 273.
48. Morens, D.M., and A.S. Fauci. 2007. The 1918 influenza pandemic: Insights for the 21st century. *Journal of Infectious Diseases* 195 (7): 1018–1028.
49. Artenstein, A.W. 2012. The discovery of viruses: Advancing science and medicine by challenging dogma. *International Journal of Infectious Diseases* 16: e470–e473.
50. Brown, *Influenza*, 63.
51. Barry, *The great influenza*, 409.
52. Barry, *The great influence*, 544.
53. Artenstein, The discovery of viruses, e472.
54. Kolata, G. 2005. *Flu. The story of the great influenza pandemic of 1918 and the search for the virus that caused it*. New York: Simon & Schuster.
55. Žižek, S. 2020. *Pandemic! COVID-19 shakes the world*, 3. New York: Polity.
56. Morens, D.M., G.K. Folkers, and A.S. Fauci. 2009. What is a pandemic? *The Journal of Infectious Diseases* 200: 1018–1021.
57. Rosenberg, *Explaining epidemics*, 281 ff.
58. Benedictow, *The Black Death 1346–1353*, 59.
59. "… for many laypeople throughout history the term *epidemic* and *contagious* were synonymous" See; Rosenberg, *Explaining epidemics*, 295.
60. Rosenberg, *The cholera years*, 133 ff.
61. See Benedictow and his use of the war metaphor; Benedictow, *The Black Death 1346–1353*.
62. The first part of the book of Barry is devoted to Warriors; Barry, *The great influenza*.
63. Benedictow, *The Black Death 1346–1353*, 231.
64. Benedictow, *The Black Death 1346–1353*, 292.
65. Shipman, P.L. 2014. The bright side of the Black Death. *American Scientist* 102 (6): 410–413.
66. Ziegler, *The Black Death*, 252 ff; McNeill, *Plagues and people*, 193.
67. "… we live within, and are inhabited by, ecologies that teem with viruses. Put simply, life on Earth has evolved in a never-ending, profoundly symbiotic interaction with co-evolving viruses." Ryan, F. 2019. *Virusphere. Ebola, AIDS, influenza and the hidden world of the virus*. London: William Collins, 205.
68. Honigsbaum, *The pandemic century*, 8.

Chapter 3
Emerging Infectious Diseases

Abstract Warnings about new pandemics have been issued for decades. Every 8 months, a new infectious disease emerges. The cautionary discourse of emerging infections focuses on preparedness. Already in 2005, the WHO urged countries to make national bio-preparedness plans. The Covid-19 pandemic, nonetheless, overwhelmed countries; policies are late and incoherent, basic materials such as tests, face masks and protective equipment are scarce. Another consequence of the emerging diseases discourse which is often not articulated by policy-makers is prevention. Once the pandemic is expanding, attention concentrates on treatment and care, as well as the search for vaccines and drugs. Efforts will be directed at mitigating the effects of the new disease rather than exploring its origins. As discussed in this chapter, human beings and the natural world are fundamentally interdependent. Humans are enveloped in a virosphere. Viruses are indispensable for the evolution and continuation of life. When the current pandemic has waned, humanity will inevitably be affected by new infectious diseases. What is necessary is more awareness of the ecological perspective on health and disease.

Keywords China · Emerging diseases · Microbial traffic · Virosphere · Wuhan

3.1 Introduction

In November and December 2019 patients with an unexplained form of pneumonia are hospitalized in Wuhan, China. At first, it is not regarded as an unusual respiratory infection during the flu season. On 31 December 2019, Chinese health officials inform WHO about 41 patients with a viral pneumonia. Most cases are related to the Huanan seafood market in the city. Worried that it might be a new form of SARS that has emerged in 2002 and caused a severe pneumonia killing hundreds of people, the authorities close the market. On 7 January 2020, the virus causing the pneumonia is identified as a novel coronavirus. Local authorities deny that the virus is transmitted from person to person. Later reports reveal that there must have been

H. ten Have, *The Covid-19 Pandemic and Global Bioethics*, Advancing Global Bioethics 18, https://doi.org/10.1007/978-3-030-91491-2_3

many more cases in Wuhan. Only persons who tested positive for the virus were counted, and only those persons could be tested if they had been exposed to the animal market or to a known patient. From 13 January 2020, the first cases begin to emerge outside of China, in Japan, Thailand and South Korea. On January 20, with the first case reported in the United States, the Chinese President recommends measures to stop the spread of the virus, which is now admittedly spreading from person to person. Three days later the authorities impose a quarantine on the city of Wuhan and a few days later in the province of Hubei. However, the biggest annual celebration, Lunar New Year, is near, and many of the 11 million inhabitants of Wuhan travel to family across the country before the lockdown is imposed. On 30 January WHO declares a global public health emergency, and it names the new disease as Covid-19 on 11 February 2020 [1].

The response to the first cases of the epidemic disease reiterates many aspects of past pandemic experiences as discussed in the previous chapter. At first, there is denial that the infection is serious. Chinese authorities emphasize that it is not like SARS but more like influenza. They also assure that there is no transmission from person to person. Physicians are instructed not to report cases to the national alert system because they do not want to disturb the economy, want people to enjoy the upcoming New Year celebrations, and continue to have a local Party Congress in January. But doctors leak via social media reports that there is dangerous infection around. Whistleblowers such as Li Wenliang (who died of Covid-19 on 7 February) are reprimanded and silenced. Journalists who reported on the first cases in Wuhan are arrested, and at least one is convicted to 4 years in jail [2]. Not only information is kept secret, also actions are delayed. The first case in China occurred probably in October or November and not in early December as reported. The cordon sanitaire imposed on Wuhan was insufficient and allowed the virus to escape before it was implemented. Furthermore, the World Health Organization, and thus the international community, was initially not involved, and left in the dark. Only in late February an international team could travel to China and investigate the epidemic. The conclusion is that earlier action that could have slowed the pandemic would have been possible. But when after 6 weeks China acted, its measures were highly effective. In mid-February, the epidemic peaked in China, and on 19 March 2020 the authorities could announce that no new locally spread infections were reported.

By the time that Covid-19 is declared a pandemic by the WHO (11 March 2020) it has affected many countries. The first cases arrive in France on 24 January, and in Germany on 27 January. On 19 February, Iran reports a serious coronavirus outbreak, and one day later cases start to emerge in Italy. Early March sees a sharp increase of cases in Spain. On 9 March 2020, all 60 million residents in Italy go into lockdown. It is estimated that on the last day of March more than one-third of the world population is under some form of lockdown. The number of infections continues to rise, from 1 million on April 2, to 5 million on 21 May, 10 million on 28 June, to almost 84 million on 31 December, and further increasing in 2021. The global death toll increases from 110,000 (10 April) to almost 1.9 million (31 December 2020). While infections are surging in Brazil and India, several countries (e.g. Italy, Spain, Iran, Germany, New Zealand and Thailand) start to ease lockdown restrictions in May 2020, hoping that the worst of the pandemic is over [3].

Since years it is known that new pandemics will arrive, although it cannot be predicted when exactly [4]. In 2003, experts estimated that since 1980 more than 35 new infectious diseases have emerged in humans; one every 8 months. Since then the list has only grown (SARS, Swine flu, MERS, Ebola, Zika). Most of these outbreaks have been localized in specific places and regions but HIV, SARS and avian flu should have been warnings that global dissemination can easily occur. In 2005, the WHO urged countries to make national bio-preparedness plans, and many countries did. It is therefore astonishing how the Covid-19 pandemic overwhelmed countries, even more so since Covid-19 is more threatening than influenza for which vaccines and antiviral agents are available. This chapter will discuss the cautionary discourse of emerging infectious diseases. One consequence of this discourse is the focus on preparedness. If the emergence of disease threats is inevitable, the best strategy is to be prepared and to take preemptive action before the pathogens become an actual threat. This response is in line with the military metaphor that dominate the 'fight' against the pandemic. It is best to survey the enemy, follow its movements, and anticipate where it will strike. The supposition is that sufficient resources are available to counter-attack and kill the invader. This is precisely the difficulty; in many cases, the infectious agent is novel and its properties are unexplored. Furthermore, if there has not been a serious threat for a long time, vigilance diminishes and preparedness plans are not updated. Prediction does not necessarily imply preparation. Another consequence of the emerging diseases discourse which is often not articulated by policy-makers is prevention. Once the pandemic is expanding, attention will naturally concentrate on treatment and care, as well as the search for vaccines and drugs. Efforts will be directed at mitigating the effects of the new infectious disease rather than exploring its origins. Ethical statements underline the need to protect vulnerable individuals, the provision of correct information and the need for international cooperation and the importance of solidarity and protection of fundamental rights [5]. What they do not mention is the necessity of primary prevention, and thus more awareness of the ecological perspective on health and disease. As discussed in this chapter, human beings and the natural world are fundamentally interdependent. Humans are enveloped in a virosphere. Viruses are indispensable for the evolution and continuation of life.

3.2 Emerging Infections

The Covid-19 pandemic that holds the world in its grip since 2020 is "an unnecessary catastrophe." [6] A new devastating pandemic has been predicted by experts for a long time. To focus attention to the fact that infectious diseases have not disappeared, a new vocabulary emphasizes that novel or previously undiscovered diseases can 'emerge' and that diseases that once affected many people and that since then have declined, can 're-emerge.' The terminology of 'emergence' has two effects. One is the suggestion that it is difficult to built on past experiences since diseases are often unknown; they present serious challenges and request innovative approaches and perspectives. The second effect is that action is suggested; the term 'emergence' relates to 'emergency.' Particularly in the case of epidemics and pandemics health care systems and societies are threatened so that drastic responses are obligated.

The starting point for the discussion about emerging infections is the 1992 report of the Institute of Medicine in the United States. The report states that "… despite a great deal of progress in detecting, preventing, and treating infectious diseases, we are a long way from eliminating the human health threats posed by bacteria, viruses, protozoans, helminths, and fungi." [7] There is no room for indifference: "It is unrealistic to expect that humankind will win a complete victory over the multitude of existing microbial diseases, or over those that will emerge in the future." [8] Similar warnings have been made by William McNeill in his study of the role of infectious diseases in shaping human history, originally published in 1976. He argues that human life has always been characterized by a precarious balance between 'microparasitism' (i.e. a range of microbes and other disease organisms) and 'macroparasitism' (i.e. primarily human beings) [9]. Civilizations have created specific disease pools, while increasing trade contacts exported diseases to other civilizations so that over time more uniform disease pools are produced. The development of agriculture, especially irrigation farming created favorable conditions for the transmission of schistosomiasis and malaria [10]. Domestication of animals led to closer contacts with animals and infection of humans by animal microbes [11]. When the experience of cholera and other contagious illnesses stimulated practices of hygiene, new diseases of cleanliness emerged (for example, poliomyelitis) [12]. The outbreak of Legionnaires' disease in 1976 showed that technologies of air conditioning, indoor plumbing and hot water systems in large buildings such as hotels and hospitals allowed a previously unknown bacterium to multiply and infect susceptible individuals [13]. The hallmark of civilizations is cultivation and transformation of the environment. Human beings have always intervened in natural ecosystems. This has often disturbed the balance between parasitisms. The point is that such interventions nowadays are more extensive and take place at a global scale. Infectious diseases will therefore remain one of the fundamental determinants of human history [14].

3.3 The Illusion of Conquest

McNeill's admonition must be interpreted against the medical background of the 1960s and 1970s when many physicians and policy-makers believed that infectious diseases are no longer a threat to human life. Medicine now has effective medication (particularly antibiotics) and modes of prevention (especially vaccination) so that these diseases can be managed. In numerous places, infectious disease departments were closed or downsized. Countries decreased investment in public health. Priority was given to biomedical technology, genetics, and care for chronic illnesses. A critical turning point was the worldwide elimination of smallpox in 1977 [15]. It highlighted the power of the germ model that regarded diseases as the result of contagion which can be eradicated.

The idea that deadly infections are no longer a threat reflected a typical Western bias, assuming that death from infections is limited to developing countries while ageing, degenerative and chronic diseases are the most significant medical challenges in the developed world. Modern civilization has been so perfected that the environment, water, and food are sanitized and hardly provide any risk of contamination. Only moving outside these safe zones presents dangers but even these risks can be mitigated with precautionary measures. The general belief is that infectious diseases such as tuberculosis and malaria are still a major cause of death in the developing world, but that further development, and especially the progress of science and technology will undoubtedly eradicate them. The rest of the world will soon follow the West in its victory over lethal infections.

The first shock to feelings of security took place in 1976. In July, an outbreak of an atypical pneumonia affected and killed mostly elderly men participating in an American Legion convention in a hotel in Philadelphia. The disease was a puzzle for some time and created panic until its causal agent was identified in January 1977. The same year there was a limited outbreak of respiratory illness among soldiers in New Jersey. Soon, the cause was identified as influenza A virus subtype H_1N_1. Since this virus was related to the one that caused the Spanish flu, it produced the fear that these so-called swine flu cases might be the beginning of a new pandemic. Medical experts and government authorities did not want to take any risks, and took rapid action to produce vaccines and immunize the entire American population, beginning in early October 1976. But in November there were no signs of an impending pandemic. In fact, a pandemic never materialized and the vaccination campaign was halted in December, especially since an increasing number of cases has been reported as adverse effects of the vaccine, after 40 million Americans had received the vaccine [16]. The fiasco of the swine flu scare illustrates that the appearance of a new virus is no reason for aggressive intervention when scientific evidence of a pandemic is not certain. The horrific memories of the past reinforced the idea that pandemics were first of all expected from influenza.

A more serious challenge emerged in 1980 with a new virus affecting the immune system. At first it occurred in healthy individuals with a homosexual lifestyle, and was therefore labelled Gay-related Immune Deficiency until it was termed Acquired Immune Deficiency Syndrome (AIDS) in 1982 when it was clear that it also affected other populations. The new disease expanded over the globe, with (at the end of 2019) approximately 76 million infected people and 32.7 million deaths since the start of the epidemic. AIDS strongly reminded scientists, healthcare professionals and policy-makers that infectious diseases have not been conquered. It made clear that death was not merely related to ageing but that mortality is associated with the human body and human modes of behavior and relations with the environment [17]. The unfolding of this new global pandemic reiterated some experiences of past pandemics. Physicians and activists who demanded a response were initially not taken seriously. Some people regarded the disease as punishment for transgressive behavior making a distinction between 'deserving' and 'innocent' victims (gay people versus recipients of infected blood). There was a significant gap between laboratory and policy; identification of the virus did not generate preventive programs aimed at changing individual behavior. It took time before knowledge of the causal mechanism translated into effective policies to prevent transmission. When medication became available, access to it was limited, especially in developing countries. Conspiracy theories flourished with denials that the human immunodeficiency virus is the cause of AIDS and with South-African President Thabo Mbeki refusing people access to antiretroviral drugs [18].

3.4 Microbial Traffic

The HIV/AIDS pandemic brought home two lessons: the world is undeniably interconnected and all global citizens are vulnerable to infectious diseases. These dimensions of globalization are combined in the concept of 'microbial traffic.' [19] Pathogens that are present in the environment, in other species or in infected humans are introduced into a new host population and then disseminate within this population. Most infections are transmissible directly or indirectly from animals to humans. An infection is said to 'emerge' when it reaches a new population, regardless of its origin. This concept of microbial traffic assumes that most emerging infections are not really new but are due to existing pathogens (which are often not known). De novo evolution of microorganisms resulting from changing genetic properties and causing a novel disease is relatively rare [20].

Microbial traffic between species is a powerful image since it highlights aspects of the contemporary global world that are responsible for disease emergence. Ecological changes due to economic development and land use (e.g. deforestation and changes in agriculture and water ecosystems) can produce schistosomiasis and Lyme disease. These changes have brought human beings in closer contact to natural reservoirs of microbes. Human demographics and behavior (e.g. migration, urbanization, war, intravenous drug use, and sexual behavior) play a role in the spread of bubonic plague, HIV and dengue. International travel and commerce have disseminated mosquitos and cholera. Technology and industry (especially food production) have resulted in bovine spongiform encephalopathy and *E. coli* infections, while breakdown in public health measures (due to economic policies or war) has led to the resurgence of tuberculosis, diphtheria and cholera in recent times. Many of these activities are related to human behavior. There are also natural causes at work, particularly climate change. Microbes themselves can evolve and adapt to new environments, leading to resistance to antibiotics.

3.5 Virosphere

The image of traffic underlines the vulnerability of human beings to microbes. Like contemporary global citizens use to travel around the globe, and traffic goods and services in extensive international commerce, micro-organisms are trafficking with an even higher speed and to a wider extent. The reason is that human beings are embedded in natural environments; they cannot separate themselves from their biological surroundings. That means that infectious diseases due to microbial trafficking cannot be eliminated because humans are not able to control the organic world [21]. However, humans often give microbes a chance to emerge. Infectious diseases are not simply natural processes but the consequence of human actions. As McNeill pointed out, their dissemination is associated with the progress of civilizations, disturbing the fragile balance between microorganisms and humans. Like human traffic, viral traffic is increasing with new and faster technologies of transportation and mobility. Infectious diseases are not merely natural disasters but man-made calamities. This understanding is important for the approach of epidemic and pandemic diseases. It is necessary to comprehend the behavior of microbes but this is not sufficient to prevent and control diseases. The science of virology can focus on the molecular biology of viruses but this will not explain the viral spread [22].

Recent developments in virological research have brought a new perspective, broader than the one motivated by the germ theory of disease. Previous studies had concentrated on viruses that infect humans, animals or plants with often harmful consequences. It became clear that viruses are not synonymous with disease; they often have a clearly positive role in the biosphere. They are a key component of the living world. First of all, they are everywhere on Earth and especially in oceans. They are more abundant than all other life forms together. Secondly, there is enormous diversity. Estimations are that more than 100 million species of virus exist but only a limited number has been identified so far [23]. The omnipresence and diversity of the viral world is expressed in the concept of 'virosphere.' Human beings are embedded in nature while nature, including viruses, is embedded in humankind [24]. The virosphere is not only around us but within us. The healthy body is inhabited by massive numbers of viruses, particularly in the colon, skin, nasal cavity, mouth and genitalia. But viral material is also incorporated in our genes. Endogenous retroviruses constitute at least 8% of the human genome, and probably 90% of the human genome consist of virus-like genes [25].

A world without viruses would not survive. The majority of viruses are not deleterious to humans. They are indispensable for the health of the biosphere and not merely agents of disease. The view that they are parasites is too limited. Recent understanding of viruses shows that they are symbionts. In some cases, they can be aggressive and parasitic but in many other cases the symbiotic relationship is mutually beneficial ('mutualism') or benefitting only one species without harming the other ('commensalism'). Viruses are packages of genetic material that are only able to replicate within a host but they evolve faster than any other biological entity, and sharing genes with their host, they are a source of new genetic material, helping organisms to adapt and survive [26]. They should be regarded as a "creative evolutionary force." [27] An often-cited example is the evolution of the placenta in mammals some 150 million years ago. It was the result of an infection by a retrovirus, inserting itself into the genome of a cell [28]. Remnants of ancient viral infections can be found in the genomes of every living organism. Viruses are an archive of all genetic information on Earth [29].

The growing awareness of an invisible ecosphere of viruses and the scientific appreciation that viruses play a vital role in the universe, presents a perspective beyond the traditional one of viruses as causative disease agents. Only a tiny portion of the virosphere affects humans (between 200 and 320 species of viruses have human beings as their host) [30]. Most viruses are bacteriophage: they infect bacteria. Better knowledge of their behavior could be used to create new therapies for bacterial infections, overcoming the problem of resistance to antibiotics. Until recently, the focus of virology has been anthropocentric. Viruses are best understood if they are important for humans, either because they lead to disease or because they affect animals (e.g. cattle, swine and chickens) and plants (e.g. tobacco, potatoes, and yeast) connected to human economic interests [31].

3.6 Viral Threats

Given the ubiquity of viruses, the question is why some viruses are so deadly and produce pandemics such as smallpox, Spanish flu, HIV and Ebola. Excessive virulence is not a successful evolutionary strategy for viruses since they can no longer reproduce themselves if their hosts die quickly. In most cases, however, the viruses are not aggressive to the natural host from which they traffick to human beings. Infected humans are therefore not necessary for their survival. Smallpox was an exception. For the variola virus, humans were the only reservoir, and this is one of the main reasons why the disease could be eradicated. In most other cases, when the virus disappears in humans, it continues to exist and evolve in its natural hosts. The epidemic is over but the virus can re-emerge in the time to come.

Between 1940 and 2004, 335 emerging infectious diseases have been reported globally. In the majority of cases (60%), these diseases have been caused by zoonotic pathogens, and almost 72% of these pathogens originated in wildlife. Emerging diseases due to such pathogens are significantly increasing with time [32]. Although the mechanisms underlying emergence are still insufficiently known, scientific studies have at least clarified two issues. One is that infectious diseases are more likely to emerge in some geographical areas, so-called emerging disease 'hotspots,' especially in tropical Africa, Latin America and Asia. These hotspots are related to wildlife biodiversity which is also increasing towards the Equator. In these areas there is expanding land-use with deforestation, development of roads and dams, logging and mining, and intensive agriculture [33]. A second issue is that some wild animal species such as bats, primates, and rodents have a higher proportion of zoonotic viruses than other groups of mammals. Particularly bats host significantly more viruses than all other species. They have been identified as reservoir hosts of recent zoonoses such as Ebola and Marburg as well as coronaviruses causing SARS and MERS [34]. Studies have shown that 9% of bats carried at least one of 91 distinct coronaviruses [35]. Viruses are not harmful to bats. Their immune system controls the viruses but when the viruses spread to humans, they are usually much more virulent.

Better understanding of the emergence of diseases has practical implications since it can focus research on specific regions and species. The idea of global emerging disease hotspots directs attention to geographical regions where new diseases are most likely to originate. This implies that potentially new viral threats can be detected with targeted surveillance. However, currently most scientific efforts of research and surveillance are directed to places (e.g. Europe and North America) from where the next deadly pathogen will most probably not emerge. Predicting future emerging disease will require a substantial reallocation of resources to hotspots in Africa, Latin America and Asia [36]. The finding that specific species are natural reservoirs of virus encourages efforts to predict zoonotic spillover with field programs to sample wildlife and identify host species which are likely to harbor the next emerging virus, and to discover novel pathogens and determine which viruses can cross species boundaries [37]. However, such field programs have been reduced in the last decennia. For example, in November 2019 the French Ministry of Foreign Affairs eliminated the position of a professional virologist at the Institute Pasteur in Laos (only to be reinstated in March 2020) [38].

3.7 The Source of Covid-19

Early January 2020 Chinese scientists isolated the virus that causes Covid-19. Since the virus has a distinctive crown of proteins, spikes, it was identified as belonging to the family of coronavirus. Gene sequencing revealed that the virus is genetically similar to the virus that caused SARS in 2002, so that the two viruses belong to the same species. In March virologists named the virus SARS-CoV-2. In January, Chinese physicians reported on the clinical features of 41 patients admitted to a designated hospital in Wuhan. Of these patients, 66% had been exposed to the Huanan seafood market; 13 were admitted to the ICU and 6 died [39]. Although the zoonotic source of the virus is not definitely known, all viruses isolated from humans so far are genetically close to coronaviruses isolated from bat populations. This finding has led to a growing scientific consensus that this virus has its ecological reservoir in bats [40]. The most likely hypothesis is that the virus crossed from these animals to humans.

The initial assumption was that the source of the outbreak was the market in Wuhan. But since the market has been closed in 1 January 2020 and thoroughly disinfected, further exploration is impossible. Furthermore, not all patients had a connection to the market. It is also not likely that bats on the market have directly infected humans. Perhaps an infected patient has introduced the virus in the market.

Another possibility is that the virus is transmitted through an intermediary host. Pangolins have been mentioned as possible intermediaries in the transmission to humans. They are the most frequently illegally trafficked animals and used as food and as basic ingredient in traditional Chinese medicine [41]. Soon, various wild and domestic animals were detected to be potential intermediary hosts, such as ferrets, minks, and cats [42].

Already in 2007, scientists in Hong Kong concluded that the presence of SARS-CoV-like viruses and other novel viruses from the large reservoir in horseshoe bats is a "time bomb." [43] China is known as a hotspot for bat coronaviruses. Several transmissions of SARS-like viruses from bats to humans had been detected in the recent past. In 2011, a novel virus had been isolated from horseshoe bats in the province of Yunnan and further studied in the Wuhan Institute of Virology. Despite warnings about the threat of coronavirus emergences, no vigorous monitoring in order to predict and prevent emerging disease has been undertaken. Now that an aggressive coronavirus has produced a pandemic, understanding how it has emerged is of crucial importance for predicting and preventing new pandemics. Future coronavirus transmissions into human populations are not merely possible but to be expected [44].

The question of the source or origin of Covid-19 has become more complicated since pandemics, as discussed in the previous chapter, are never merely microbiological or medical events. They also have social and political dimensions. Initially, most scientists agreed that Covid-19 has its origin in a natural source, pre-adapted to humans and becoming a virulent agent in this host or an intermediary before zoonotic transfer to humans. It can also be the case that the virus jumped to human beings and developed its virulent capacities after zoonotic transfer. Reports that the virus already circulated in November 2019 seem to confirm this scenario, acknowledging that many people can be infected but not showing any symptoms. That Covid-19 has a natural origin corresponds to the experience that most emerging infectious diseases start with a spillover from nature. The difficulty is that scientists so far have not found a virus in bats or other animals with the same genetic characteristics of SARS-CoV-2. Recently, three viruses have been found in bats in North Laos that are more than 95% identical to SARS-CoV-2 [45]. While the Wuhan market was a superspreading event, the virus might therefore have originated outside of China, or at least in Southeast Asia which is a hotspot of diversity for related viruses, leaving it still a mystery how it or its progenitor travelled to Wuhan [46].

The issue where and how the virus originated is not simply a neutral fact. It is related to questions of culpability and blame; references to the 'Chinese virus' are used to blame Asian people, and to denounce their supposed lack of hygiene and exotic food habits. On the other hand, the Chinese government emphasizes that the virus first occurred outside China and was imported via contaminated food products. Whatever the scientific theories, a politically inspired discourse on the origin has emerged: SARS-CoV-2 is an artificial construct. The suspicion is that the virus is fabricated in a Chinese laboratory and has intentionally or accidentally been released from this setting. Although genetic analyses have shown that the new virus is not man-made and that viral escape from a laboratory is unlikely, these speculations are widely disseminated [47]. They are part of a set of conspiracy theories blaming China for the origin of the pandemic. These theories are actively promoted by American officials, especially in the previous Administration, emphasizing Chinese responsibility for the disease. The effect of the theories is to undermine the credibility of science and to reduce the willingness of individuals to adhere to public health measures and to engage in behavior that can mitigate the spread of the virus [48]. On the other hand, the Chinese authorities after initially concealing the emergence of an epidemic, later strongly endorse the natural origin of the epidemic in order to promote a favorable (global) public opinion on how they have dealt with the pandemic and to promote the idea that, like anybody else, Chinese citizens are victims of a new disease. At the same time, they not only emphasized scientific knowledge but also propagated conspiracy theories, for example that the virus has originated in the United States [49]. However, the secretiveness of the Chinese government, the suppression of information, and the unwillingness to cooperate with an in-depth investigation of the origins of the pandemic have enhanced the mystery of how Covid-19 could spread around the world. Although the scientific community as well as the WHO continue to advance the need for further exploration of its origins, almost 2 years into the Covid-19 pandemic it is still an enigma where and how the outbreak emerged [50].

While it is important to reconstruct where and how SARS-CoV-2 emerged, the focus on a specific geographical location is too restrictive. In this chapter it is argued that new infectious diseases emerge within particular ecological constellations; economic development, exploitation of land, expanding agriculture, urbanization and migration have reduced the natural habitats of pathogens and increased microbial traffic to humans. China is one of the world's most rapidly growing economies, and one of the prime examples of modern industrialization and global capitalism [51]. It has seen massive rural-urban migration and heavy pollution of the environment. It is the world's largest producer of meat and poultry. Many wildlife species are now industrially farmed and produced for the commercial market. The country has also developed a sophisticated scientific and technological research community, and has applied neoliberal policies in many sectors of society, including privatization of hospitals, operating for profit. In all these aspects, China is embedded within and often at the forefront of global capitalism and neoliberalism. Understanding the emergence of a new infectious disease is therefore more complex than identifying a wet market, a laboratory, an animal, or food product. It should take into account that such emergence is related to human activities, specifically global capitalist practices that are not unique for China. In order to prevent the further emergence of pandemic diseases, an ecological perspective is required that examines the conditions and structures that promote microbial traffic. Such perspective also reveals that the strenuous efforts of the Chinese state and society to control the pandemic with biomedical and technocratic interventions and surveillance conceal these structural conditions because threats are not prevented or eliminated but primarily made 'governable' [52].

3.8 Conclusion

The SARS-CoV-1 virus that caused the SARS pandemic in 2002 was a dangerous virus with a high fatality rate (10%). It disseminated from human to human by contact, but only when people had symptoms [53]. The SARS-CoV-2 virus, though less lethal, is more dangerous since a large number of infected people have no symptoms, so that it can spread rapidly without any visible symptoms. The Covid-19 pandemic highlights the difference between infection and disease. Infection occurs when a pathogen has trafficked to a host and is multiplying within the host. It is not necessarily or not yet associated with signs of disease. Historically, a disease was recognized on the basis of symptomatology while the causative agent was unknown or detected much later. Nowadays, symptoms of disease and microbes are determined simultaneously. In the Covid-19 pandemic the causal virus is discovered very quickly. The tragedy is that such knowledge does not lead to fast and adequate responses. This chapter examines how despite all predictions and warnings, epidemic and pandemic diseases usually present surprises. Really 'new' viruses are extremely rare; they usually come from closely related pre-existing ones, and they have therefore antecedents and relatives. The main concern should be how an existing virus in a host species would be able to cross over into humans, articulating viral traffic rather than viral origin. Since humans are embedded in a virosphere, they cannot reject the ultimate reality that there is always a risk for a lethal infection. This does not imply fatalism. There is a need for policies and research strategies to understand and recognize viruses that exist in nature giving us some advance warning. What kind of viruses are there and under what conditions are they introduced and disseminated in human beings? The first step is therefore effective global surveillance, identifying the conditions that promote viral traffic. It is also argued in this chapter that infectious diseases have wider environmental and social causes, and are not just the effect of a specific pathogen. These diseases emerge because the environments in which pathogens reside are disturbed. Programs of surveillance and prevention early in the chain of emergence, close to their source, can permit early intervention and measures that limit the spread of disease as soon as possible. However, as the subsequent chapters will show, public policies are generally indifferent to the point of departure of pandemics, and mainly interested in their point of arrival. Once the disease is expanding, the focus is not on emergence but on emergency, i.e. what is happening after it has occurred.

References

1. Mackenzie, D. 2020. *Covid-19. The pandemic that never should have happened, and how to stop it*. London: The Bridge Street Press.
2. Davidson, H. 2020. Wuhan Covid citizen journalist jailed for four years in China crackdown. *The Guardian*, December 28.
3. Horton, R. 2020. *The Covid-19 catastrophe. What's gone wrong and how to stop it happening again*. Cambridge: Polity.

4. Wolfe, N. 2011. *The viral storm. The dawn of a new pandemic age*, 15–16, 237 ff. New York: St. Martin's Press; Khan, A.S. 2020. *The next pandemic. On the front lines against humankind's gravest dangers*. New York: PublicAffairs.
5. An example is: UNESCO. 2020. *Statement on Covid-19: Ethical considerations from a global perspective*. Paris: UNESCO.
6. Mackenzie, *Covid-19*, 225.
7. Lederberg, J., R.E. Shope, and S.C. Oaks, eds. 1992. *Emerging infections: Microbial threats to health in the United States*, 27. Washington, DC: National Academies Press.
8. Lederberg, et al. *Emerging infections*, 32.
9. McNeill, W.H. 1998. *Plagues and people*, 24. New York: Anchor Books.
10. McNeill, *Plagues and people*, 62 ff.
11. "Most and probably all of the distinctive infectious diseases of civilization transferred to human populations from animal herds." McNeill, *Plagues and people*, 69.
12. Because immunizing poliomyelitis infections in early childhood were less common in middle-class homes, children from such more pristine homes were more at risk to develop serious complications than those in poor immigrant neighborhoods. Honigsbaum, M. 2020. *The pandemic century. One hundred years of panic, hysteria, and hubris*. New York: W. W. Norton & Company, 6.
13. Honigsbaum, *The pandemic century*, 187 ff.
14. More knowledge and better organization "cannot cancel humanity's vulnerability to invasion by parasitic forms of life." McNeil, *Plagues and people*, 295.
15. The WHO certified in 1979 that smallpox has been eradicated. See: Hopkins, D.R. 2002. *The greatest killer. Smallpox in history. With a new introduction*. Chicago/London: The University of Chicago Press.
16. Honigsbaum, *The pandemic century*, 175 ff; Kolata, G. 2005. *Flu. The story of the great influenza pandemic of 1918 and the search for the virus that caused it*. New York: Atria Paperback.
17. Rosenberg, C.E. 1992. *Explaining epidemics and other studies in the history of medicine*, 278 ff. Cambridge: Cambridge University Press.
18. Honigsbaum, *The pandemic century*, 225 ff.
19. Morse, S.S. 1991. Emerging viruses: Defining the rules for viral traffic. *Perspectives in Biology and Medicine* 34 (3): 387–409; Morse, S.S. 1995. Factors in the emergence of infectious disease. *Emerging Infectious Diseases* 1: 7–15.
20. The importance of new variants is best demonstrated in influenza. Lederberg et al., *Emerging infections*, 42–43.
21. "…infectious diseases are … the price we pay for living in an organic world." Morse, *Emerging viruses*, 387.
22. Morse, *Emerging viruses*, 405.
23. In April 2019, researchers had identified 195,000 species of virus; Goodman, J.R. 2020. Welcome to the virosphere. *New Scientist* 245 (3264): 41; Only approximately 7,000 species have received an official name. But nowadays with new methods and techniques (especially metagenomics) increasing numbers of viruses are identified; Zimmer, C. 2020. Welcome to the virosphere. *New York Times*, March 24.
24. Suttle, C. 2005. The viriosphere: The greatest biological diversity on Earth and driver of global processes. *Environmental Microbiology* 7 (4): 481–482; Ryan, F. 2020. *Virusphere. Ebola, AIDS, influenza and the hidden world of the virus*. London: William Collins, 3; Rohwer, F., and K. Barott. 2013. Viral information. *Biology and Philosophy* 28 (2): 283–297.
25. Hamilton, G. 2008. Welcome to the virosphere. *New Scientist* 199 (2671): 40; Zimmer, C. 2021. *A planet of viruses*. Chicago/London: The University of Chicago Press.
26. Goodman, *Welcome to the virosphere*, 42.
27. Hamilton, *Welcome to the virosphere*, 38.
28. Ryan, *Virusphere*, 207 ff.
29. Suttle, *The viriosphere*, 481.
30. Zimmer, *Welcome to the virosphere*; Rodrigues, R.A.L., A.C. dos S.P. Andrade, P.V. de M. Boratto, et al. 2017. An anthropocentric view of the virosphere-host relationship. *Frontiers in Microbiology* 8: 1673.

31. Rodrigues, et al. 2017. *An anthropocentric view of the virosphere-host relationship.*
32. Jones, K.E., N.G. Patel, M.A. Levy, A. Storeygard, D. Balk, J.L. Gittleman, and P. Daszak. 2008. Global trends in emerging infectious diseases. *Nature* 451 (7181): 990–993; Wolfe, *The viral storm*, 64, 101.
33. Allen, T., K.A. Murray, C. Zambrana-Torrelio, S.S. Morse, C. Rondini, M. Di Marco, N. Breit, K.J. Olival, and P. Daszak. 2017. Global hotspots and correlates of emerging zoonotic diseases. *Nature Communications* 8 (1): 1–10.
34. Olival, K.J., P.R. Hosseini, C. Zambrana-Torrelio, N. Ross, T.L. Bogich, and P. Daszak. 2017. Host and viral traits predict zoonotic spillover from mammals. *Nature* 546 (7660): 646–650.
35. Burki, T. 2020. The origin of SARS-CoV-2. *The Lancet Infectious Diseases* 20 (9): 1018–1019.
36. Jones, et al. *Global trends in emerging infectious diseases.*
37. Olival, et al. *Host and viral traits predict zoonotic spillover from mammals.*
38. Schlegel, T. 2020. *Didier Sicard: "Il est urgent d'enquêter sur l'origine animale de l'épidémie de Covid-19".*
39. Huang, C., Y. Wang, X. Li, et al. 2020. Clinical features of patients infected with 2019 novel coronavirus in Wuhan, China. *Lancet* 395: 497–506.
40. WHO. 2020. *Origin of SARS-CoV-2.*
41. Lam, T.T., M.H. Shum, H.-C. Zhu, et al. 2020. Identifying SARS-CoV-2 related coronaviruses in Malayan Pangolins. *Nature* 583 (7815).
42. Kumar, O.R.V., B.S. Ramkumar, B.S. Pruthvishree, et al. 2020. SARS-CoV-2 (Covid-19): Zoonotic origin and susceptibility of domestic and wild animals. *Journal of Pure and Applied Microbiology* 14 (suppl 1): 741–747.
43. Cheng, V.C.C., S.K.P. Lau, P.C.Y. Woo, and K.Y. Yuen. 2007. Severe acute respiratory syndrome coronavirus as an agent of emerging and reemerging infection. *Clinical Microbiology Reviews* 20 (4): 683. See also: Menachery, V.D., B.L. Yount Jr., K. Debbink, et al. 2015. A Sars-like cluster of circulating bat coronaviruses shows potential for human emergence. *Nature Medicine* 21: 1508–1513.
44. Morens, D.M., J.G. Breman, C.H. Calisher, et al. 2020. The origin of Covid-19 and why it matters. *American Journal of Tropical Medicine and Hygiene* 103 (3): 955–959.
45. Mallapaty, S. 2021. Laos bats host closest known relatives of virus behind Covid. *Nature* 597: 603.
46. World Health Organization, *WHO-convened Global Study of Origins of SARS-CoV-2: China part*, 7; Mallapaty, S. 2020. Where did COVID come from? WHO investigation begins but faces challenges. *Nature* 587: 341–342; Mallapaty, S. 2020. Meet the scientists investigating the origins of the Covid pandemic. *Nature* 588: 208; Lei, R., R. Qiu, and P. Jia. 2021. WHO-China report on Covid: Important step forward, more to be done. *The Hastings Center*, April 9.
47. Andersen, K.G., A. Rambaut, W.I. Lipkin, E.C. Holmes, and R.F. Garry. 2020. The proximal origin of SARS-CoV-2. *Nature Medicine* 26: 450–452.
48. Bolsen, T., R. Palm, and J.T. Kingsland. 2020. Framing the origins of Covid-19. *Science Communication* 42 (5): 562–585.
49. Lemus-Delgado, D. 2020. China and the battle to win the scientific narrative about the origin of Covid-19. *Journal of Science Communication* 19 (05).
50. This issue will be further discussed in Chapter 5 (paragraph 5.4.2).
51. Zhang, L. 2021. *The origins of Covid-19. China and global capitalism*. Stanford: Stanford Briefs.
52. Zhang, *The origins of Covid-19*, 21.
53. Abraham, T. 2007. *Twenty-first century plague. The story of SARS*. Baltimore: Johns Hopkins University Press.

Chapter 4
Diverging Policy Responses

Abstract The current pandemic has reactivated ancient metaphors (especially military ones) but also initiated a new vocabulary: flattening the curve, social distancing, lockdown, self-isolation, and sheltering in place. In this chapter, policy responses to Covid-19 are examined. The epicenter of the first wave of Covid-19 is China and other Asian countries, then moving to Europe and North America, and next to Latin America. This movement in principle leaves time for preparation and exchange of experiences so that policies can be adapted and refined. In practice, however, learning effects are slow and limited, and policy responses diverse and heterogeneous. The question is why such varieties occur. All countries are faced with similar challenges, particularly the need for testing, shortages of protective equipment, and lack of coordination and coherence. Nonetheless, every country is inventing the wheel and experiences in other countries are hardly taken into account. More astonishing is that even exposure to the first wave of the pandemic does not much improve policy responses to subsequent waves.

Keywords Cooperation · First wave · Lockdown · Pandemic policies · Policy strategies · Netherlands · Shortages · Testing

4.1 Introduction

Two months after the outbreak of Covid-19 in China, Europe and North America became epicenters of the pandemic. Italy was the first country in the Western world where the pandemic exploded. Like in most countries, the events in China were first denied. Since Italy had not faced any serious outbreaks in recent time, it had no experience in dealing with a massive epidemic. It was also unprepared, although a national plan against pandemics had been issued in 2005. But this plan was simply forgotten and its recommendations (for example, stocking personal protective equipment) were never implemented. The first two cases in Italy concerned Chinese tourists on January 31. Direct flights from China were stopped and a state of health emergency was declared. No further action was taken. Health protocols focused on

H. ten Have, *The Covid-19 Pandemic and Global Bioethics*, Advancing Global Bioethics 18, https://doi.org/10.1007/978-3-030-91491-2_4

control of symptomatic patients who had been in contact with China. Things changed with the discovery of the first local cases (without any connection to China) in Codogno in Lombardy on February 20. The region ordered a lockdown in some municipalities. On March 8, Lombardy and other provinces in northern Italy (with 16 million people) were locked down. A few days later national stay at home orders were promulgated, and all non-essential industries and businesses halted. Only on May 4 the national lockdown was relaxed, lifted on May 17, while all mobility was allowed on June 3. The policies to mitigate the transmission of the virus were therefore incremental. A national lockdown on March 22 became unavoidable when the hospital system was completely overwhelmed. A few days later, the country counted more than 74,000 confirmed cases and 7500 deaths. Shocking images of Italian care facilities circulated in the world media. The majority of hospital beds were occupied by Covid patients, emergency rooms crowded, patients lying in corridors, doctors having to decide who would have ventilator treatment, and many health professionals ill and dying because of lack of protective equipment [1].

Italy was not only the first country in the West ravaged by coronavirus, it was also more severely affected than other countries. At the end of March, 9.8% of the total population was presumably infected (compared to 3.0% in France and 0.72% in Germany; only somewhat later Spain was more infected with 15%) [2]. When the lockdown was lifted in May, Italy had reported almost 230,000 cases and 32,876 deaths (data on May 24), while China had reported 92,015 cases and 4547 deaths. The reasons for the high mortality in Italy are not clear. It is suggested that one explanation is the older population (23% of Italians is over 65). Also, the risks in nursing homes for older adults (often with multiple comorbidities) were underestimated. For example, the political authorities of Lombardy appealed to nursing homes to admit patients with less severe forms of Covid in order to create space in hospitals [3]. Furthermore, there was a desperate lack of protective materials for health professionals. In general, the country was unprepared for a disaster of this scale. Policies were incoherent and slowly adopted and implemented. It is surprising to see that many patterns of previous pandemic experiences are repeated [4]. Initially there is a failure to recognize the magnitude of the threat and to organize a systematic response. Policy-makers are skeptical and do not listen to experts warning for a catastrophe. In February many public events take place in Lombardy such as agricultural fairs and a soccer match bringing together thousands of people. Measures are only partially and gradually enforced, while the virus is spreading exponentially. Policies follow the virus rather than preventing it. The decision to close northern Italy produces a massive exodus to southern Italy, spreading the virus to untouched areas. Hospitals are unprepared and only slowly reorganized, creating separate Covid and non-Covid care tracks. Preventive materials are lacking. A coherent system of testing, contact tracing and tracking movements of people is absent. At the same time, regions that have the responsibility for health care, take different approaches. The region of Veneto has much less cases and deaths because it adopts a proactive approach with early extensive testing of symptomatic and asymptomatic cases, tracking of potential positive cases, testing everybody near these infected persons, and a strong emphasis on home diagnosis and care with monitoring and protecting healthcare providers and other potentially exposed workers. In Lombardy the approach to testing is more conservative, focused only on symptomatic cases, with limited investment in tracing, home care, and protection of healthcare professionals [5].

Soon after the outbreak in Italy, the pandemic expands over Europe which becomes the epicenter in March. The question is whether the Italian experiences have been instructive for the policies in other countries. Italy itself has not learned from the lessons from China where stringent containment measures had reduced new cases by more than 90%. In Italy, it is evident that the disease severely affects the elderly and people with underlying conditions such as diabetes, cardiovascular disease, cancer and obesity. The mean age of patients who died is 81 years. Between 9% and 11% of infected patients need intensive care. Approximately 20% of the healthcare professionals are infected and some die. Ventilators, hospital beds, protective equipment but also doctors and nurses are in short supply [6]. The Italian experience also shows that the effect of the pandemic is substantially underestimated since a large number of people is infected and not tested. The official numbers of infected cases and deaths are not representative because people die outside of hospitals, untested and untreated. People also die from other causes since regular care is suspended. During March, April and May the excess mortality (above the normally expected mortality) is 44,000, and thus higher than the officially registered deaths from Covid-19 [7]. Italy moreover reiterates, although late, the lessons from China that the adoption of restrictive measures to limit virus diffusion is effective. However, it also illustrates that policy errors and delays can turn the pandemic into a disaster. Incoherent and disconnected actions of national and regional governments, each with their own bodies of experts (primarily doctors, virologists and epidemiologists), delay the design and implementation of effective interventions, while best practices (such as in Veneto) are not shared [8]. Finally, the experiences in Italy highlight significant ethical challenges. Physicians have to work below the normal standard of care. They must cancel and postpone surgery and procedures, leaving other categories of patients without adequate care and treatment. Often, protective materials are not available so that care providers have to decide whether or not to take serious risks or abandon care. They frequently watch patients die because ventilatory support cannot be provided. Patients often expire in complete isolation and cannot see their friends and family. Since ventilators are scarce, allocation decisions must be taken, often using age as a decisive criterion [9].

In this chapter, policy responses to Covid-19 will be examined. In contrast to natural disasters, pandemics do not hit everybody within a particular area at the same time and with similar severity, but progress gradually and come in waves expanding in some regions and declining in others. The epicenter of the first wave of Covid-19 is China and other Asian countries, then moving to Europe and North America, and next to Latin America. This movement in principle leaves time for preparation and exchange of experiences so that policies can be adapted and refined. In practice, however, learning effects are slow and limited, and policy responses diverse and heterogeneous. The first paragraph of this chapter will discuss how policies developed within the Netherlands, showing that within countries uncertainties and inconsistencies are prevalent. The next paragraph will analyze policy differences between countries. This analysis will raise the question why such varieties occur. Possible answers will be explored in examining three global challenges in subsequent paragraphs: the need for testing, shortages of protective equipment, and lack of coordination and coherence. In fact, every country is inventing the wheel and experiences in other countries (notably China and Italy) are hardly taken into account. More astonishing is that even exposure to the first wave of the pandemic does not much improve policy responses. Experiences with the disease have led to better medical insights and treatments; virologists obtained more sophisticated knowledge of the virus, while numerous efforts are started to develop vaccines. But in summer 2020 many countries assume that the worst is over, and relax their stringent measures, with the inevitable result that in the autumn a second wave of the disease occurs. The resurgence of the virus illustrates the persistent controversy between public health and social and economic interests, as well as between individual and collective approaches.

4.2 Variability of Policies Within Countries; The Example of the Netherlands

On January 18, 2020 the National Institute for Public Health and the Environment (RIVM), a Dutch research institute acting as an independent agency of the Ministry of Health, Welfare and Sport, and the main organization providing advice and information of the new coronavirus to the government, issues a statement that there is a small chance for the coronavirus to appear in Europe. Since there are no direct flights from Wuhan to the Netherlands, the risk of infection is minimal. The Institute has an Outbreak Management Team (OMT) operating when there is an epidemic disease in the country. On 24 January this team has its first meeting and recommends to make the coronavirus disease a reportable disease. It does not advice to start entry screening at airports. The general idea is that if the virus is spreading in the country it will be limited to a handful of cases since the virus is not contagious and municipal health services know what to do: contact tracing and isolation of infected persons. In February, the virus is already spreading undetected. Many people are on vacation, also in Northern Italy, or celebrating Carnival (23–25 February). With the experiences in Italy, some experts in the southern provinces warn that there might be capacity problems in healthcare but RIVM experts maintain that the country is sufficiently prepared and that containment will be successful, based on experiences with known infections such as influenza. The severity of the outbreak as watched in other countries is comparable to the flu season. They also advise against the use of face masks, and enormous quantities of personal protective equipment are flown to China, to help it cope with the pandemic. Testing for people without symptoms is regarded as useless. However, on 21 February the first patient with confirmed Covid is admitted to the ICU in Gorinchem, only later found to be tested positive for the virus. At the end of the month, municipal health services in the south again warn that testing and tracing capacity is insufficient due to previous budget cuts and insufficient numbers of personnel. Around the same time, the OMT issues a second report, concluding that it may be too late for containment; it indicates that there might be shortages of test materials. In March, policies shift from containment to mitigation, aiming at maximum control rather than eradication of the virus in the population [10]. In the beginning of the month, when 23 cases of infection become known, especially in the southern part of the country, the ministerial crisis management committee first meets. The RIVM advises people in the south to limit contacts if they have symptoms. The government announces as extra measure that people should no longer shake hands. After the WHO has declared a pandemic, the RIVM urges that it is no longer obligatory to report suspected cases but only confirmed cases. Public opinion polls reveal that 60% of the population thinks that the media exaggerate the threat of the virus [11].

On March 12 when 121 confirmed cases and the first death are reported, the government in a press conference announces that meetings of more than 100 people are forbidden, that citizens are encouraged to work from home and to stay home if they have symptoms. Persons with mild symptoms will no longer be tested. The option to test healthcare workers is rejected by Parliament. Policy measures become more stringent a few days later: schools, children day care centers, restaurants, gyms and schools are closed. The rule to keep 1½ meter distance is introduced. In this stage containment is rejected because it requires a lockdown with deleterious economic and social effects. The official policy is based on obtaining herd immunity. While waiting for vaccines and treatments, the only thing is to hope that population immunity will be built. This strategy requires that 50% of the population is infected. Since the expected death rate is 1–2% this will lead to 85,000–170,000 deaths. On March 20, the King makes a television address stipulating that the virus cannot be stopped, only its circulation can be delayed. The same day, a ban on visiting elderly homes is issued, and a few days later a so-called 'intelligent lockdown' is announced. Instead of a strict lockdown targeted measures are taken: schools and universities are closed as well as contact businesses (e.g. hairdressers and dentists) but shops and markets remain open as long as people maintain distance. Experts argue that a test, trace and isolation strategy (for example used in Singapore) is impossible because it requires a large number of tests and has a high risk of later re-introduction. It is clear that testing capacity is insufficient and the Minister of Health promises to increase this capacity (from 4000 test per day to at least 17,500).

In April 2020 the peak of the pandemic is reached with more than 1000 cases each day, and the highest mortality (234 deaths on April 8). Hospitals and ICUs can barely cope; patients are transferred to other provinces and to Germany. Regular care is postponed or cancelled. Although RIVM experts still defend the herd immunity strategy, it is no longer mentioned as a goal of policy measures. The goal now is protecting the vulnerable as well as healthcare capacity. In May, the number of patients in ICUs is going down. The government announces a timetable for relaxation of measures, although some experts argue that a longer lockdown is necessary together with intensive testing and tracing. Use of face masks in public transport is made mandatory while experts explain that this policy decision is not based on scientific evidence. On May 11, primary schools, children day centers, libraries, and contact professions are allowed to open (masks not obligated). Beginning of June high schools, restaurants, pubs, theaters and museums reopen for a maximum of 30 visitors.

Mid-May a study is published estimating excess mortality for the previous 9 weeks at almost 9000. Since 5634 confirmed Covid deaths has been reported, it means that probably 37% of Covid deaths are not reported as such (or that non-Covid patients died because regular care has been postponed). The testing rate is still low (with a maximum average of 5729 per day in April and May). Guidelines are released to expand the contact tracing capacity of the municipal health services. While major efforts are made to expand the number of ICU beds, it becomes clear that 60% of personnel working in elderly homes and home care lack protective equipment. In May limited outbreaks occur in slaughter houses and schools. In June, infections are discovered in breeding farms for minks. On June 5, the government launches the coronavirus dashboard providing information about ICU admissions, people tested positive, estimation of the number of infected people, and the estimated reproduction rate of the virus. The same day the ministerial crisis structure is abandoned. The OMT is meeting for the last time. People are advised not to go on holiday to countries with a high infection rate. At that time, the daily number of confirmed cases goes down to 209, with 13 deaths (the number of deaths will go to zero in Mid-July). The first wave of the pandemic is over and people can relax and enjoy the summer holiday.

The Netherlands has one of the highest mortality rates per hundred thousand inhabitants in Europe in Spring 2020 due to Covid-19. Analysis of the crisis management exposes some familiar characteristics [12]. The threat of the virus is not taken seriously; it is compared with the seasonal flu and viewed as a Chinese problem. Authorities articulate that the virus is not contagious. Early warnings about healthcare capacity problems are ignored. Testing practices are deficient. Experts are amazed about the stringent measures taken in China and Italy. There is a failure of imagination, underestimating the threat. A lockdown is regarded as an archaic and autocratic approach from the history of plague and Spanish flu [13]. Mass events such as Carnival, soccer matches and winter holidays can continue. Experts and policy-makers generally assume that the country is well prepared, and that the first priority is to protect the economy and prevent harm to social life. In hindsight, February is a lost month. After initial delay, the authorities are surprised about the speed with which the virus is disseminating. In March, the conclusion is that it is too late for containment. Defeatism settled in and no serious effort is undertaken to stop the virus. Given the speed of transmission, testing and contact tracing are no longer thought feasible, although the experience in Germany shows otherwise. For some time, the strategy is aimed at herd immunity but this is no longer articulated since antibody studies demonstrate that only a very limited number of people show immunity. The assumption is that the virus will disappear after some time without much intervention so that a painful choice between public health and economic well-being can be avoided. Rather than a total lockdown as in China, Italy, and also Spain and

France, the Netherlands introduces a partial lockdown with the least possible restrictions of the freedom of citizens and businesses.

4.3 Variability of Policies Between Countries

Europe became the epicenter of the pandemic because of the inertia of governments that underestimate the rapidity of the spread of the disease. At first very limited measures are taken, and more stringent and widespread measures delayed. Health ministers in most countries repeat that everything is under control. The same message is initially propagated by the Director-General of WHO: the international spread of the disease appears to be minimal and slow. During the first 14 days after the first confirmed cases, the increase of infected persons is similar in China and Europe but the number continues to rise exponentially in Europe when it decelerates in China. Within hundred days since its emergence the virus has frozen international travel, confined most human beings to their homes, and extinguished economic activity. In March 2020, more than 170 countries are affected by the virus. France and Spain join Italy in imposing lockdowns. Later in March India and South Africa go into lockdown. In April some European countries start to relax restrictions. At that time, the United States is the country with most cases of Covid infections, followed by Brazil, India and Russia. Early June, New Zealand is the first country to return to pre-pandemic normality.

From the beginning of the pandemic, the World Health Organization repeats the message of tracing, testing and treating. Already early January it publishes a package of guiding documents for infection prevention and control, as well as laboratory testing. The recommendations emphasize the traditional public health methods: active case finding, immediate testing and isolation, contact tracing and rigorous quarantine of close contacts. Many countries ignore the advice of WHO [14]. However, these methods are followed by Asian countries. They had worked well during the SARS epidemic of 2002. Person-to-person transmission could be interrupted through isolation, quarantine at home or in designated facilities, social distancing and community containment. These measures were mandatory, rigorously applied, restricting the movements of everyone, and enforced with checkpoints and legal penalties [15]. Countries in East-Asia with the recent experience of SARS had expanded their capacity to deal with infectious diseases. They had created organizational and administrative structures to manage a crisis with a centralized policy structure. For example, Taiwan established a National Health Command Center in 2004 [16]. They also had made the healthcare system more resilient with designated hospitals with extra beds for patients with infectious diseases. Singapore, to mention another example, built a national center for infectious diseases, completed in May 2019. Countries also developed capacities for testing and extensive contact tracing. Moreover, they held regular exercises so that procedures and skills were trained [17].

Rigorous lockdowns, forced quarantine, travel restrictions and massive testing allowed China to effectively control the coronavirus disease. Countries close to China took less draconic measures but they acted quickly with travel restrictions, quarantines, self-isolation, social distancing and heightened hygiene [18]. Singapore cancelled flights from Wuhan, and placed all travelers from affected areas in mandatory quarantine in special facilities. The city-state intensified contact tracing and surveillance to identify as many cases as possible. Testing capacity was rapidly scaled up, drive-through testing stations were set up, so that all suspected cases could be tested. Testing and medical care for all citizens was free. The use of face masks was strongly recommended, and the government distributed masks to every household. Schools were not closed and major social distancing measures were not implemented. Although the number of cases increased, there was no exponential growth [19]. Taiwan used home quarantine enforced with strict penalties. Borders were closed and individuals were tracked on the basis of their recent travel history. Face mask production was rapidly increased and masks were distributed among the population. Schools were only closed for 2 weeks. However, there was not large-scale testing [20]. Hong Kong did not close its borders but imposed travel restrictions, requiring mandatory self-quarantine for people arriving from mainland China, and rapidly established special quarantine facilities. Schools were closed for a limited time. Almost everyone used face masks in public. South Korea had experienced in 2015 an outbreak of the Middle East Respiratory Syndrome (MERS, also caused by a coronavirus). It quickly expanded diagnostic capacities and extensively tested, traced and isolated suspected cases. Domestic testing kits were quickly developed and approved. A lockdown was not imposed so that restaurants and businesses remained open [21].

In the above countries the pandemic was taken seriously from the beginning. The threat was rapidly recognized and the response was quick, even before the first case was diagnosed. The countries had built adequate capacities and management structures prior to the emergence of the crisis because of recent experiences with SARS and MERS. Although only South Korea had a major surge of cases, the countries managed to control the pandemic relatively well. They are therefore often presented as positive examples of disease management, especially when the epicenter has moved to Europe and North America. While it is tempting to compare the approaches of different countries in order to identify best practices, comparison is questionable because of uncertainties regarding the behavior of the virus and differences between countries. The above countries for example make extensive use of mobile phone location data and face recognition technology to track movements of people, making private data easily accessible. They also execute a centralized approach in policy making with mandatory quarantine, extensive surveillance, and harsh penalties in case of disobedience [22]. Populations generally have a high level of trust in public institutions and compliance with governmental measures is high. It is argued that in Asian countries social distancing is normal and most people are used to wear masks in public [23]. People are aware that they were 'at war' with the virus and they support the collective effort to fight back [24]. These characteristics do not apply in other countries so that the stringent approaches applied in East Asia will not be feasible in other parts of the globe. What can be concluded perhaps are two things. One is that early detection and rapid responses have a notable effect and flatten the curve of the pandemic. The other is that health systems that are not only focused on hospital care but also on community-centered public health have more resilience. The experiences with previous coronavirus infections motivated the above countries to build a wider set of capacities in order to cope with a future pandemic. Rapid responses and focus on public health seem to be related to public attention and concern. In Italy and the Netherlands media interest in the new corona disease is initially low. In Singapore and South Korea after the first domestic case, public attention increased rapidly. South Korea imposed domestic restrictions 11 days after the first case, and Singapore after 15 days. Public attention was relatively low in the United Kingdom and the United States; first restrictions in the UK were ordered after 45 days, in the USA after 34 days. In Europe, public attention increased fast in Switzerland, Denmark and Ireland with first actions taken in respectively 3, 8 and 9 days after the first infectious case. Other countries waited more than a month before taking action, usually after the first death (Belgium 35, France 36, Spain 37 and Sweden 41 days after the first patient was infected) [25].

Assessments later in the year have become more nuanced. In Singapore the number of infections significantly increased, although the fatality rate is rather low. The focus of detection of infectious cases has for a long time been on citizens, ignoring the conditions of foreign workers who are crowded in dormitories. This led to the emergence of large clusters of infection among these workers. In South Korea, infections rapidly increased in August after several major outbreaks. A significant problem for policies modelled on the experiences with SARS is that the Covid-19 virus is infectious before symptoms are noted. Public health measures against SARS were successful because most patient were symptomatic and thus easily identifiable so that they could be isolated. Because SARS-CoV-2 behaves differently, symptom-based detection is not sufficient to control transmission [26].

4.4 Policy Strategies

A basic challenge is to determine which policies and strategies work best, especially in view of the expectation of subsequent waves. Information is scarce, and studies are recent. The problem is not only that responses are different among countries but the implementation varies as well as the compliance of the population with policy measures [27]. Despite the lack of scientific evidence, governments feel obliged to take measures, particularly since the experiences in China and Italy are more and more reported in the media. Managing and controlling the pandemic is not primarily regarded as a sign of scientific expertise but first of all of political competence and government effectiveness. This is reflected in the profiling of some countries as exemplary. East-Asian countries are often presented to Europe as examples of disease control. Within Europe, Germany is lauded for its effective management. The country has a large number of infected cases but a much lower fatality rate than neighboring countries. This is attributed to an early response with widespread testing, detecting also milder cases within non-vulnerable groups. Unlike other countries, testing laboratories in Germany have been expanded across the country so that diagnosis does not rely on a few central laboratories. Due to early and widespread testing the case fatality rate in Germany is much lower than in other European countries [28]. Another example is Sweden that applies a relaxed mitigation strategy. Instead of a lockdown, citizens are recommended to practice social distancing, to work from home and to avoid large gatherings. Testing is limited and stringent measures are not taken [29]. New Zealand is also acclaimed for its approach. It quickly implements the existing pandemic influenza plan and institutes border-control policies. A few weeks after the first patient is diagnosed with Covid-19, the country switches to an elimination strategy and imposes a stringent lockdown (March 26) The pandemic is declared over 103 days after the first case, and public life returns to near normal [30]. The irony, however, is that in the second half of 2020, exemplary countries such as Germany and Sweden suffer from a huge increase of infections and deaths after the summer relaxation of public health measures. In December 2020, the number of deaths in Sweden is higher than in all neighboring Nordic countries together. After the Swedish king criticized public health policies, the authorities finally took restrictive measures, and advised the public for the first time to wear face masks [31].

The political dimension of policy strategies is furthermore reflected in the repeated and vexing question why countries do not learn from each other. Why is the West not listening to China? Chinese publications in January point out that the coronavirus is more infectious than seasonal flu. In February, several studies from China, Hongkong and Japan indicate that the virus is transmitted before any symptoms occur, so that massive testing is necessary. But this information is not used in most Western countries. The first inclination of policy-makers is to consider the epidemic as a Chinese problem. It is suggested that China is withholding information, and that this is the reason for the late response in Western countries. While this has been the case in early January, increasing information is available since then. Even if the information is available, it is not trusted because the country is ruled by a totalitarian regime, using propaganda and falsified statistics. The measures taken, though the lockdown in Wuhan was effective, are regarded as the typical response of an authoritarian state. Many ethical questions are also raised about China's policies [32].

The variety of policy responses, and the aversion to learn from other countries illustrate that policies are guided by specific normative frameworks that are characteristic for countries and regions. Faced with the pandemic threat, many countries demonstrate an attitude of exceptionalism [33]. They assume that their country is so special that there is no urgency to learn lessons from other countries because the response needs to be adapted to the social and cultural climate of the particular national setting, and furthermore that the country is so different that experiences elsewhere cannot be applied because it is better prepared, has a more resilient health care system, and efficient health services. This attitude is particularly prevalent in wealthier countries with the most advanced healthcare systems. Generally, poorer countries have taken stricter measures than richer countries. It seems that the coronavirus disease is especially affecting affluent countries and regions. Lombardy, for example, one of the wealthiest regions in Europe, was first and most severely harmed. Covid-19 is hyperbolically labelled as the 'Ebola of the rich.' [34] Although Egypt, Morocco and South Africa have many cases, most African countries are doing much better than expected. Although the predictions of the impact of the pandemic in this Continent were dismal, Africa in the first wave has a lower number of cases and deaths than Europe, Asia and the Americas. It is difficult to explain this difference. The virus arrived rather late but most African governments acted swiftly and introduced a range of public health measures even before the first cases were diagnosed. Public support for the measures was high. Countries were also prepared because of the experiences with HIV and Ebola outbreaks as well as recent polio vaccination campaigns, so that community health systems were well developed. The African Centre for Disease Control and Prevention, established in 2017, did an excellent job in coordinating testing and medical supplies. At the same time, populations are much younger than in Europe (average age is 19 compared to 43 in Europe), so that less people became ill. Perhaps that the climate was also a favorable factor [35].

That normative frameworks are guiding policies can be recognized in different types of approaches [36]. Restrictive strategies as used in China apply stringent limitations on the freedom of movement and business. They are executed by a strong central government that massively mobilizes resources, like in a state of war. Permissive strategies employ measures with minor, or at least as minimal as possible limitations of freedom. They are reluctant to impose lockdowns and rely on behavioral advice and voluntary compliance of citizens, appealing as in the Netherlands to the individual responsibility of citizens for social distancing and healthy behavior. These strategies are used in the European Union and the United States. A third type of strategies, applied for example in South Korea, Singapore and Taiwan is hybrid. They combine stringent measures such as mandated quarantine and intensive contact tracing with liberal approaches to regulating movement and business. These types of strategies exemplify the conflict between individualistic and collectivistic values which always exist in public health. Infectious diseases affect individuals but they cannot be controlled by individual action, only by community responses, i.e. when the majority of individuals adhere to specific regimes such as wearing masks, keeping physical distance, and washing hands. Individual freedom therefore needs to be balanced against public health. In communitarian societies, more people are used to give priority to collective interests than in liberal societies. Autocratic regimes can impose rigorous lockdowns, police individual behavior, and use modern technology to monitor citizens [37]. But in liberal societies such interventions will not be acceptable, or only for a very limited period of time. In many countries, forced lockdowns and isolation of patients in designated facilities are legally impossible and will require special legislation. Electronic surveillance of citizens will not be possible because of concerns about privacy.

The challenge for liberal societies is that aggressive measures seem to work, at least in the first wave. Restrictive strategies significantly reduce the number of infectious cases. In relatively short time China manages to contain the pandemic, blocking the transmission of the virus by radical measures and locking down society and the economy [38]. Such a containment strategy is no longer possible in Western countries since they acted late when the virus was already widespread. The permissive strategy could not prevent that the number of infections, hospitalizations and deaths was climbing. Many countries (for example Italy and Spain) therefore moved from a permissive to a restrictive strategy with incremental measures which gradually limited individual freedom. While staying at home was only an advice in the Netherlands, it was enforced in Italy and France with police and fines. The initial aim of mitigation is combined with efforts of containment. As long as there are no effective medicines and vaccines available, the main interventions are aimed at slowing the disease and reduce its exponential growth so that healthcare systems are not overwhelmed. The combination of measures indeed delays the spread of the outbreak ('flattened the curve') and mitigates the consequences: cases are spread over a longer period of time. With the emergence of more contagious variants of the virus, the balance between permissive and restrictive measures had to be reconsidered (see the next chapter).

Countries have implemented a range of measures to halt the spread of the coronavirus disease: case identification, isolation of infected persons, contact tracing, testing, staying at home orders, quarantining at home or at special facilities, social distancing, promoting the wearing of facemasks and telework, cancellation of public events and mass gatherings, closure of schools, universities, workplaces, libraries and museums, travel restrictions, airport screening, and border controls. It is unclear to what extent these measures have been successful, and what specific measures were most effective. Effectiveness is usually measured on the basis of data on the number of infected cases, hospitalized patients and deaths. While the policy interventions are heterogeneous across countries, taken together the measures have been effective in flattening the pandemic curve. What seems most effective is avoiding human contacts and mobility. Studies of the effect of public health measures on the number of infections during the first wave show that limiting gatherings to ten people or less, particularly inside, and closing schools and universities is most effective. Massive testing and contact tracing have less effect, unless implemented early [39]. It is estimated that changed social behavior has reduced the transmission rate in Italy by 45% [40]. Restrictive measures have probably prevented the death of 38,000 Italians [41]. Rapid and rigorous intervention is most effective. An example is New Zealand where stringent containment measures were applied quickly; these measures likely reduced the number of fatalities by over 90% [42]. On the other hand, the United States could have avoided more than half of cases and deaths (as of May 3) if it had implemented the same control measures 1 week earlier [43].

Studies of policy measures tend to focus on lockdowns. While the effect of individual interventions is difficult to determine, the effect of a lockdown is identifiable. Scholars argue that lockdowns have been very effective, leading to 81% reduction of transmissions, averting millions of deaths [44]. A report of the OECD concludes that the proportion of persons in a population contracting the disease (i.e., the attack rate) can be reduced by at least 40% with a package of comprehensive measures

[45]. But findings are contradictory. Other studies indicate that full lockdowns, border closures and high rates of testing are not associated with reduced numbers of critical cases or mortality per million inhabitants [46]. The problem with this focus on lockdowns is twofold. First, the concept of lockdown itself is unclear. It may refer to a partial lockdown, primarily implementing physical distancing, or to a complete lockdown, sealing off cities and areas so that residents cannot travel or move around, sometimes imposing a curfew, and suspension of all non-essential services. Many measures are in between these two extremes, emphasizing staying at home, working from home, prohibiting public gatherings, and reducing meetings at home or in public to a small number of (related) people. The second problem is that other factors have an important effect. Morbidity and mortality are influenced by the age of the population as well as the prevalence of comorbidities (for example, obesity). The implications is that older populations are more severely affected. An interesting finding is that countries and regions with a higher per capita GDP have more reported cases and deaths [47]. It also seems that in the initial phase of the pandemic, the virus spreads more rapidly among people of higher social status. Only in later phases, usually once lockdowns have been implemented, people of lower social status are more exposed and affected. The explanation is that the first group has more diverse social contacts and networks, and is more mobile, so that initially they are at the center of the viral spread. These findings are remarkable since it is commonly assumed that socioeconomic factors such as high unemployment rates and wealth inequalities are associated with an increased number of cases. While it is true that people of lower social status are most burdened after the novel virus is spreading, people of higher social status may import the virus and provoke its initial spread [48].

Furthermore, an observation discussed above is that the effectiveness of measures is enhanced if they are implemented quickly. Countries that acted early did better, independently from the stringency of the policies. When measures are late, very stringent rules had only a limited effect on the medical outcomes (e.g. in Spain and Belgium) [49]. Finally, the effect of policy measures depends on the compliance of the population. When a significant number of people do not adhere or no longer follow the policies, the effect of lockdown measures will diminish.

The general findings are that lockdown measures help with the management of the pandemic; they are flattening the curve and reduce pressure on healthcare systems. The above considerations, however, make it doubtful how the effects can be explained. Faced with a crisis it is attractive to ascribe effectiveness to policy decisions, especially when they imply a drastic measure such as locking down the society. Whether or not effective, a lockdown is problematic for various reasons. It has short-term effects while it is not clear what the effect is in the longer term. Relaxation of restrictive measures will possibly lead to a growing number of infections, so that in the future, alternation of stringent measures and relaxation will be required. Lockdowns raise the issue of sustainability. They are seriously impacting society and economy so that it is questionable how long they can be maintained. People do not like to spend life in social distancing, and after some time 'lockdown fatigue' is setting in. Lockdowns also cause anxiety, depression and loneliness. They increase risks for domestic violence and child abuse.

Even if they can be morally defended with an appeal to the priority of public health and the effort to save human lives, the ethical balancing with economic and social well-being does not disappear, and will be increasingly discussed the longer the lockdown exists. Lockdowns can finally reinforce existing inequalities. They are difficult to implement for poor people who cannot work from home or have essential jobs in cleaning, delivery and food production. Lockdowns sometimes exacerbate health inequalities. This was first observed in China when the virus spread quickly in hospitals in Wuhan and infected many health care workers. Resources in the city were limited and materials, equipment and medical workers were transferred to Wuhan. Outside of the province of Hubei many hospital beds were empty so that more lives would have been saved when patients would have been relocated, like later Dutch, French and Italian patients were moved to Germany [50].

These problematic aspects of lockdowns suggest a curious conclusion. The use of the expression 'lockdown' promises a firm and drastic intervention. Because of their severity, lockdowns in Western countries are postponed as long as possible in order to avert the dire social and economic consequences, but eventually they become unavoidable.

Countries and regions that have managed to contain the pandemic (e.g. South Korea, Hong Kong and Taiwan) did not impose a nation-wide lockdown. Many countries apply a lockdown because they have missed the opportunity to intervene early (e.g. Italy and the Netherlands). Lockdown orders are executed with ambivalence. In the USA, coordination among federal, state and municipal governments is poor; some states declare stay-at-home orders while others do not. Orders are not strictly enforced while there is no leadership at the federal level. Emphasis is on individual behavior, not addressing the collective problem that is posed by the pandemic [51]. Whatever their policies, most countries face global challenges because they are not prepared for a disaster of this magnitude.

4.5 Global Challenges

When the pandemic unfolded, countries are confronted with similar challenges. Testing capacities are insufficient and it takes a lot of time and effort to improve them. Essential materials such as protective equipment are in short supply and they cannot be provided to vulnerable groups, notably healthcare workers. These challenges highlight lack of coordination and cooperation, and aggravate lack of leadership.

4.5.1 Testing Capacity

The World Health Organization has repeatedly emphasized that testing, isolation and contact tracing are the backbone of the response to the pandemic. Its Director-General stressed the same message again and again: test, test, test. All patients suspected for Covid-19 should be tested as well as the persons who have been in contact with them [52]. Nonetheless, many countries have not followed these recommendations. The rationale for testing is twofold. Without testing there is not a clear picture about how many people are infected and how the virus is disseminating across a territory. Testing is furthermore the basis for containment through contact tracing and isolation of patients. While some countries (e.g. South Korea and Germany) resort to extensive testing, other countries perform only a small number of tests (for example in Spring 2020 the average number of tests in the Netherlands is 400 per million inhabitants, compared to 2000 in South Korea) [53]. Limited testing is the result of lack of preparation. Germany already had an extensive network of laboratories while France and the US had a limited number of accredited public laboratories and the United Kingdom relied on a few centralized laboratories [54]. Germany could also launch a diagnostic protocol for the detection of the virus in January [55]. The test kits developed and distributed by the CDC in the United States in early February proved to be unusable because of a faulty reagent. Since the government has ordered that only US-made kits can be used, the supply of test kits was insufficient for some time. Limited testing is furthermore the deliberate result of policies. In the initial stage of the pandemic, the Netherlands for example test only people with symptoms; health care professionals are not tested; during a short period, there is only voluntary testing of people entering the country, even from infected areas. In Germany on the other hand there is preventive testing of everybody when there is an outbreak in schools, abattoirs, and nursing homes. South Korea applies massive testing with sometimes ten thousand people tested in outbreaks after public demonstrations and church ceremonies. The country has sufficient testing capacity since the government had encouraged companies to produce tests. Lack of testing in many countries is finally the consequence of scarcity of materials such as test kits, swabs, glassware, and reagents. Because of this scarcity, authorities (e.g. in France and the Netherlands) argue that systematic testing is not needed. Test systems (PCR machines) are also different, and specific materials are required, provided by the producers of the systems. Since the shortage of materials is a universal problem, countries are competing in a global market, unless they themselves manufacture the items [56].

Insufficient testing implies that it is not clear how many people are infected. The only remaining indicator for measuring the effect of policies is the number of corona patients in intensive care units. Another implication is the impossibility of contact tracing. Because the virus is too rapidly disseminating in the population, countries such as the Netherlands and France decide that tracing cases is no longer a priority. In the United Kingdom the policy of test, trace and isolate is stopped on March 12. The approach is to watch and wait [57]. Contact tracing starts with a positive test result and it is most effective when it is done quickly. But in many countries it usually takes a long time (more than 48 h) before the test can be done, while the result also can take longer than 48 h, sometimes 10 days [58]. A major problem is that the virus is often transmitted before persons have symptoms; up to 40% of people are infectious but never develop symptoms [59]. The focus on symptomatic cases is therefore insufficient to control the pandemic. Testing only in hospital settings and missing milder or asymptomatic cases in the community also implies that there are many more infections than officially reported.

4.5.2 Shortages

In 2009, when the swine flu pandemic emerged, France has one billion face masks in stock. This reserve was not renewed with the argument that the masks could be quickly ordered in China. In 2020, France like many other countries faces serious shortages of masks. Not only are there no stocks but because of the lockdowns fabrication and transportation are halted. Globally, China produces 50% of facemasks [60]. But when it imposed a facemask wearing policy in the country, the normal production of 20 million masks per day was not sufficient, and China imported more than 2 billion facemasks in January and February. Several European countries in this stage donated protective materials such as face masks to China. These early experiences should have alerted other countries. In March, not only facemasks are in short supply but all protective materials such as gloves, goggles, gowns and face shields. The WHO warns on March 3 that these shortages endanger healthcare workers [61]. When doctors and nurses got ill, the health system itself cannot cope with increasing demand for care. A serious ethical challenge surfaced: how can health professionals take care of infected patients if they cannot protect themselves against the disease? Can they be morally required to perform their work if they risk their own health and life? Similar ethical questions emerge when it becomes evident that the number of hospital beds, ICU equipment and medicines is insufficient in many countries. A major ethical debate concerns the problem of selecting which patients will qualify for intensive treatment when resources are insufficient to treat all patients who need them. Experiences in Italy show that the catastrophic scenario of rationing and triage is no longer unrealistic [62]. These ethical challenges will be further explored in following chapters.

Important in the context of this chapter is the role of policies, firstly by neglecting and underestimating the need for equipment and protective materials, and secondly by addressing the shortages. Already in February 2020, the WHO had warned for imminent worldwide shortages. Nonetheless, the Minister of Health in the Netherlands assures in the same month that a large-scale purchase of face masks and protective materials is not necessary [63]. A few weeks later, when it becomes clear that there are serious shortages, it is troublesome to solve them because shortages are global. The Ministry of Health engages in panic buying and negotiates with unreliable parties while prices are surging. Also, the population starts hoarding, not only face masks and disinfectants but also food and toilet paper. Addressing the shortages is a huge challenge. At first, there is little coordination; each health facility is trying to obtain as much material as possible. At the end of March, the Dutch government establishes a national consortium for medical resources so that provision is better coordinated. After shipments of surgical masks are stolen from hospitals, the French state requisitions all stocks and production of masks to distribute them fairly. Manufacturing of masks and other equipment is increased in countries in order to diminish the dependency on imports from abroad. People are encouraged to make masks at home. In May there is finally a sufficient number of face masks available. Although there are a few companies in the world that manufacture ventilators, production is rapidly ramped up with massive government support, although the shortage of trained healthcare workers is more difficult to address [64].

The experience of shortages demonstrates the downside of globalization. For many essential materials countries have become dependent on a few low-cost production sites. For example, 80% of the active pharmaceutical ingredients for drugs used in the United States are produced in China and India [65]. China is the largest producer of face masks and eye protection [66]. When global supply chains are disrupted due to lockdowns and export bans, countries face immediate problems. They are associated with lack of redundancy: there are limited stockpiles and excess supplies. Because market thinking has dominated healthcare, maintaining adequate inventories of personal protective equipment is regarded as expenditure that should be minimized. Governments are hesitant to interfere, have not maintained supplies, and are initially at least, reluctant to require private companies to manufacture protective equipment and ventilators [67]. Another concern is that globalization implies inequality and that the pandemic reinforces inequalities. Already limited resources become even scarcer in low-income countries [68]. That developing countries are hit hard is for example illustrated in Brazil, Colombia and Ecuador: 8 out of 10 care professionals have to perform high-risk procedures without adequate personal protective equipment [69]. What is striking is the lack of global solidarity. Countries block the export of protective materials to secure the provision for their own population. Ordered shipments of protective equipment are intercepted and delivered to countries which pay more. In the global competition high-income countries secure more materials than low-income countries. Coordination to make sure that equipment goes to the areas that are hit hardest is absent.

4.5.3 Lack of Coordination and Cooperation

One of the most worrying aspects of the pandemic is the lack of a coordinated, global response. When the corona disease emerges, there is "no global leadership, no willingness to cooperate… Instead there was inattention, rivalry and accusation." [70] Rather than reinforcing global cooperation and solidarity, the United States and some other countries blame WHO for the global dissemination of the virus. Most countries prefer to address the viral threat on their own; in this 'Alleingang' they do not take any lessons from the experiences of foreign countries. Pandemics are distinct from other disasters because they come in waves. Because not all parts of the world are affected at the same time, there are opportunities for acquiring experience and knowledge of the virus, its spread, and effective public health measures. As discussed earlier in this chapter, the will to learn from the experiences of others was limited. It seems that most countries, thinking and acting for themselves, mirror the mentality of nationalistic populism, rejecting globalization and denying that the virus is a global threat. This mentality is self-destructive since the pandemic is a global phenomenon that can only be overcome through collective action. What is lacking from a bioethical perspective is solidarity. This is not in the first place a matter of economic support. At a more fundamental level, it expresses the experience of being together in the same ordeal. At this level, solidarity is clearly manifested within neighbourhoods, cities and countries. While politicians in numerous countries repeat the message that we are all in this together, their first concern is their own citizens, although there are examples of transnational solidarity, such as the treatment of patients from France, Italy, and the Netherlands in Germany and Luxemburg.

Assessing the policy responses in the first half of 2020, three lessons can be drawn. First is the role of the state. Preliminary data suggest that countries with centralized government institutions did somewhat better in managing the pandemic (for example France compared to Belgium) [71]. In many instances, while policies to contain and mitigate the disease are initially delegated to regions and provinces within a country, when transmission cannot be halted, the state has to step in with national measures. Shortages of essential resources could not be solved at local or regional levels but required a strong coordinating role of the central government. In the Netherlands for example, testing capacity could only be expanded after the establishment of a national coordination structure [72]. The role of central government is furthermore evident in efforts to address the economic impact of policies, launching financial support packages to support businesses and individuals.

The second lesson is the importance of leadership. In a crisis that cannot be quickly resolved and that requires drastic interventions in individual and social life, good leadership acts quickly and aggressively. But leadership is not just management. It requires nurturing a sense of shared mission (controlling the spread of the virus in order to save lives and economic recovery), but also a willingness to listen to scientific experts, developing creative responses, informing and educating the public, uniting the population and mobilizing collective efforts as well as empathic understanding for the distress and suffering of people [73]. Otherwise, the population will not trust the authorities and compliance with policy measures will diminish. Autocratic and populist leaders generally did a bad job: they denied the seriousness of the virus for a long time, delayed policies, and did not implement stringent policies [74].

The third lesson concerns international cooperation and the role of international organizations. Many calls for global collaboration have been launched with the basic argument that the virus does not respect borders [75]. Despite these calls, a coordinated global response has not been forthcoming. Countries are primarily concerned of their national interest. Some countries undermine global cooperation, trying to secure as fast as possible supplies of potential treatments and vaccines through bilateral deals and bypassing multilateral organizations [76]. The result is that policies are different across countries and sometimes across regions within a country. For example, in the European Union cross-border regulations have created labyrinths. Travelers from Spain must go into quarantine in Belgium and the Netherlands, but not in France. Since June 2020 the internal borders between countries are open again, but it is onerous to travel because of different regulations. In September the borders between countries are closed once more. Countries have different color codes; guidelines for mandatory quarantine differ and are often not controlled. Some countries require negative Covid tests but it differs how recent the test must be. Though many argue that better coordination is needed with the same rules for travel and color coding, no harmonization has been adopted by member states.

One area that is remarkable is science. The pandemic has boosted international cooperation of scientists and researchers, primarily in clinical medicine, virology, epidemiology and the development of antivirals and vaccines. The genome sequence of Covid-19 is immediately shared by Chinese scientists, so that the virus can be studied across the world. Research data are rapidly published, supported by the provision of open access by many publishers, so that experiences and knowledge can be widely exchanged. An example of collaborative activity is the Solidarity Trial launched by WHO with the aim of clinically testing treatments and enrolling patients from more than 30 countries [77]. Scientific collaboration can build on previous inspiring examples such as the intensive worldwide collaboration during 13 years to eradicate the smallpox virus [78]. Following the SARS outbreak in 2003, it took 20 months to develop a vaccine for human trials; for the Zika virus it took 6 months. Already in March 2020, four vaccine candidates for Covid-19 are tested in humans while more than hundred vaccines are in various stages of development. In December, the first Covid vaccine is approved for use.

4.6 Conclusion

This chapter examines how SARS-CoV-2 expanded over the world in the first half of 2020, and how countries initiated and implemented policies to contain or mitigate the disease. Without a cure or vaccine, policy measures available are the traditional ones used in previous pandemics. Policies are heterogeneous: within countries they are not consistent, and between countries there are substantial differences. As soon as the viral threat is recognized, responses are needed while scientific data are scarce because the virus is novel, so that decisions must be made in conditions of uncertainty and ambiguity about the impact of various measures. While movement restriction and physical distancing are highly effective, they bring substantial damage to economic and social life but also human rights.

Although a wide range of public health interventions can be applied (from mandatory quarantine and testing to voluntary appeals to distancing and masking), in hindsight it is clear that countries that focus on specific preventive strategies and implement them consistently and strictly, are most successful [79]. It is also clear what went wrong: the threat was not taken seriously, testing slow, contact tracing inadequate, quarantining and isolating infected persons deficient, expert recommendations confusing, and strong national responses missing [80].

Whereas the pandemic is a global challenge, policy responses are primarily local. Rather than scientific evidence, though initially limited and provisional, policies reflect political strategies that articulate normative frameworks, emphasizing different positions in regard to fundamental values such as individual freedom and privacy versus common interest, personal versus social responsibility, public health versus economic well-being, saving lives versus social interaction. All countries face global challenges: testing capacity is insufficient and shortages of essential medical materials arise. It takes time to resolve both challenges because international cooperation and coordination is wanting. It is recognized that Covid-19 particularly affects vulnerable populations and exacerbates existing health inequalities; therefore, the value of solidarity is frequently invoked but in practice, like other basic principles of global ethics, is it put aside by concerns with national interests.

While the perspective of global bioethics could have directed attention to global threats, as well as to the necessity of preparedness and prevention, as argued in the first chapter, the lack of such perspective not only made that in many countries the risk of emerging infectious diseases was not taken seriously and that crucial delays occurred in responding to the virus. Global bioethics also articulates that humans and the natural world are interconnected and interdependent. Infectious diseases are not a threat to some countries or regions, but a menace for all inhabitants of the globe which can only be overcome by global cooperation. Experiences with Covid-19 are therefore a reason to rethink the significance of global bioethics for dealing and coping with a pandemic disaster. Pandemics are not just biological or medical events that appeal to science for guidance and management; they are calamities because they have social, political and economic effects that impact on the fundamental values of societies. For learning to deal with a pandemic disease, scientific knowledge of the virus and the disease is not sufficient.

References

1. Capano, G. 2020. Policy design and state capacity in the Covid-19 emergency in Italy; if you are not prepared for the (un)expected, you can be only what you already are. *Policy and Society* 39 (3): 326–344; Carvalho, A.C.C., and A. Kritski. 2020. Learning from the Italian experience in coping with Covid-19. *Journal of the Brazilian Society of Tropical Medicine* 53: e20200199.
2. Flaxman, S., S. Mishra, A. Gandy, et al. 2020. Estimating the number of infections and the impact of non-pharmaceutical interventions on Covid-19 in 11 European countries. Imperial College Covid-19 Response Team. *Nature* 584: 257–261.
3. Carvalho and Kritski, Learning from the Italian experience in coping with Covid-19; Boccia, S., W. Ricciardi, and J.P.A. Ioannidis. 2020. What other countries can learn from Italy during the Covid-19 pandemic. *JAMA Internal Medicine* 180 (7): 927–928.

4. Pisano, G.P., R. Sadun, and M. Zanini. 2020. Lessons from Italy's response to coronavirus. *Harvard Business Review*, March 27.
5. Capano, Policy design and state capacity in the Covid-19 emergency in Italy, 337; Carvalho and Kritski. Learning from the Italian experience in coping with Covid-19; Pisano, Sadun, and Zanini. Lessons from Italy's response to coronavirus.
6. Remuzzi, A., and G. Remuzzi. 2020. Covid-19 and Italy: What next? *Lancet* 395: 1225–1228.
7. Alicandro, G., G. Remuzzi, and C. La Vecchia. 2020. Italy's first wave of the Covid-19 pandemic has ended: No excess mortality in May, 2020. *Lancet* 396 (10253): e27–e28.
8. Capano, Policy design and state capacity in the Covid-19 emergency in Italy, 336 ff.
9. Rosenbaum, L. 2020. Facing Covid-19 in Italy – Ethics, logistics, and therapeutics on the epidemic's front line. *New England Journal of Medicine* 382 (20): 1873–1875; Nacoti, M., A. Ciocca, A. Giupponi, et al. 2020. At the epicenter of the Covid-19 pandemic and humanitarian crises in Italy: Changing perspectives on preparation and mitigation. *NEJM Catalyst Innovations in Care Delivery* 1–5.
10. Platform Containment Nu. 2020. Covid-19 in the Netherlands. *A time line.*
11. Stokmans, D., and M.L. Adriaanse. 2020. Hoe Nederland de controle verloor. De corona uitbraak van dag tot dag. *NRC*, June 20–21.
12. Boin, A., W. Overdijk, C. van der Ham, J. Hendriks, and D. Sloof. 2020. *Covid-19. Een analyse van de nationale crisisresponse*. Leiden: The Crisis University Press.
13. Korteweg, N. 2020. Ik maak me het meeste zorgen dat mensen de maatregelen niet vasthouden op vakantie. *NRC*, July 5.
14. Pollock, A.M., P. Roderick, and B. Pankhania. 2020. Covid-19: Why is the UK government ignoring WHO's advice? *British Medical Journal*, March 30.
15. Wilder-Smith, A., and D.O. Freedman. 2020. Isolation, quarantine, social distancing and community containment: Pivotal role for old-style public health measures in the novel coronavirus (2019-nCoV) outbreak. *International Journal of Travel Medicine*: 1–4; See also: Abraham, T. 2007. *Twenty-first century plague. The story of SARS*. Baltimore: The Johns Hopkins Press.
16. Huang, I.Y. 2020. Fighting Covid-19 through government initiatives and collaborative governance: The Taiwan experience. *Public Administration Review* 80 (4): 665–670.
17. Woo, J.J. 2020. Policy capacity and Singapore's response to the Covid-19 pandemic. *Policy and Society* 39 (3): 345–362.
18. Cowling, B.J., and W.W. Lim. 2020. They've contained the coronavirus. Here's how. *New York Times*, March 13.
19. Lee, V.J., C.J. Chiew, and W.X. Khong. 2020. Interrupting transmission of Covid-19: Lessons from containment efforts in Singapore. *International Journal of Travel Medicine*: 1–5; Woo. Policy capacity and Singapore's response to the Covid-19 pandemic.
20. Steinbrook, R. 2020. Contact tracing, testing, and control of Covid-19; learning from Taiwan. *JAMA Internal Medicine* 180 (9): 1163–1164; Wang, C.J., C.Y. Ng, and R.H. Brook. 2020. Response to Covid-19 in Taiwan. Big data analytics, new technology, and proactive testing. *JAMA* 323 (14): 1341–1342; Huang. Fighting Covid-19 through government initiatives and collaborative governance: The Taiwan experience.
21. Shokoohi, M., M. Osooli, and S. Stranges. 2020. Covid-19 pandemic: What can the West learn from the East? *International Journal of Health Policy and Management* 9 (10): 436–438.
22. For example, in Japan a mandatory lockdown is not allowed by the law. See: Tashiro, A., and R. Shaw. 2020. Covid-19 pandemic response in Japan: What is behind the initial flattening of the curve? *Sustainability* 12 (13) 5250.
23. Tashiro and Shaw. Covid-19 pandemic response in Japan.
24. Huang, Fighting Covid-19 through government initiatives and collaborative governance: The Taiwan experience.
25. Aksoy, C.G., M. Ganslmeier, and P. Poutvaara. 2020. *Public attention and policy responses to Covid-19 pandemic*. Bonn: IZA Institute of Labor Economics.
26. Gandhi, M., D.S. Yokoe, and D.V. Havlir. 2020. Asymptomatic transmission, the Achilles' heel of current strategies to control Covid-19. *New England Journal of Medicine* 382 (22): 2158–2160.

27. Gibney, E. 2020. Whose coronavirus strategy worked best? Scientists hunt most effective policies. *Nature* 581 (7806): 15–16.
28. Shokoohi, Osooli, and Stranges. Covid-19 pandemic; Deen, K. 2020. Dit kan Nederland leren van de Duitse aanpak van de coronacrisis. *Trouw*, September 22; Stafford, N. 2020. Covid-19: Why Germany's case fatality rate seems so low. *British Medical Journal* 369.
29. Baldwin, P. 2021. *Fighting the first wave. Why the coronavirus was tackled so differently across the globe*, 69 ff. Cambridge: Cambridge University Press.
30. Baker, M.G., and N. Wilson. 2020. Successful elimination of Covid-19 transmission in New Zealand. *New England Journal of Medicine* 383 (8): e56; Wilson, S. 2020. Pandemic leadership: Lessons from New Zealand's approach to Covid-19. *Leadership* 16 (3): 279–293.
31. Orange, R. 2020. As Covid death toll soars ever higher, Sweden wonders who to blame. *The Guardian*, December 20.
32. Vervaeke, L. 2020. Waarom luisterde het Westen niet naar China? *De Volkskrant*, September 7, 6–7; Lei, R., and R. Qiu. 2020. Report from China: Ethical questions on the response to the coronavirus. *The Hastings Center*, January 31; Műller, O., G. Lu, A. Jahn, and O. Razum. 2020. Covid-19 control: Can Germany learn from China? *International Journal of Health Policy and Management* 9 (10): 432–435.
33. Baldwin, *Fighting the first wave*, 75.
34. Nacoti, Ciocca, Giupponi, et al. At the epicenter of the Covid-19 pandemic and humanitarian crises in Italy, 3.
35. Tangwa, G.B., and N.S. Munung. 2020. Covid-19: Africa's relation with epidemics and some imperative ethics considerations of the moment. *Research Ethics*: 1–11; Soy, A. 2020. Coronavirus in Africa; Five reasons why Covid-19 has been less deadly than elsewhere. *BBC News*, October 7; Makoni, M. 2020. Covid-19 in Africa: Half a year later. *The Lancet Infectious Diseases* 20 (10): 1127.
36. Travica, B. 2020. *Containment strategies for Covid-19 pandemic*, May 18.
37. Frey, C.B., C. Chen, and G. Presidente. 2020. Democracy, culture, and contagion: Political regimes and countries' responsiveness to Covid-19. *Covid Economics* 18: 222–238.
38. Kupferschmidt, K., and J. Cohen. 2020. Can China's Covid-19 strategy work elsewhere? *Science* 367 (6482): 1061–1062.
39. Haug, N., L. Geyrhofer, A. Londei, et al. 2020. Ranking the effectiveness of worldwide Covid-19 government interventions. *Nature Human Behavior* 4: 1303–1312; Brauner, J.M., S. Mindermann, M. Sharma, et al. 2020. Inferring the effectiveness of government interventions against Covid-19. *Science* 371 (6531): eabdg338.
40. Gatto, M., E. Bertuzzo, L. Mari, et al. 2020. Spread and dynamics of the Covid-19 epidemic in Italy: Effects of emergency containment measures. *Proceedings of the National Academy of Sciences of the United States of America* 117 (19): 10484–10491.
41. Flaxman, S., S. Mishra, A. Gandy, et al. 2020. *Estimating the number of infections and the impact of non-pharmaceutical interventions on COVID-19 in 11 European countries*. Imperial College London. (30-03-2020).
42. Deb, P., D. Furceri, J.D. Ostry, and N. Tawk. 2020. *The effect of containment measures on the Covid-19 pandemic*. IMF Working paper.
43. Pei, S., S. Kandula, and J. Shaman. 2020. Differential effects of intervention timing on COVID-19 spread in the United States. *medRxiv preprint*.
44. Flaxman, Mishra, Gandy, et al. *Estimating the number of infections and the impact of non-pharmaceutical interventions on Covid-19 in 11 European countries*.
45. OECD. 2020. *Flattening the Covid-19 peak: Containment and mitigation policies*.
46. Hopman, J., and S. Mehtar. 2020. Country level analysis of Covid-19 policies. *EClinicalMedicine* 25: 100500; Chaudhry, R., G. Dranitsaris, T. Mubashir, J. Bartoszko, and S. Riazi. 2020. A country level analysis measuring the impact of government actions, country preparedness and socioeconomic factors on Covid-19 mortality and related health outcomes. *EClinicalMedicine* 25: 100464.

47. Chaudhry, Dranitsaris, Mubaskir, Bartoszko, and Riazi. A country level analysis measuring the impact of government actions, country preparedness and socioeconomic factors on Covid-19 mortality and related health outcomes.
48. Berkessel, J.B., T. Ebert, J.E. Gebauer, T. Jonsson, and S. Oishi. 2021. Pandemics initially spread among people of higher (not lower) social status: Evidence from Covid-19 and the Spanish fly. *Social Psychological and Personality Science* 11: 20098. The authors argue that the same is true for the Spanish flu pandemic which mainly affected people younger than 40 years. They point out their findings apply to the first wave of the Covid pandemic; it remains to be seen whether the same diverging patterns exists for subsequent waves.
49. Migone, A.R. 2020. The influence of national policy characteristics on Covid-19 containment policies: A comparative analysis. *Policy Design and Practice* 3 (3): 259–276.
50. Liu, L. 2020. Sustainable Covid-19 mitigation: Wuhan lockdowns, health inequities, and patient evacuation. *International Journal of Health Policy and Management* 9 (10): 415–418.
51. Ren, X. 2020. Pandemic and lockdown: A territorial approach to Covid-19 in China, Italy and the United States. *Eurasian Geography and Economics* 51 (2): 162–183.
52. This is known as the WHO mantra of FTTIS: find, test, trace, isolate and support. WHO. 2020. *Critical preparedness, readiness and response actions for Covid-19: Interim guidance.*
53. Boin, Overdijk, Van der Ham, Hendriks, and Sloof. *Covid-19*, 45.
54. Moatti, J.-P. 2020. The French response to Covid-19: Intrinsic difficulties at the interface of science, public health, and policy. *Lancet Public Health* 5 (5): e255; Editorial. 2020. The Covid-19 testing debacle. *Nature Biotechnology* 38 (653).
55. Corman, V.M., O. Landt, M. Kaiser, et al. 2020. Detection of 2019 novel coronavirus (2019-nCoV) by real-time RT-PCR. *Euro Surveillance* 25 (3): pii=2000045.
56. Rajan, S., J. Cylus, and M. McKee. 2020. Successful find-test-trace-isolate-support systems. How to win at snakes and ladders. *Eurohealth* 26 (2): 34–39.
57. Beale, R. 2020. Short cuts. *London Review of Books*, May 21.
58. Kretzschmar, M.E., G. Roshnova, M.C.J. Bootsma, et al. 2020. Impact of delays on effectiveness of contact tracing strategies for COVID-19: A modelling study. *The Lancet Public Health* 5: e452–e459.
59. Parshley, L. 2020. The magnitude of America's contact tracing crisis is hard to overstate. *National Geographic*, September 1; Moghadas, S.M., M.C. Fitzpatrick, P. Sah, et al. 2020. The implications of silent transmission for the control of Covid-19 outbreaks. *Proceedings of the National Academy of Sciences of the United States of America* 117 (30): 17513–17515.
60. Wu, H., J. Huang, C.J.P. Zhang, Z. He, and E. Ming. 2020. Facemask shortage and the novel coronavirus disease (Covid-19) outbreak: Reflections on public health measures. *EClinicalMedicine* 21: 100329.
61. WHO. 2020. *Shortage of personal protective equipment endangering health workers worldwide.*
62. Vincent, J.-L., and J. Creteur. 2020. Ethical aspects of the Covid-19 crisis: How to deal with an overwhelming shortage of acute beds. *European Heart Journal: Acute Cardiovascular Care* 0 (0): 1–5; Craxi, L., M. Vergano, J. Savulescu, and D. Wilkinson. 2020. Rationing in a pandemic: Lessons from Italy. *Asian Bioethics Review* 12: 325–330.
63. Boin, Overdijk, Van der Ham, Hendriks, and Sloof. *Covid-19*, 68 ff.
64. Iyengar, K., S. Bahl, R. Vaishya, and A. Vaish. 2020. Challenges and solutions in meeting up the urgent requirement of ventilators for Covid-19 patients. *Diabetes & Metabolic Syndrome: Clinical Research & Reviews* 14: 499–501.
65. Gustafsson, L. 2020. Covid-19 highlights problems with our generic supply chain. *The Commonwealth Fund*, May 7.
66. Ramney, M.J., V. Griffeth, and A.K. Jha. 2020. Critical supply shortages – The need for ventilators and protective equipment during the Covid-19 pandemic. *New England Journal of Medicine* 382 (18): e41.
67. Cohen, J., and Y. Van der Meulen Rodgers. 2020. Contributing factors to personal protective equipment shortages during the Covid-19 pandemic. *Preventive Medicine* 141: 106263.

68. McMahon, D.E., G.A. Peters, L.C. Ivers, and E.E. Freeman. 2020. Global resource shortages during Covid-19: Bad news for low-income countries. *PLoS Neglected Tropical Diseases* 14 (7): e0008412.
69. Martin-Delgado, J., E. Viteri, A. Mula, et al. 2020. Availability of personal protective equipment and diagnostic and treatment facilities for healthcare workers involved in Covid-19 care: A cross-sectional study in Brazil, Colombia, and Ecuador. *PLoS One* 15 (11): e0242185.
70. Horton, R. 2020. *The Covid-19 catastrophe. What's gone wrong and how to stop it happening again*, 37. Cambridge: Polity Press.
71. Desson, Z., E. Weller, P. McMeekin, and M. Ammi. 2020. An analysis of the policy responses to the Covid-19 pandemic in France, Belgium, and Canada. *Health Policy and Technology* 9 (4): 430–446.
72. Boin, Overdijk, Van der Ham, Hendriks, and Sloof. *Covid-19*, 78 ff.
73. Wilson, Pandemic leadership, 279–293; Fraser, M.R. 2020. Leading in the Covid-19 crisis: Challenges and solutions for state health leaders. *Journal of Public Health Management and Practice* 26 (4): 380–383.
74. Åslund, A. 2020. Responses to the Covid-19 crisis in Russia, Ukraine, and Belarus. *Eurasian Geography and Economics* 61 (4–5): 1–14.
75. Mohamed, K., E. Rodriguez-Roman, F. Rahmani, et al. 2020. Borderless collaboration is needed for Covid-19 – A disease that knows no borders. *Infection Control & Hospital Epidemiology* 10: 1–2; Editorial. 2020. Communication, collaboration and cooperation can stop the 2019 coronavirus. *Nature Medicine* 26: 151; Perez, G.I.P., and A.T.B. Abadi. 2020. Ongoing challenges faced in the global control of Covid-19 pandemic. *Archives of Medical Research* 51: 574–576.
76. Editorial. 2020. Global collaboration for health: Rhetoric versus reality. *Lancet* 396: 735.
77. Momtazmanesh, H., D. Ochs, L.Q. Uddin, et al. 2020. All together to fight Covid-19. *American Journal of Tropical Medicine and Hygiene* 102 (6): 1181–1183.
78. Heymann, D.L., and A. Wilder-Smith. 2020. Successful smallpox eradication: What can we learn to control Covid-19? *Journal of Travel Medicine* 27 (4): 1–3.
79. Baldwin, *Fighting the first wave*, 101.
80. Lewis, T. 2021. How the U.S. pandemic response went wrong – And what went right – During a year of Covid. *Scientific American*, March 11.

Chapter 5
Diverging Facts and Values

Abstract This chapter analyzes the resurgence of Covid infections since the second half of 2020. Many countries again are not sufficiently prepared for new waves of the pandemic. While most policy-makers emphasize that their decisions are based on scientific evidence, the fundamental problem is that evidence about effective policy measures is insufficient. Policy decisions need to be taken in a context of uncertainty and unknown risks. Major controversies exist regarding the origin and characteristics of the virus, symptomatology, infectiousness, transmission, distancing, masking, immunity, vulnerability, and the role of animals. In all these cases, scientific facts are lacking, weak, questionable, or contested so that in practice policy decisions are based on value judgments. Interpreting the available evidence and proposing what should be done often involves an ethical point of view.

Keywords Controversies · Distancing · Masking · Precautionary principle · Risk rituals · Second wave · Testing · Uncertainty · Vulnerability

5.1 Introduction

The argument of this chapter is that the diverging policy responses, described in the previous chapter, should be interpreted against the background of a supposed distinction between facts and values. It is often stated by scientists and policy-makers that policies should be based on facts. Most governmental authorities repeatedly emphasize that their decisions are based on scientific evidence. However, the novelty of the virus implies that science is hardly able to produce and corroborate factual evidence, – at least it takes time to produce evidence. An example is the use of face masks. In some countries (e.g. Belgium and Germany) they are mandatory in all public interactions, while in other countries (e.g. the Netherlands) for a long time they are not recommended and physical distancing is emphasized. Scientific evidence concerning the protective value of masks is disputed. Similar controversies concern the question how the virus is disseminated through the air, via droplets or

H. ten Have, *The Covid-19 Pandemic and Global Bioethics*, Advancing Global Bioethics 18, https://doi.org/10.1007/978-3-030-91491-2_5

aerosols. The answers determine policies of distancing and use of ventilation and air-conditioning but science is not able to provide definitive conclusions. Another example is immunity. Does infection with SARS-CoV-2 result in immunity or not; and if so, is it permanent or temporary? This question has implications for the argument of herd immunity and for the potential effectiveness of vaccination. It is also unclear how and when this coronavirus actually emerged. According to some scientists, the virus had already been circulating long before the first outbreak. Perhaps it will behave like the previous SARS virus and suddenly disappear. These factual controversies underscore the limitations and uncertainties of evidence-based policies. Actually, policies are primarily based on values. While they all focus on the common good of public health, their implementation is determined by other values, especially individual autonomy and liberty, and economic growth and survival. In many countries in the Global South, the choices are limited; without work, no income, so health security is not an option for many people.

This chapter starts with examining the resurgence of Covid infections in the second part of 2020. In the summer many are hoping that the pandemic is receding. But the virus has not disappeared and comes back as soon as human caution is diminishing. Again, many countries are not sufficiently prepared for this expansion of the disease. Medical care, however, has somewhat improved and shortages of basic resources have been addressed. The fundamental problem is the insufficiency of evidence about effective policy measures. Nonetheless, policy decisions need to be taken in this context of uncertainty and unknown risks. This issue will be examined in the third paragraph of this chapter. But first attention will be given to the persistence of Covid. Almost everybody expected that with the start of vaccinations in 2021, infectious cases would decline, whereas on the contrary new waves of infections occurred, even in countries where the majority of the population has been vaccinated. Subsequently, major controversies will be analyzed. They concern the origin and characteristics of the virus, symptomatology, infectiousness, transmission, distancing, masking, immunity, vulnerability, and the role of animals. In all these cases, scientific facts are lacking, weak, questionable, or contested so that in practice policy decisions are based on value judgments. Interpreting the available evidence and proposing what should be done often involves an ethical point of view.

5.2 The Second Wave

On November 24, 2020 a new milestone is reached with globally 60 million infections and 1.416.112 deaths. Since the end of May, the number of infections across the world is continuously rising, while in Europe the number is declining and many countries relax restrictive measures. Since the end of July, however, confirmed cases start to increase across the world (from 16.234 on June 1 to 263.639 on November 1). In fact, the pandemic follows the same patterns as during its first wave. Cases first start to rise in East-Asian countries (from 682 on April 1 to 100.276 in September, and since then declining). In October, Europe again is the epicenter of the pandemic. At the end of November, one person dies in Europe from Covid-19 every 17 s. In the Americas, the number of confirmed cases is continuously rising (from 78.959 on June 1 to 165.857 on November 1) [1]. Already early August, simulation models predict a second wave in October and November in most European countries with a higher number of infected cases than in the first wave [2]. Many experts and health professionals expect a second peak of the coronavirus. A survey of doctors by the British Medical Association in September 2020 shows that 86% of respondents believe that a second peak is likely or very likely in the next 6 months [3]. At that time the number of reported infections is already rising, not only in England but in other European countries as well. In the Netherlands, more people test positive in the second week of August than in the entire month of July, possibly because infected persons return from holidays. After the summer vacation period, primary and secondary schools are re-opened. Since May public opinion has changed. Warm weather and impatience with restrictive measures make it difficult to enforce measures. Demonstrations and sometimes violent protests against public health measures become common in many countries. However, in September, with the continuing growth of cases and the rise of hospital admissions, new measures are announced, first in specific regions but inevitably later in the entire country [4]. Early October, the Netherlands is one of the major epicenters of the pandemic with more infected cases than in the United States, Brazil and most other European countries. Restrictive measures are unavoidable, and later that month curfews are imposed in cities and regions in Belgium, France, and Germany, followed by national lockdowns in Ireland, the Czech Republic and France. In the Netherlands, policy responses again are hesitant; the main concern is to avoid too much economic damage. The imposition of a partial lockdown in November (closing restaurants but not shops, amateur sports but not schools) initially flattening the curve but at the end of the month the number of infected people explodes and hospitalizations increase. Stringent measures are unavoidable, and finally on December 15, the government decided to have a 'hard lockdown' for 5 weeks, closing schools and most shops, and prohibiting gatherings of more than 2 people. Even that is not turning the trend, allegedly because of the spread of the British variant of the virus, forcing the government to introduce, for the first time since the Second World War, a curfew. This controversial measure, initially provoking extensive protests and riots, is eventually prolonged until the end of April 2021.

At the time that most European countries report daily record numbers of infected cases, higher than in the first peak, the second wave is declining in East Asian countries. New Zealand never has a second wave. The situation is most dramatic in the United States. In March and April daily cases grow exponentially (around 35.000 per day), remain relatively stable until June 15 (with approximately 20.000 cases per day), grow again until July 16 with new record highs (more than 75.000 daily cases), as more states have reopened businesses. When more measures are taken, and universities and colleges switch to online teaching, cases diminish until September 7 (to around 25.000 cases) and since then grow exponentially until more than 180.000 daily cases on November 13. That day, the country has 11 million infected patients, and a quarter of a million deaths.

In contrast to other countries, the fact that the number of infected people in the United States remains relatively high problematizes the notion of 'wave.' It seems that the country is still struggling with the first wave, while others argue that it is already in the third wave. European countries assume that after the decline of cases in the summer the worst is over, until unexpectedly, infections start to multiply and a 'new wave' can be recognized. In reality, the virus has never disappeared. Lockdowns are lifted while the virus is still circulating. After the stringent policies in the first part of the year, people lose fear of the virus and they enjoy the holiday season. Policy-makers hope that limited and targeted measures, mainly at regional and local levels, can prevent a resurgence of the pandemic. Local interventions presuppose an effective test and trace system, and adherence to social distancing and self-isolation but these counter-measures fail. Regardless of the specific measures taken, in all European countries the number of infected patients starts to rise after the summer. This synchronicity of the surges is difficult to explain. Perhaps it is due to the similarity of public health measures: lockdowns are lifted more or less at the same time, and schools open with the beginning of the academic year at the end of August. Perhaps the weather plays a role with the onset of colder temperatures in September. Also, some failures of policies in the first wave are repeated in many places: warnings of experts are ignored and they are accused of alarmism; fatigue with restrictive measures affects the populations [5]. Furthermore, there might be basic problems. One is the inability to understand exponentiality. While linear growth means that a quantity increases by a constant amount for each unit of time, a virus spreads more rapidly because each infected person infects multiple other persons. There is thus no constant rate of change but the rate of increase accelerates when the number of infected persons grows. With Covid-19, the number of cases is doubling every 3–4 days. Exponential growth will initially not be recognized since the increase of cases is limited and slow. If interventions are delayed because the numbers are relatively low, the pandemic will soon be out of control because the rate of increase is underestimated. The bias concerning exponential growth may explain why compliance with safety measures such as wearing masks, handwashing, and self-isolation is initially low in the beginning [6]. Another basic problem is the idea that public health and economy can be combined, i.e. that lives can be saved while simultaneously the economy can be spared. Policy-makers are reluctant to intervene drastically in order to protect economic interactions. But a choice is unavoidable. Economies are seriously damaged and contracting but the first priority should be early and stringent interventions to slow the transmission of the virus. Such interventions "set the stage of an eventual economic recovery" [7]. Lockdowns and investment in public health will pave the way for eventual economic recovery. Hesitation between efforts to contain the virus and rebuild the economy will make matters worse [8]. Desperate situations call for desperate remedies, and in the end, for most policy-makers saving human lives comes first.

After 1 year, SARS-CoV-2 has infected more than 84 million people and caused 1.850 million deaths. The daily number of confirmed Covid cases worldwide continues to grow until January 2021, declines until mid-February, and rises again since then. While vaccinations have started at the beginning of the year, growth of new cases decreases in some countries such as the USA, UK and Israel but many countries face a new surge of cases. Particularly Brazil and India experience wide-spread devastation due to more aggressive mutations of the virus. Rather than diminishing, Covid infections multiply in 2021. In the first half of that year, the number of confirmed cases doubled (177 million by mid-June) while more people died from Covid than in the entire previous year (bringing the total number of deaths at 3.8 million by mid-June) [9].

The resurgence of Covid-19 in the second part of 2020 demonstrates that many countries have not used the summer to better prepare for new waves of the pandemic. Massive testing and contact tracing continue to be problematic in a number of countries. Policies remain vacillating between strict control and relaxation. Many people have the feeling that they are back to square one, and that there is no perspective of an exit strategy. At the same time, some things have changed. The death rate has declined. Mortality of people in intensive care units is lower because care experiences have improved and some medication (particularly steroids) has proved to be effective. The time during which patients need intensive care is also shorter than in the first wave. The average age of infected people is lower, although the case fatality rate is still high for older people (especially those older than 80 years) [10]. Hospitals are better equipped and organized to maintain regular care. Since care givers have become familiar with the course of the disease, there is less fear and anxiety. The number of hospital beds, especially in intensive care units, has expanded. Scarcities are addressed and stocks of protective materials are adequate, so that more attention can be given to care and elderly homes. Although experiences are growing and numerous research projects have been initiated, controversies persist about the best approaches against the virus because scientific evidence is limited and inherently uncertain.

5.3 Continuing Waves

There is no formal definition of a pandemic or epidemic wave. In fact, it is a metaphor indicating a pattern of ups and downs. Graphs of the number of infected and deceased people over time, clearly show peaks and declines. For example, the Spanish flu had three distinct waves during 1918–1919. Looking at the global situation of Covid deaths, the pandemic thus far has four waves, with peaks in April 2020, January, April and August 2021, while the number of confirmed cases only shows three waves (in January, in Spring and late Summer 2021) [11]. The difference might be due to limited testing so that cases are underreported. But waves also affect countries and continents at different times. The first wave in Europe for example is in April 2020, and in Africa in July of that year. Europe and the Americas have until now experienced four waves, Africa and South-East Asia three. Within countries there are major differences. The Netherlands is now facing a sixth wave, rising in October 2021 (after waves in April, October and December 2020, and April and July 2021). Israel has had four waves (February and September 2020, January and

September 2021). In both countries, each subsequent wave includes more confirmed cases of infection. In hindsight, the first wave which made such political and emotional impact is relatively insignificant compared to the waves that followed. China is an exception; it experiences no waves after the initial one in February 2020, but nonetheless several ripples with a low number of cases in 2020 and 2021, despite drastic public health measures after detection of a few cases.

The continuing surges of the virus infection are attributed to the emergence of new Covid variants. The Delta variant, first identified in India in December 2020, and twice more contagious than previous variants rapidly becomes the predominant variant in most countries in 2021 (with a reproduction number of 3.5–4). Policymakers and scientists had expected that increasing vaccination rates will bring down the number of infections and hospitalizations but this expectation has been jeopardized by the new variant. Israel, one of the first countries to start vaccinations early 2021, is confronted with the highest surge of cases in September 2021, necessitating reintroduction of public health measures, such as mandatory face masks in public spaces. The Delta variant is especially raging in countries with low vaccination rates (for example India and Indonesia which were confronted with overwhelmed healthcare systems) but also in countries where most people are vaccinated, it can infect everyone. Even vaccinated people can transmit the virus, although they have a much lower risk of serious illness and death. This is not true for unvaccinated persons who are now disproportionately at risk and in many countries the majority of patients in the ICUs.

The continuing waves of Covid-19 raise two questions. First is that it puts a damper on the optimism regarding vaccination. This is generally considered as the preferred way to end the pandemic. In countries where most people are vaccinated, the virus does not disappear. This is certainly due to the behavior of the virus that is continuously evolving, but also related to human behavior. After vaccination, humans are not invulnerable, and relaxation of public health measures and providing a preferred status to vaccinated citizens, should therefore be reconsidered. The second question is whether and when we can say that the pandemic has ended. As long as the virus is not controlled globally, and the majority of the world population not vaccinated, is will have multiple opportunities to mutate, and develop variants that are more transmissible and that diminish or circumvent the immune protection afforded by vaccines. Against this backdrop, eradication (as happened with smallpox) is wishful thinking, but also elimination is not realistic. Perhaps it is possible to eliminate the virus in specific regions and countries with a high vaccination rate but then Covid-19 will continue to be epidemic in other parts of the world, especially since the virus has animal reservoirs from which it can be transferred to humans. The most likely scenario for the future of the pandemic is cohabitation. We have to learn to live with the virus since the most lethal phase of the pandemic is over, and the virus will become endemic. From time to time there are small outbreaks but with lower hospitalizations and deaths than until now, and we need regular vaccinations [12]. In this scenario the viral threat has not disappeared but the risks have become more tolerable. Although the disease is still there, its presence and effects are acceptable as part of 'normal' life. There is no return to pre-pandemic life since a certain level of death and illness due to Covid-19 will be included in the new order of life. Declaring that the pandemic has ended is therefore not a medical or epidemiological but political and ethical decision.

5.4 Evidence, Uncertainty and Risk

In March 2020, John Ioannidis published a critical article, arguing that policies against Covid-19 are based on unreliable data [13]. There is no solid evidence on how many people have been infected. Because of limited testing, the majority of infections are missed. It is uncertain what exactly is the risk of dying from Covid-19. The WHO estimations that the fatality rate is 3.4% are probably wrong and a reasonable estimate for the general population in the United States varies from 0.05% to 1%. If true, the fatality rate of Covid-19 will be lower than for seasonal influenza and common cold (caused by other coronaviruses). Because data is uncertain, it is unknown whether extreme measures of social distancing and lockdowns will be effective, while their social and economic consequences are also unknown. Ioannidis' publication not only emphasizes uncertainties and lack of evidence but also points to an ethical challenge: how do we know whether policies are beneficial or harmful? And if we don't know, how are actions that impact the life of all people justified?

If factual information is absent or deficient, and there is urgency or pressure to act because of the perceived threat, policy decisions are only to a limited extent based on scientific evidence, and more often determined by the weighting and balancing of values: saving individual lives, preventing illness, shielding vulnerable populations, protecting social life, education and culture, salvaging employment and the economy. Against this backdrop, dealing with uncertainty in pandemic times fosters specific ethical concerns. Although many other ethical challenges will be discussed in subsequent chapters, five issues will be important in connection to uncertainty: the reliance on quantitative data, the promotion of specific habits, the application of the precautionary principle, the role of experts, and the impact of social media.

5.4.1 The Numbers Game

In most countries, the population is confronted with daily statistics about the number of infected patients, hospitalizations and deaths. This information has two functions. First, it provides insight into the evolution of the pandemic, visualizing the rise and decline of case numbers, and thus the impact of the disease within society, as well as the effect of policy measures. It shows an image of the severity and impact of the virus in the national context and also allows comparisons between different countries. Second, it is a means of making the population aware of the threat, providing estimations of risks, and thus an incentive to adhere to the policy measures. However, the daily reports are also causing horror, and creating anxiety and fear.

Reliance on quantitative data suggests that there is at least some way to assess the effectiveness of measures, opening up the possibility to relax and return to a more normal situation. This sense of security is false. Since many people are not tested, particularly not asymptomatic people who can be infectious, the number of cases is underestimated; many more people are infected than the number of infected cases that is reported. Case data is dependent on testing policies, and these are different among countries, and changing within countries over time. It is the same for the number of deaths; more people are dying than reported. Mortality data is differently reported in countries and over time; early deaths may have been missed, and delays in reporting are usual. People with mild symptoms dying at home are not recognized as confirmed cases, and therefore not included in the statistics. Furthermore, people with other conditions who died because regular care is postponed due to Covid are not counted in the pandemic statistics [14]. For example, patients died of heart attacks or strokes because they could not receive timely and appropriate care. Furthermore, an unknown number of people died because of the economic consequences of the lockdowns.

Policy-makers increasingly use the reproduction number (R_t) as a measure for the transmission of the virus. The assumption is that it predicts the course of the pandemic and that it can guide the planning of policy interventions. This number indicates how many people are infected by one infected person. Usually one such person infects 2–3 other people. Only when the R is below 1, the disease will slowly extinguish. In the early stages of the pandemic, the reproduction number was 3.87. In Italy it ranges from 2.76 to 3.25 [15]. The spread of the virus depends on its transmissibility, the number of contacts of infected persons, and the duration of contagiousness. The reproduction number is an estimation used by disease modelers. However, estimations differ even if they are based on similar data. Data depends on accurate and timely reporting of cases, and are affected by changes in case definition, and sufficient or insufficient testing and tracing capacity [16]. The reproduction number does not necessarily reflect what goes on in the real world. Like the other numbers disseminated to the public, they are a means to control uncertainty. They evoke the failure or success of containment and mitigation policies, and embody the hope to return to normal. Behind the statistics of lethality and morbidity, however, are real people. The focus on numbers not only hides uncertainty but also distress and anxiety [17].

5.4.2 *Risk Rituals*

Initially, responses to the pandemic are based on previous experiences. The terminology of 'emerging infections' suggests otherwise. It implies unpredictability and discontinuity with the past; risks and threats are not a simple repetition of the past.

In Asian countries, Covid is compared with the SARS outbreak in 2002, in Western countries with influenza. When faced with the increasing impact of Covid-19 in China, experts perceive that the risk is low that a similar situation will occur in Europe. The images of the devastating impact in Italy where Covid rapidly overwhelms the healthcare system, change the perception of risk, especially in the general public. Panic buying and hoarding at the start of the first wave leads to shortages in grocery stores as well as of protective materials such as masks. This reflects the fear and uncertainty among numerous people [18].

Perception of risk is not equivalent to objective threat. Evaluating hazards to which people might be exposed, means interpreting the world in terms of what might be harmful or beneficial. This interpretation is based on experiences, values, beliefs and emotions, as well as social and cultural context [19]. Scientific evidence is only one of the contributing factors to risk perception. While evidence indicates that Covid-19 has a lower fatality rate than SARS and MERS, research in Spring 2020 shows that risk perception is high across countries in Europe, Asia and North America [20]. Since Covid has a high transmission rate compared to other coronavirus diseases, there is widespread fear of contagion. The threat is invisible and the virus unknown. As a novel disease, it is unfamiliar, and by many perceived as uncontrollable. It is even more frightening since there is no treatment and vaccination. Numerous studies of previous outbreaks show that risk perception is important for the implementation of policy measures. The perception that the disease is serious and that one can be susceptible to infection, and thus vulnerable, is associated with the adoption of protective behavior [21]. Measures against the pandemic are basically aimed at changing behavior so that risk perception will be crucial for the adoption of these preventive measures. High risk perception possibly explains the compliance with policy measures in the first wave of Covid-19. But nobody can imagine the future, and when the disease persists longer than expected, risk perception declines.

Feelings of uncertainty and perception of risk often motivate the development of ritualistic behavior to allay risks. Mask wearing for example has become a risk ritual in Japan, providing wearers with a sense of control when facing uncertainties [22]. Such rituals are not primarily a personal choice but a response to social pressure because the risk is collective, although they provide a sense of personal control. They are precautionary behavior that cannot simply be assessed in terms of efficacy. When the state of uncertainty is prolonged, as in the case of Covid-19, such ritualistic behaviors can develop in many countries [23].

5.4.3 The Precautionary Principle

The opinion of Ioannidis, mentioned above, that lack of good data does not justify rigorous measures such as lockdown, is countered with the argument that precisely because the facts are unknown, there is a reason to be cautious [24]. A prudent approach is not to wait until estimates and evidence are clear and convincing. This argument is based on the precautionary principle. The principle is formulated in international treaties and declarations. It is used in two senses: negative, as the duty to abstain from acting when the potential exists for serious and irreversible consequences, and positive as the obligation to take measures to avoid potential, severe and irreparable harm, in the absence of scientific certainty. A famous example of the latter is the removal of the handle of a water pump by John Snow during the cholera epidemic in London in 1854. There was only weak evidence at the time that the spread of cholera was causally linked to the water pump.

The precautionary principle is defined as: "When human activities may lead to morally unacceptable harm that is scientifically plausible but uncertain, actions shall be taken to avoid or diminish that harm" [25]. Harm is morally unacceptable when it is threatening to human life or health, and serious and irreversible. Applied in the context of Covid-19 the principle implies that action that may reduce risks, even when scientific certainty is lacking, should be used to prevent infection. In the early phases of the pandemic, several countries did not recommend or even disadvise the use of face masks, arguing that benefits are scientifically not proven, while harms may possibly occur. However, using the precautionary principle in its positive sense, it can be supposed that while current evidence of benefits from wearing a mask is sparse and contested, and while there is no hard evidence for harms and negative effects of masking itself, the potential harms of not wearing them may be severe and irreparable. Given these uncertainties, dismissing mask use may potentially and plausibly result in harms which can be avoided by action. Masking is furthermore a low-cost intervention that is easy to apply with potentially a large impact on the pandemic [26]. There is therefore an ethical imperative of applying the precautionary principle, and to wear face masks [27]. This is not simply a risk-benefit analysis since harms, risks and benefits are all scientifically uncertain, at least at this stage of the pandemic. Later, when more robust evidence is acquired about the benefits of masking, a different situation emerges.

5.4.4 The Role of Experts

The pandemic has highlighted the role of scientific experts in policy decision-making. In most countries, politicians are advised by scientific and medical officers. They often argue that all decisions are based on the best scientific evidence. When the pandemic lasts longer, the role of experts becomes increasingly contested. One criticism is that the range of expertise is limited. Most advisory bodies include physicians, epidemiologists and virologist but not psychologists, behavioral scientists, sociologists, let alone ethicists. The result is a technocratic approach based on mathematical disease models rather than real-world evidence, and concerned primarily with hospital capacity, particularly intensive care beds. A broader perspective could have avoided the catastrophe in nursing and elderly homes, and enhanced attention to diminishing compliance with restrictive measures [28].

Experts work from existing frames of knowledge. They proceed on the basis of historical precedent, experiences from previous pandemics such as the Spanish flu (1918) and the Mexican swine flu (2009). Although pandemics have been predicted, what was expected first of all was influenza. However, existing knowledge can be wrong when applied to an emerging disease. Prior to Legionnaires' disease the assumption was that water towers and air conditioning systems are not risky for hotel guests. Initially, experts assumed that Ebola does not circulate in West Africa, and that Zika is a relatively harmless disease carried by mosquitos. For some time, the official policy in the Netherlands and other countries is that Covid patients are only infectious when they have symptoms such as cough and fever. Expert opinion also is that testing should be very selective and only applied in the hospital setting. The reluctance to learn from colleagues in other countries has already been mentioned in the previous chapter. Although early information from China is available, many experts question the reliability of the data. Officially there are 2548 deaths in Wuhan as a result of the outbreak. But in social media people report that there are at least 5000 urns delivered to families. People who died at home, are not included in the official statistics. Asymptomatic cases are not calculated, and only later included in the statistics. The real death toll in Wuhan could have been ten times higher than the official number [29]. These instances of political manipulation of information undermined the credibility of reports and articles of Chinese scientists in established journals so that real-time experiences are less influential than historical precedents.

Controversies and uncertainties are common in science. In the context of the pandemic, policy-makers as well as scientific advisory bodies are concerned to avoid mass panic. At the same time, they are optimistic about future solutions such as widespread testing and the arrival of vaccines. Differences of opinion are not welcome, and scientific disagreement is discouraged. Deliberations of advisory bodies are secret, so that the various arguments and justifications for recommendations are not apparent. Initially, there is a high level of self-assurance among experts. Later, alternative teams of experts are operative, publicly questioning the recommendations of the official expert bodies that advice governments. An example is the so-called Red Team in the Netherlands which is promoting more stringent measures and more extensive testing, especially in schools and nursing homes. Moreover, discrepancies emerged between scientific recommendations and policy decisions. In the Netherlands, schools and restaurants are closed in the first wave against the advice of experts. In general, expert advices in numerous countries are not unanimous. Politicians on the other hand use the scientific advice to justify the measure they want to implement anyway. Countries where experts have the initiative in policy-making do not perform better than those where politicians have the leading role [30].

5.4.5 The Impact of Social Media

A final issue is public communication. Political decisionmakers use to inform the population, often in regular press conferences, flanked with scientific experts. Communication that is trustworthy not only relies on expertise but also on honesty and transparency [31]. Uncertainties need to be disclosed. Besides stating what is known, it should be discussed what is not known, what is investigated and explored, and how evidence is changing. But in many cases, such disclosures are considered as confusing, and therefore not provided [32]. The Covid pandemic illustrates that nowadays ways of communication have drastically changed, with social media as the platform of choice. From the beginning, people have used media such as Twitter, Facebook, Instagram, You Tube, and Whatsapp to share stories and experiences with the pandemic. Public health measures have increased the use of the media so that people could stay connected for work, education and communication despite the need of physical distancing. Social media is also used by public authorities to disseminate urgent information and recommendations. Scientific information is quickly exchanged, so that the latest medical and research information is widely distributed [33]. Thanks to social media messages, the world was first alerted that a new infectious disease had emerged in Wuhan in December 2019, before the Chinese authorities informed the World Health Organization, and prohibited such messages [34].

In the course of the pandemic, however, attention shifted from the positive to the negative dimensions of social media. They are crucial communication tools for health information but also of a wide variety of opinions, perceptions and attitudes which contents cannot be controlled, so that misinformation and false information are rapidly disseminated through what has been called an 'infodemic'. It is difficult to distinguish accurate and inaccurate information [35]. Social media has developed into a major source of false information and conspiracy theories (see Chap. 8). When in times of uncertainty and threat the need to seek information is growing, even a few minutes of exposure to Covid news can have negative consequences for human wellbeing [36].

5.5 Controversies

"I've studied 100 diseases. Covid is the strangest one I have seen in my medical career," remarks British epidemiologist Tim Spector [37] SARS-CoV-2 is a highly unpredictable virus. It presents a wider range of symptoms than previously recognized. Patients can have complaints for a long time. The pathology is not well understood and potentially any organ can be affected. Covid-19 really is a new disease that cannot be compared with the seasonal flu. Numerous dimensions of the infection are not known. Covid-19 makes more people ill at the same time than the usual influenza, and many experts agree that it is more serious than this seasonal illness. Covid-19 as well as influenza cause respiratory disease, and both viruses are transmitted in a similar way but Covid-19 has more severe consequences: 15% of patients require oxygen and 5% need ventilation. Covid-19 is more deadly: approximately 1% of infected people die, while the fatality rate for influenza is less than 0.1% [38]. Lethality depends on age and gender. In the Netherlands, the probability to die for people younger than 35 is 0.004%, for 70 years old people 2.5% and for those older than 85 years 28.3%. The probability to die is 1.5 times higher for men than for women [39]. The novelty of the SARS-CoV-2 virus implies that hard facts are scarce. Many controversies therefore exist since scientific information is uncertain, ambiguous and slowly developing.

5.5.1 The Origin of the Virus

As discussed in Chap. 3, a politicized controversy concerns the source of the virus. It is not clear where and how the virus emerged and trafficked to humans. Such knowledge is important to prevent future pandemics, especially enabling prevention at the source. It can be the case that the virus was already circulating for some time before the 2019 outbreak happened, in China or elsewhere. For the Spanish flu it is argued that the virus circulated between 1900 and 1915 and for some unknown reasons became lethal in 1918 [40]. For SARS-CoV-2 four possible scenarios are explored: direct spillover from animals to humans; spillover through an intermediate host; introduction through frozen foods; and escape from a laboratory [41]. After almost a year of negotiations a WHO mission of experts is allowed to visit Wuhan in January 2021 [42]. The mission concludes that it is possible/likely that the virus jumped from an animal, most probably a bat, to humans. It is more likely that the virus first infected another animal (perhaps a pangolin or mink) and then infected a human being. It is also possible that the virus is imported on the surface of food packaging or in the food. The mission report finally concludes that it is extremely unlikely that the virus escaped from a laboratory [43]. After the report is released, 14 countries criticized it, arguing that a full and independent investigation has not been possible. China has not provided key information such as data of the early Covid cases. The mission team was prohibited to contact community members. The team was not really independent since half of the team members were Chinese scholars. The Director-General of the WHO admitted that the investigation was not extensive enough, and that all scenarios should be further explored [44].

In October 2021, the WHO undertook a new effort to identify the origin of the pandemic, launching the Scientific Advisory Group for the Origins of Novel Pathogens (SAGO) to explore the origin, not only of SARS-CoV-2 but of all future emerging pathogens [45]. This initiative came after the possibility of a laboratory leak has been more seriously considered [46]. Such leaks are not uncommon. The last person dying from smallpox in the world (in 1978) was a medical photographer working in the same building as the smallpox laboratory at the Medical School of the University of Birmingham in the United Kingdom [47]. Over the years, many dangerous pathogens have escaped from laboratories across the world [48]. In September 2021, the US intelligence community released a report to find the origins of SARS-CoV-2. It investigated the two main options: spillover from nature, and a laboratory leak but was inconclusive. The only conclusion was that the virus was not genetically engineered and developed as a biological weapon [49].

5.5.2 The Characteristics of the Virus

Like all viruses, SARS-CoV-2 is mutating but compared to the influenza, HIV and Ebola viruses, mutations are limited. Already in September 2020, researchers have identified more than 12,000 mutations. Because the virus has infected millions of people, multiple variants are circulating but what the effect of mutations is for the contagiousness and potential danger of the virus is initially not clear while there can also be implications for the effectiveness of antiviral strategies and vaccines [50]. In December 2020, a new virus variant (Alpha) spreading faster than the initial pathogen is discovered in the United Kingdom, at that time confronted with a surge of Covid cases. This finding creates a sense of panic, in the United Kingdom where vaccinations has just started, as well as in other countries that immediately prohibited travel from the UK just before the Christmas holidays, while all countries imposed more stringent public health measures [51]. Soon, other variants are identified in South Africa and Brazil [52]. The healthcare system in South Africa is overwhelmed because of the dissemination of a virus variant (Beta) that is not only more transmissible but also evades the immune system since antibodies and vaccines are less effective against them. The Gamma variant is widely circulating Brazil (and detected in Japan in January 2021), ravaging the city of Manaus that already had experienced a serious Covid outbreak the previous year, and where more than 70% of the local population was assumed to be immune [53]. This variant is not only the most transmissible of all variants, infecting younger people, but is also more lethal. It is perfectly capable of evading immunity from earlier coronavirus infection, so that many reinfections occur. Variants can only be detected with genetic sequencing but this is initially hardly done in most countries, except in Denmark and the United Kingdom that have extensive programs of genomic surveillance of SARS-CoV-2 [54]. When other countries start or expand this surveillance more variants are discovered [55]. Most concerning is the Delta variant, discovered in India at the end of 2020. It caused a second wave in that country in Spring 2021 with an extremely high number of cases. This variant of SARS-CoV-2 is more transmissible and causing more serious and different symptoms than pre-existing variants. In June 2021, it has spread to 74 countries, necessitating countries such as the United Kingdom to delay relaxation of public health measures. It also seems that one vaccine dose is less effective against this variant while two doses are strongly protective, at least for some time [56]. Nonetheless, it becomes clear that the Delta variant is able to infect vaccinated persons ('breakthrough infection'), and in some cases make them ill so that they need hospitalization. In Autumn 2021, the Delta variant itself mutated into a new variant which could be even more transmissible than the original Delta variant. When variants are detected it usually is too late to contain them; the viruses have already migrated to other countries [57]. The finding that highly contagious variants such as Delta can infect fully vaccinated people has tempered the optimism that vaccination can end the pandemic, and has reiterated the continuing need for preventive and public health measures.

When Covid-19 emerged and many characteristics of the virus were unknown, two prognostic models are used. Some experts assume that the best-case scenario is that the virus mutates and dies out, like human coronaviruses such as SARS and MERS have disappeared. Another scenario is that the virus will not go away. Waves of the pandemic will continue at least until 2024. Outbreaks of coronavirus will arrive every winter. The virus must be eliminated worldwide to end the pandemic, and this seems impossible. The experience with new virus variants makes the second model more likely but it is still uncertain since the outcomes depend on unknown factors such as the duration of immunity to the virus, the policy choices of governments and the behavior of individuals, and the effectiveness of vaccines [58]. Although it is important to know the characteristics of the virus, for example to determine the risk of severe illness or reinfection, this in itself does not determine the dangerousness of the virus. SARS-CoV-2 can be compared to similar viruses as more or less dangerous, but 'danger' is a value judgment depending on estimations of risks in specific circumstances, making the virus more dangerous to some people than others, for example because of social inequality. The experience with new variants strongly articulates the global dimension of the pandemic and public health policies. The enormous number of infections across the world provides many opportunities for the virus to mutate. As long as public health measures are not applied everywhere and the majority of the world population is not vaccinated, new variants will continue to emerge, creating risks for other people, and making the gains of vaccination questionable [59]. Failing or denialist policies in some countries such as Brazil jeopardize efforts to mitigate the impact of the pandemic in all other countries.

5.5.3 Symptomatology

Initially, the clinical picture of Covid-19 is well described but now it seems clear that at least 15 different types of symptoms can be recognized [60]. For example, in critically ill patients the respiratory muscles particularly in the diaphragm can be affected causing persistent dyspnea and fatigue [61]. It also becomes clear at the end of 2020 that the virus may cause neurological symptoms in one out of every seven people infected. These symptoms can be serious such as stroke, encephalitis and cognitive decline [62]. While it is generally assumed that children are less vulnerable to Covid infections, and have milder symptoms or no symptoms at all, sporadically a serious condition is associated with Covid-19 in children [63]. It is obvious that everybody can be infected but older people and persons with underlying conditions are more at risk. After infection the incubation period is 3–5 days; on average symptoms appear after 5–7 days, mostly within 10 days. It can last a week before the symptoms are so serious that hospital care is needed. In the Netherlands 2% of ill persons are hospitalized; some of these patients need intensive care. In the first wave, about 25% patients of patient with serious pneumonia are admitted to the ICU, in the second wave, about 20%. Approximately 24% of patients in the ICU have died during the first wave, the majority older than 80 years. Why men have higher risks is unknown; women and men are equally infected but men are more frequently hospitalized. Furthermore, there are uncertainties about the long-term effects of Covid-19. Many patients discharged from hospital and recovering from Covid-19 report a variety of symptoms 6 months later (most commonly fatigue, muscle weakness and sleep difficulties) [64]. These complaints can also be related to hospitalization and intensive care treatment, and it takes time to fully recover. However, it became clear that anybody after infection, even asymptomatic persons and those with mild illness, can experience symptoms lasting weeks or months after being infected. This so-called 'long Covid' is manifested with a range of symptoms such as tiredness, headache, brain fog, shortness of breath, muscle pain, and depression [65]. It is estimated that one in ten patients do not feel well after 3 weeks and suffer from long-term symptoms [66]. Better insights into the symptomatology of Covid infection is important since it not only facilitates improved care and treatment but also dispels some myths about the disease, for example, the argument that it is comparable with (and even less harmful than) the seasonal flu.

5.5.4 Infectiousness

It took some time to acknowledge that the virus is transmitted from person to person but soon it became clear that infection takes place through direct contact and inhalation of respiratory droplets from a coughing and sneezing patient. Infected people can transmit the virus to others, 1 or 2 days before they have symptoms. They are most infectious in the first 5 days after development of symptoms. It is estimated that it takes 7–13 days to develop symptoms. A substantial number of infected persons are not experiencing symptoms, and they will 'silently' infect other people. Initial studies conclude that 81% of all infections are asymptomatic but that was at the time when fever was not considered a criterion for Covid. Now, calculated numbers are lower but still varying from 50% to 17% [67] A famous example is the Italian town of Vo'; during the lockdown the majority of its population was tested; more than 40% of those who tested positive were without any symptoms. Another example is an outbreak on a French aircraft carrier: nearly 50% of positive tested persons had no symptoms. Similar experiences are observed in nursing homes, prisons, cruise ships, and universities [68]. While it is argued that symptomatic people are the most infectious because their viral load is higher whereas asymptomatic persons present a lower risk but can transmit the virus for an extended period, sometimes longer than 14 days, these findings are contested since other studies show that both groups of persons have a similar viral load that is peaking in the first week of infection. Probably, asymptomatic patients clear the virus faster and are infectious for a shorter period of time [69]. The problem is that there is lack of clarity because 'asymptomatic' is not clearly defined. Some patients who are asymptomatic at the time of testing develop symptoms later, and are in fact pre-symptomatic. It is furthermore unclear whether infection without any symptoms may lead to long-term damage to at least some organs in some people. While data and studies are imperfect, it is assumed that asymptomatic persons play a significant role in the progression of the pandemic but there is no consensus about how much they actually contribute to transmission [70].

The uncertainties regarding infectiousness have implications for policies. People who do not experience symptoms but are infectious may adhere less to the public health measures than people who are ill. They will thus contribute more to the spread of the virus than patients. If it is the case that the majority of viral transmissions take place very early in the disease course, testing, contact tracing and self-isolation should start immediately after symptoms are experienced. Testing policies are different: some focus on patients with symptoms or specific categories such as healthcare workers, patients admitted to hospital or those having been in close contact with infected patients. Other policies make testing available to asymptomatic people. In the Netherlands, the official policy at first is that testing people without symptoms is not effective. In the United States, the CDC in August 2020 issues guidelines that persons without symptoms do not need to be tested. Previously, the agency has stated that those who has been in close contact with an infected person should be tested even when symptoms are absent. In September, the guidelines are reversed to the initial position.

Some scholars argue that widespread testing of persons without symptoms is the cornerstone of policies to reduce transmission of Covid-19 [71]. If there is a substantial number of silent carriers of the virus symptom-based testing and isolation will be insufficient to control outbreaks [72]. Frequent testing of the entire population of a country will improve the management of the pandemic. It will enhance the predictability of the disease and also shorten the interval between policy changes and observable effects in terms of cases and hospital admissions, so that there is better guidance for the modification of restrictive measures. But it requires an enormous increase in the extent and frequency of testing. This can be done for a small country such as Switzerland. With 1.5 million test per day, it will be possible to test every resident every 5–6 days. It will also require that positive persons will go into strict quarantine. In this way the exponential growth of the pandemic can be prevented and lockdowns will not be necessary, allowing the economy to reboot [73]. In October–November 2020 Slovakia started to test the entire adult population (approximately 4 million people) with 80–90% of the people participating. Participation is voluntary but anybody who does not take part must self-isolate together with his or her household for 10 days; substantial fines will apply if the quarantine is broken. If the test is negative, a certificate valid for 7 days is provided, necessary to access public offices, banks and workplaces. The program with weekly testing of almost all residents, organized by the Slovak army, decreased the prevalence of Covid cases by 58% [74]. Similar proposals to frequently test every citizen have been made for other countries [75]. The promise is that in this way the pandemic can be ended rapidly and lockdowns avoided. Those citizens subjected to regular testing will be granted more freedom than the untested who remain barred from social interactions.

Massive testing is nowadays possible because a new type of test is available. The usual molecular test (PCR test) determines the amount of virus in the body. It is the most reliable test because it is very specific: with a negative result, it is more than 99% certain that a non-infected person has no virus. The sensitivity is lower: a number of non-infected persons receive a positive result (estimated at 15%). It is argued that the test is too sensitive since it is also positive when the viral load is low, and the person is therefore not so contagious. All positive tested persons need quarantine and contact tracing while for many it will not be necessary. This raises an ethical concern: how harmful is it when a person has to isolate when in fact it is not necessary? During the lockdown almost everybody needs to remain at home, now with wide-spread testing only a limited number of people, but there is better surveillance of the spread of the virus and the number of people that potentially infect others is limited. The introduction of rapid antigen tests after the summer of 2020 changes the debate. They can provide results in 15–30 min. The PCR test takes time, trained personnel, infrastructure and logistics. Rapid tests do not require laboratories. However, they are less sensitive so that cases are missed and it is unclear how many people are really negative. A review of studies of these test concludes that they work best in people with symptoms (identifying four out of five infected cases) but in asymptomatic cases they identify only half of all Covid-19 infections [76]. Even more easier are self-tests that can be performed by individuals themselves at home or elsewhere. They come into the market in the beginning of 2021 and are widely advocated before participating in social life, such as schools, sports, and visits to other people. They allow quick clarification for individuals about their health status. From a public health point of view, self-testing removes the responsibility for testing from laboratories and health professionals to individual citizens. This shift may complicate reporting of test results as well as contact tracing and quarantining since the interpretation and reporting of results is entirely in the hands of individual users. Positive results require confirmatory PCR testing, false positive results lead to unnecessary self-isolation, false negative results may give an inappropriate feeling of security and thus may increase transmission of the virus, while a true negative result is reassuring but does not preclude infection the next day [77].

There are doubts about the effectiveness of massive testing, especially with PCR tests [78]. It is a huge social effort while the goal is not clear. Apparently, testing is used for diagnostic purposes, i.e. to know who has Covid-19. Identification of infected persons is helpful for modification of policies and for protection of people at high risk, giving priority testing to care providers and school teachers. In that case, 'vital groups' have to be identified. But this diagnostic use is questioned since the symptoms of Covid are specific and can be diagnosed without using any specific test. Many patients with milder symptoms stay at home, and are not regularly tested. Others argue that PCR tests are necessary for epidemiological purposes: to help to ascertain how the virus is behaving within a population, to find people with active infection, and provide means to disrupt transmission. The tests are then used for population screening rather than individual diagnosis. This epidemiological goal is questioned because there are alternative approaches such as testing of sewerage water, surveillance systems in general practice, and the number of hospital admissions that provide information about the quantity and distribution of infected people. The introduction of rapid tests changes the discourse: they can be diagnostic (although less reliable) as well as epidemiological instruments. They provide real-time evidence, going beyond the assumptions of mathematical modelling [79].

It is critical to remember that testing is part of a chain of public health measures: after a positive test extensive tracing of sources and contacts needs to be done, with strict quarantine of infected people, while recommendations for social distancing and hygiene need to be implemented by everybody since a negative test is only a snapshot result. If there is a large number of infections, contract tracing is impossible to do. Patients, especially when they are young, have many contacts, and in most countries the number of personnel to do the tracing is insufficient. The municipal health services in the Netherlands wrongly assumed that a positive patient has 3–4 contacts, thus expecting that the tracing will require 5 h. The large number of cases and contacts made it impossible to logistically execute the tracing. Contact tracing is furthermore less effective when there is a time lag between infection and symptoms, and when there is a gap between having the test and receiving the result [80]. In 70% of the cases in the Netherlands it was not possible to identify the source of the infection. If identification was possible, it turned out that most infections took place between housemates (54%), family members (9%), friends (8%) and at work (9%). Less infections occurred in restaurants (6%), sports clubs (3%) and among travelers (6%) [81]. If rapid testing is readily accessible, it may potentially improve contact tracing since the result are quickly known. But it will only strengthen one part of the chain. Even if tracing capacity is sufficient, not all contacts can be identified, while infected people need to be quarantined or isolated.

The controversies around testing illustrate the ethical issues at stake. With PCR testing, at least some non-infected people will needlessly stay in quarantine, while with rapid antigen testing more positive cases are missed so that a number of infectious people participate in social life. What kind of harm should be taken more seriously? The shifting aims of testing raise different concerns. If tests are used as a screening mechanism, an ethical framework for mass testing is necessary [82]. Such framework should imply information of participants about the purpose, limitations, and uncertainties of testing, explain that it is an offer, and clarify how data will be used [83]. In practice, it is not transparent if and how such ethical requirements are applied. In Slovakia, people are not mandated to undergo testing but non-participation is made unattractive because certificates of being tested are issued and requested in various contexts. In South Korea, people in close contact with confirmed cases are in mandatory 14-day quarantine, and can only be released after mandatory diagnostic testing [84]. Proposals for universal repeated testing in the United Kingdom emphasize that testing should be voluntary but at the same time it is coupled with penalties for breaching quarantine after a positive test result. In fact, the population will be separated in positives and negatives, while the last category will receive an identification card to resume normal life [85]. Such proposals make it questionable whether participation is really voluntary. Two categories of citizens are created, the tested and the untested, with the first allowed to participate in social activities while the second continue to be isolated and 'locked down.' This classification may lead to stigmatization and discrimination on the basis of health status. The urge to expand testing and particularly rapid mass testing seems driven by the political wish to reopen the economy as soon as possible. Economic considerations and the rejection of the idea of an immediate lockdown were important motives to start population screening in Slovakia. Rapid self-testing in the Netherlands is heavily subsidized and promoted by organizations of employers and businesses. Testing in this perspective can become a technology-oriented approach to the pandemic, prioritizing speed over accuracy, and suggesting that infectiousness can be controlled, while neglecting other significant determinants of transmission. Negative test results have a reassuring and liberating effect, and may make individuals less cautious, thus making it harder to maintain and implement public health measures.

5.5.5 *Transmission*

The common idea is that the virus spreads from person to person through droplets from the airways of infected people when they sneeze, cough, cry, talk or sing. These respiratory droplets fall down within 1.5 m. This finding is the basis for physical distancing measures. The virus can also be transmitted through contact with infected persons which is the basis for the advice of frequent hand washing. Findings that the SARS-CoV-2 survives on plastic and stainless-steel surfaces for up to 72 h, and for example can persist on the touch screen of mobile phones for at least 28 days at room temperature as well as on paper currency, led to concerns about transmission through so-called 'fomites' (i.e. objects or materials which transmit infection) [86]. People started to clean groceries and parcels, and payments with paper money were no longer allowed. Authorities recommended cleaning and disinfecting surfaces. Sales of sanitizer and disinfectants boomed. Later, researchers argue that these findings are based on laboratory experiments, and that they are not applicable in ordinary life situations [87]. The risk of transmission through inanimate surfaces in real life is minimal. To reduce virus transmission, it is more important to apply physical distancing than disinfection; people rather than surfaces should be the major concern [88].

A controversy has arisen about the question whether the virus is transmitted through fine droplets (aerosols). Such droplets are produced through respiration and can be floating in the air for some time. Viable virus has been isolated from aerosols generated in hospital rooms of Covid patients, 2–4.8 m away from the patients [89]. Airborne transmission may explain infections in conditions where people are close together for a prolonged time such as households, restaurants, gyms, and schools [90]. It also explains the role of superspreading events: a small number of people with a high viral load can infect a disproportional number of other people through close contacts for example during funerals, marriages and choir singing, especially if people gather indoors. In March 2020, 102 of the 130 members of the Amsterdam Concertgebouw choir fell ill and one singer died of Covid-19, although physical distancing had been practiced [91]. Clusters of Covid infections have been identified in meat-packing plants, student dormitories and elderly homes, where people are crowded together in often poorly ventilated conditions. Contact tracing in such cases is less effective since it usually does not trace backwards in time, identifying the source that has infected the patient during 14 days before symptoms emerge.

Policy recommendations about airborne transmission are not clear. It is generally assumed that the risk of infection through aerosols is negligible. The WHO in early stages indicates that the disease is not airborne. A large number of scientists appeal to the WHO to revise its position [92]. In July 2020, the organization grudgingly accepts that aerosol transmission may happen, although further studies are needed. In the Netherlands, the RIVM maintains that under normal circumstances, there is minimal risk of spreading the virus through small droplets. Even if there is evidence of aerosol transmission it is not evident how frequently it happens and how serious the risk of infection is. Prevention should be based on the assumption that all droplets, large or small can lead to infection at close proximity so that the usual measures of distancing and hand washing should be sufficient. This conclusion is contrary to the point of view that airborne transmission requires additional measures such as better indoor ventilation, assuming that distancing and hand washing are not enough to prevent infection, certainly not indoors. These suggestions are rejected by policy-makers because airborne transmission may cause public fear and will require disproportional efforts in regard to the minimal risk. Perhaps, it is also an unwelcome reminder of the miasma theory of the past.

The debate on viral transmission illustrates the ethical aspects of risk assessment. When the role of fomites and aerosols in the spread of the virus is not certain and hard evidence is lacking, then it can be argued that relevant measures (such as cleaning surfaces and ventilating) deliver only minimal benefits compared to the efforts that should be undertaken to eliminate infectiousness. On the other hand, an appeal is made to the precautionary principle. If harm occurs, it is serious, and in some cases, lethal; all attempts to reduce the risk are therefore beneficial, even when the evidence is insufficient.

5.5.6 Distancing

One of the most common policy recommendations is physical distancing. It is based on the observations that respiratory droplets may transmit the virus. The chance to be infected is highest when people are close to each other for at least 15 min. The risk of infection decreases with the distance taken from an infected person: it is reduced by 82% with a physical distance of 1 m; every extra meter may half the risk of infection [93]. The specific distancing rules differ among countries: in Japan and the United Kingdom people are instructed to keep 2 m distance from each other, in Germany, the Netherlands and Australia 1.5 m, and in France, Norway, Vietnam and Singapore it is 1 m; in Sweden distancing is advised but a specific measure it not provided. Scholars, however, have argued that distancing rules are based on outdated science. The assumption is that respiratory droplets come in two sizes with large droplets falling and small droplets floating in the air. But droplets exist in a continuum of sizes. The virus may travel more than 2 m through coughing and sneezing, while transmission also depends on the crowded setting, indoor airflow patterns, and exposure time [94]. The almost universal introduction of distancing seems to reflect value judgments about what is safe rather than scientific evidence. Politicians in the United Kingdom suggesting that the 2 m rule can be reduced, were severely criticized. Avoiding close contacts is problematic from a human and social point of view. Some countries have introduced the idea of 'social bubble:' a small group of people can have close contacts with each other outside their household while practicing distancing with others outside the group. A corona bubble approach is applied in New Zealand, Australia and Belgium. In the last country, each household can invite up to four guests to their home; when the number of infections rises, the number of guests per family is reduced to one close contact [95]. In the United Kingdom three households are allowed to create a 'Christmas bubble.' The idea is to ameliorate the need for social interaction and to help families and friends to come together without distancing and masks. It allows vulnerable and isolated persons to engage in social connections and to cope with the stress of the pandemic.

From an ethical point of view, interaction and intimate contact are essential for human beings. Shaking hands welcomes people and establishes connections, while hugging is an almost universal expression of emotion and affection. The meaning of both gestures has now reversed, and they are interpreted as biohazards. Physical distancing and lockdowns are antithetical to the relationality that is fundamental for human existence. Isolating people produces stress and loneliness but it is not merely a source of mental troubles but itself an existential threat [96]. In this perspective, rigorous application of anti-pandemic measures has serious implications for the human dimensions of life. For example, the stringent prohibition of visits to people in elderly and care homes during the first wave of the pandemic, deteriorated the quality of life of many. Even more dehumanizing were the conditions of seriously ill patients who died alone, and were often buried quickly without ceremony and family. Especially now that the pandemic last longer, maximum risk reduction should not be an absolute priority. Risks can never be reduced to zero, and some risks should be accepted in order to make human life more bearable. Policies should be flexible and find compassionate approaches which can perhaps also increase compliance and reduce corona fatigue.

5.5.7 *Masking*

The issue of face masks is one of the most controversial challenges in the pandemic. In several countries it has become a political statement. Wearing a mask is regarded as a muzzle, an instrument to restrict individual liberty or a manifesto, a sign of solidarity and concern with fellow citizens. Facemasks have become the most visible symbol of the pandemic. Policies regarding masking have been very different. Although in some countries face masks are used from the beginning, in many other countries they are initially not deemed essential. Later, their use is incrementally advised, first in circumstances where physical distancing is less feasible (e.g. in public transport), subsequently in the public domain (especially indoors, e.g. supermarkets, museums, and restaurants). Policies in the Netherlands have been confusing. In the first stage of the pandemic masks are dis-advised (emphasizing that their effect is not proven, and that they even might have negative effects). Somewhat later they are required only in public transport. Since October 2020 the government urgently advises them, and finally (in December) they are made mandatory in public indoor spaces, while in other countries such as Germany they are obligatory since the summer. These changing policies are mainly the result of pressures from society and businesses. Voluntary advise is difficult to implement, putting the burden of enforcement on businesses and leading to polarization and disputes with customers. There are also increasing problems in hospitals with aggression against care providers. In countries such as the United States with many protests against mask use, the public is initially discouraged from wearing masks. There is fear that there will be an increasing shortage of masks especially for healthcare professionals. In April 2020, the CDC reverses its position and states that face covering is advisable. The agency strengthens its advice to wear masks in November. An overwhelming majority of the American public (75%) say that they always wear a face mask when they leave home [97]. International guidelines are confusing. The WHO first opposes the use of masks by healthy people, and recommends medical masks for patients with symptoms but it changes position in June 2020, advising the use of non-medical masks in public spaces because updated information has shown pre-symptomatic and asymptomatic transmission. At the same time, it underlines that there is no direct evidence on the effectiveness of universal masking, and it details a long list of potential harms and disadvantages [98]. Changing policy advices may produce confusing messaging to the general public.

Reluctancy to recommend masking is based on lack and insufficiency of scientific evidence. Nonetheless, many studies indicate that the use of face masks is associated with a large reduction in the risk of infection. They especially diminish the transmission of virus by asymptomatic and pre-symptomatic persons who are not aware that they may infect others; infected persons using masks transmit viral droplets less far. In order to prevent transmission all healthy persons should wear a mask [99]. Studies furthermore show that masks reduce the dose of the virus to which somebody is exposed; they prevent infection and lessen the severity of symptoms if somebody gets infected [100]. Using masks therefore is not only an altruistic act but also self-protective. Other studies, however, have different outcomes. A randomized clinical trial for example concludes that surgical masks do not protect wearers against infection, but do prevent transmission to others [101]. This conflicting evidence points to two different justifications for the use of masks: source control (preventing transmission of the virus to others) and personal protection (protecting the wearer against infection). Experience with mask use exists in hospital settings, thus most research focuses on the extent to which masks protect the wearer. What is needed from a public health perspective is knowing whether and how covering the face is protecting others from droplets emitted by the wearer. These two situations are different, and the requirements for masks differ; even with home-made face coverings large droplets are blocked before they are aerosolized [102].

The opposition to masks focuses on the lack of proof of effectiveness. Hard evidence is lacking because there are no clinical trials comparing people with and without masks (the above-mentioned study is criticized because it only relies on self-reporting). It is not clear what the effect of masks is in addition to distancing and hygiene. Studies assume that mandates to wear masks are implemented and that people wear them correctly. For example, in Spain during the second wave, cases rise while there is a national mandate to wear masks. It is also argued that there are very different types of masks with divergent filtering capacity [103]. N95 or FFP2 respirators are the standard of care for professionals working in Covid wards, and their effectiveness is proven within this context but for surgical masks and face coverings in the community this is unclear. Another type of argument concerns the use of masks. They are often ineffective because they are not correctly applied and removed, not changed regularly, and not disposed of safely so that wearers can be exposed to higher risk of infection. Another argument against masks articulates that they provide a false sense of security; with a mask people are less cautious to follow the other measures of distancing and hygiene. Policy-makers are concerned that the focus on masks will undermine their measures. Whatever the evidence, masks will not be protective on their own but only in combination with other measures. Since many factors determine viral transmission, it is not possible to obtain scientific evidence by comparing countries where mask-wearing is mandatory and other countries where masks are not used. Whether face masks prevent the transmission of the virus depends of various factors such as the quality of the mask and its adequate use. A final argument, particularly in the first wave, is that masks, given the existing shortage, should be reserved for healthcare workers.

Protagonists argue that masks are used since some time in healthcare settings. There is a long history of efforts to evaluate the efficiency of masks for surgeons and many attempts to improve them [104]. Face coverings were also commonly employed during the Spanish flu pandemic. A 2011 review of 67 studies on the effectiveness of physical interventions to limit or reduce the spread of respiratory viruses (e.g. influenza) concludes that wearing masks is effective in containing epidemics and pandemics [105]. Earlier, Wu and colleagues had already demonstrated that the use of masks was strongly protective during the SARS epidemic in 2003 [106]. In fact, there is no evidence that mask users neglect physical distancing out of a sense of security. There is no increase of risk compensation behavior at the population level, even if perhaps some individuals have riskier behavior [107]. The general public can be trusted to behave responsibly. Better use of masks can be learned through information and education campaigns. The argument of shortage should be an incentive to manufacture, and this is indeed what many people in many countries have done, – fabricating home-made cloth masks so that face coverings have multiplied in communities, sometimes making them into a fashion item. The conclusion of these arguments is that official recommendations in the first period of the pandemic not to wear masks in public have failed to apply the precautionary principle; they are in fact unethical since they could have prevented morbidity and mortality of Covid-19 [108]. A modelling study points out that universal mask use in the United States could have saved almost 130,000 lives during the winter of 2020–2021 [109].

From an ethical perspective it is furthermore interesting to note that for a long time, masking is approached from a medical-technical point of view. The mask is regarded as an object, a medical device that is evaluated in terms of medical benefits and harms, particularly in terms of effect on the spread of the infection. Masks need to comply with specific standards for quality and use. The main concern therefore is efficacy and reliability. First of all, they are requested for a specific purpose, i.e. to protect healthcare professionals. Masks are a matter of personal and professional responsibility; it is up to the individual healthcare worker to protect oneself and in doing so to protect others. The primary beneficiary of masking is the individual wearer rather than the population. However, this medical narrative is one-sided. Masks never work by themselves; their effects depend on human behavior in the real world. The use of masks is perceived differently across the globe. In East Asian countries masks are commonly used in public life, especially during the flu season. In Japan, masks are adopted since the Manchurian plague outbreak (1910–1911) and the Spanish flu [110]. As barrier against the inhalation of invisible pathogens, they provide a sense of control amidst threats and uncertainties. Masks have become a symbol of personal responsibility, a form of selfcare and personal expression of concern with other people [111]. The symbolic meaning of masks in Western countries is different. Covering the face produces anonymity and discomfort, difficulties in communication (especially for deaf people); they make people expression-less and disconnected from their surroundings. In a healthcare setting, masks have a mixed message, expressed in the seventeenth century image of the plague doctor showing a bizarre mask with a pointed beak. The contemporary mask is less frightening: it shows that doctors will not abandon their patients and have overcome their own fear of contamination. At the same time, the intimacy of the relationship between doctor and patient is barred; both look at each other behind a veil [112]. Outside of the context of healthcare, masks are a symbol of illness; they identify somebody as infected and may provoke discrimination and stigmatization. In the beginning of the pandemic, people of Asian ethnicity in some European countries are indeed stigmatized for wearing masks; they are viewed as threatening carriers of disease.

A pandemic is a time of fear and confusion. When the infectious disease is more disseminated, and an increasing number of people cover their faces, the symbolic meaning of masking changes. Face coverings are a visible signal that threat is in the air, and they indicate that the threat is taken seriously. When they are no longer only recommended for ill people, and used for source control, the discriminatory effect is reversed. Now people without masks can be stigmatized [113]. Wearing a mask is a form of politeness and civility. A mask symbolizes social solidarity and respect for others. By covering their faces, people show that they are not indifferent to the health of other citizens. Besides, they feel empowered; they can play a role in stopping the pandemic. Masking demonstrates that health is not an individual concern but a public good: everybody's behavior contributes to the health of everybody else. Though perhaps uncomfortable and inconvenient, masks can be compared to seatbelts and condoms. They manifest that something is done, and that social responsibility is exercised. Covering the face in this perspective is a moral act of concerned citizenship [114].

Against this backdrop, the mask is more than a medical device. It has a social function as a health ritual, independent from the discourse of effectiveness [115]. The mask provides individuals a sense of control in the face of uncertainties, even if its effect is questionable [116]. Wearing a mask has a moral meaning, symbolizing a rule of conduct and civilized behavior in dealing with risks.

5.5.8 Immunity

Infections activate the immune system and make people resistant to future infections with the same microorganism. Other coronaviruses confer short-term immunity (2–3 years), so that people get reinfected later [117]. Several cases of reinfection with SARS-CoV-2 have now been described, usually occurring between 2 weeks and 4 months, but it is unknown how often this happens. Immunity against the virus is not clear, but it seems to offer protection for at least a year [118]. Some infected people do not seem to produce antibodies while in others, especially with mild Covid, antibodies are only detectable for a limited time [119]. But this does not mean that there is no immune response, since antibodies are only one part of the immune system. A Danish study of patients tested positive in the first and second waves in 2020 concludes that natural immunity occurs in 80–83% of patients younger than 65 years, but only in 47% of older patients [120]. How long immunity against the virus will last, is unsure. This uncertainty complicates the ethical debates about herd immunity and immunity passports which was important in the first year of the pandemic. Epidemiologists estimate that 60–70% of the population would have to be immune in order to accomplish group immunity. The idea to let the virus run its course in order to rapidly diminish the number of people vulnerable to the virus has raised significant ethical issues but it becomes unfeasible when people surviving an infection are not immune or only for a short time [121]. The suggestion to use positive antibody tests as 'passports' to allow individuals to return to work or travel faces similar challenges. A positive test of SARS-CoV-2 may not be associated with a positive antibody test, while the existence of antibodies does not guarantee that a person cannot be reinfected, particularly not, if immunity is short-lived. Immunity passports thus provide a false sense of security while they are also ethically problematic since their implementation may result in discrimination, stigmatization, and increasing inequity [122]. Finally, the short duration of immunity after natural infection and the possibility of reinfection, especially when new virus variants develop and spread, indicate that it may not be possible to eradicate the virus, even not with a vaccine, and more so since it is circulating in animal reservoirs. The virus will stay with us and nobody will be invulnerable. The only long-term approach then is prevention of transmission [123].

The debate about immunity becomes particularly important after the start of vaccinations in 2021. It is expected that vaccines against Covid-19 will induce more effective and longer lasting immunity than natural infection. Indeed, licensed vaccines generate a robust immune response, protecting against serious illness, hospital admission, and death [124]. Currently, there are two uncertainties that are now subjected to many studies, while knowledge is evolving [125]. One concerns how long immunity will last after vaccination. Antibody levels are steadily declining after vaccination with a growing risk of breakthrough infections in fully vaccinated people. Since October 2021, some countries have therefore started to offer a third dose of vaccination, mostly to high-risk persons. The second uncertainty is whether vaccines will be protective against circulating viral strains and new variants. The effectiveness of vaccines against recent variants, particularly the highly infectious Delta variant seems to be reduced, threatening especially people who are older and have underlying health conditions. As long as transmission is not controlled (through vaccination of the majority of the world population and continuous implementation of public health measures), resistant viruses will continue to emerge.

5.5.9 Vulnerability

SARS-CoV-2 is not affecting everybody in the same manner. Some people are more susceptible than others. While some are exposed to the virus, they are not infected. The young have usually less risk to be infected than adults. The severity of the disease can differ also. Some people such as men and persons with underlying health conditions have more serious symptoms than others. Racial and ethnic minority groups have increased risk of getting sick, being hospitalized and dying from Covid-19 [126]. People in poor neighborhoods are more vulnerable to the disease, while recurrent Covid outbreaks are described in nursing homes, slaughterhouses, and prisons [127].

It is not clear how these different vulnerabilities can be explained. One type of explanation focuses on biological factors. For some time, it is assumed that blood type A+ is a predisposition that makes people more susceptible to Covid, but later the evidence is assessed as extremely weak [128]. Another finding is that increased risk is related to a particular segment of human DNA with genes that are inherited from Neanderthals (60,000 years ago) [129]. A different type of explanation emphasizes social and economic conditions, especially social inequality. Factors such as discrimination, limited access to healthcare, unsafe work settings, low paid and less stable jobs, and crowded housing are associated with more Covid-19 infections, hospitalizations and deaths [130]. A curious finding in some studies is that among positive and hospitalized patients, the number of smokers is significantly lower than in the general population [131]. Some scientists conclude that smoking or nicotine uptake have a protective effect. In France, some advice the use of nicotine patches. In China, there is a surge in tobacco consumption during the pandemic. The WHO rejects the claim that tobacco or nicotine can reduce the risk and argues that smokers have a higher risk of contracting Covid (a respiratory disease) than non-smokers [132]. Later research revealed that these studies have methodological flaws and present incomplete data, while in several cases the authors have financial links to the tobacco industry which is actively promoting the use of smoking against Covid-19. One of the papers is formally retracted by the journal for unreported conflicts of interest [133].

5.5.10 Role of Animals

Another controversy is related to animals. While it is assumed that the coronavirus originated in animals (particularly bats), it is not clear what role animals play in the transmission of Covid-19. Evidence suggest that the virus can also spread from infected people to animals such as cats, dogs, and rabbits. There are reports that lions and tigers in a New York zoo have been infected [134]. In several countries, minks in farms have been detected with SARS-CoV-2. Studies report that in these farms the virus likely jumped from minks to humans [135]. Because of the risk of zoonotic transmission, in some countries (e.g. the Netherlands and Denmark) millions of minks at infected farms have been culled. Moreover, mutations of the virus in infected animals, could impede the efficacy of vaccines. When domestic animals such as cats and dogs can be infected by close contact with infected humans, there are worries that they might transmit Covid-19 to other humans. The general assumption is that currently there is no evidence for such transmission. Covid-19 is the result of person-to-person transmission. However, CDC recommends pet owners to restrict interaction of their pets with people outside their household: cats should be kept indoors, and dogs should be walked practicing physical distancing rules but masks should not be used [136]. Pets might potentially transmit the virus to humans, although there is no evidence, and thus precautionary measures must be taken (as the example of minks may demonstrate) [137].

5.6 Conclusion

Past experiences tell us that a pandemic is not just a medical and scientific challenge. Covid-19 illustrates that a widespread infectious disease is a multidimensional event, disrupting social life, damaging the economy, reinforcing political tensions, testing cultural customs and traditions, and provoking ethical questions. Even if it is addressing these various dimensions of the pandemic, and not only focused on medical, virological and epidemiological concerns, scientific knowledge will not be sufficient to control the challenges of the disease. Not only is it necessarily tentative and limited because the disease is novel, but it is also inherently uncertain so evidence will always be contested since that is the nature of critical scientific inquiry. At the same time, policy decisions cannot wait until sufficient evidence is available. Choices and compromises have to be made, perhaps reluctantly and hesitantly. Underlying these decisions are value judgments about what is best to do.

As this chapter demonstrates, policy measures must be coherent and persistent, otherwise the infections will multiply and a new wave of cases will emerge. In the face of uncertainty, several mechanisms have developed to cope with anxiety, ambiguity and unpredictability. The general public is bombarded with quantitative data about infected, hospitalized and deceased people. Politicians frequently refer to the reproduction number as a reliable indicator for the extent of viral transmission. They deliver positive messages about fast-tracked medicines, and express the hope that vaccines are soon available. The public is almost every day reminded that it is relatively simple to stop or mitigate the spread of the virus with distancing, hand washing, use of masks, testing, and isolating persons who test positive or have been exposed to the virus. It is cautioned that nobody is invulnerable to SARS-CoV-2. In the population certain risk rituals are evolving, transforming the meaning of social interactions. While shaking hands used to be a sign of courtesy and friendliness, it is now regarded as lack of concern, and even hostility. Working at home is no longer viewed as reclusiveness and laziness.

Dealing with uncertainty and the evolving coping mechanisms assume an estimation of which risks are acceptable and how much safety we want, not only for ourselves but also for others. Many people are cautious because they do not want to infect their relatives and friends. Many avoid larger gatherings and do shopping in the early morning to avoid crowds. These behaviors are based on the precautionary principle. When evidence is weak or lacking, human conduct is guided by the moral desire to avoid potential and serious harm. In many essential dimensions of the pandemic factual information that can help us to navigate our social life is not or not yet available, as the controversies discussed in this chapter illustrate.

Accordingly, policies and behavior are directed by moral considerations rather than scientific evidence. At the level of policies, decisions are necessary to balance various values. Even if human life and health are primary considerations, policy interventions have an impact on social well-being, mental health, and economic and cultural life. When a disease is pandemic, individuals are not able to resist the threat by their own means; a collective response is required that engages as many people as possible, and only the state is able to formulate and enforce such response. This raises another set of ethical considerations: how far can individual autonomy be restricted, curbing human rights such as freedom and privacy? In many countries, appeals are made to responsibility and rationality, but what are the options if people do not change their behavior? Are there any possibilities between respecting individual choice and strict enforcement of policy measures? Answers to these questions differ, as shown in this and the previous chapter. In most countries, doctors are obliged to report cases and deaths of Covid-19 so that an accurate overview at the national level can be obtained. Efforts to influence and modify individual behavior, however, vary from mandatory testing and quarantine to voluntary cooperation and implementation. At the level of individual behavior, the specific character of an infectious disease is manifested. Patients have a double role as victims as well as vectors of the disease. The first role requires respect for individual autonomy, the second concerns the public good. Since a substantial number of transmissions is from asymptomatic persons who feel well, in fact everybody can be a vector. This finding complicates the usual concept of disease; do carriers of the coronavirus have the disease (and are thus patients) although it is not manifesting itself? In that case, everyone is a patient or a potential patient, and from a public health perspective, the appeal to individual responsibility will be insufficient. Individual health is inherently connected to everybody's health, and each person has responsibility for the health of other citizens, at least to prevent harm to others. This is exemplified in the operations of testing and contact tracing; they are not primarily done to clarify one's health status but to identify other people at risk. A positive source person has the responsibility to cooperate in tracing contacts. From this point of view, individual interests cannot be disconnected from social responsibility and solidarity. Controlling harm at the source will benefit the population.

In mainstream bioethics, the individual is at the center of ethical concern, as it used to be in ordinary times. But an infectious disease stretches this concern beyond the individual perspective. It highlights that individuals are embedded in the physical, cultural and social environment. That signifies that a broader ethical perspective is required, which will be elaborated in subsequent chapters.

References

1. World Health Organization. 2021. *WHO coronavirus disease (Covid-19) dashboard.*
2. Cacciapaglia, G., C. Cot, and F. Sannino. 2020. Second wave Covid-19 pandemics in Europe: A temporal playbook. *Scientific Reports* 10: 15514.
3. BMA Survey. 2020, 14 September.
4. Looi, M.-K. 2020. Covid-19: Is a second wave hitting Europe? *British Medical Journal* 371: m4113.
5. Spinney, L. 2020. Why is Europe yet again at the Centre of the coronavirus pandemic? *The Guardian*, 2 November.

6. Banerjee, R., J. Bhattacharya, and P. Majumbar. 2020. *Exponential-growth prediction bias and compliance with safety measures in the times of COVID-19*. Bonn: IZA Institute of Labor Economics.
7. IMF. 2020. *World economic outlook. A long and difficult ascent*. Washington: International Monetary Fund, October, xvii.
8. IMF, *World economic outlook*, 23.
9. World Health Organization. 2021. *Coronovirus (Covid-19) dashboard.*
10. Ledford, H. 2020. Why do Covid death rates appear to be falling? *Nature* 587: 190–192.
11. World Health Organization. 2021. *Coronovirus (Covid-19) dashboard.*
12. Kofman, A., R. Kantor, and E.Y. Adashi. 2021. Potential Covid-19 endgame scenarios. Eradication, elimination, cohabitation, or conflagration? *JAMA* 326 (4): 303–304.
13. Ioannidis, J.P.A. 2020. A fiasco in the making? As the coronavirus pandemic takes hold, we are making decisions without reliable data. *STAT*.
14. Flaxman, S., S. Mishra, A. Gandy, et al. 2020. Estimating the number of infections and the impact of non-pharmaceutical interventions on Covid-19 in 11 European countries. Imperial college Covid-19 response team. *Nature* 584: 257–261.
15. Flaxman, S., S. Mishra, and A. Gandy, et al, Estimating the number of infections and the impact of non-pharmaceutical interventions on Covid-19 in 11 European countries; Remuzzi, A., and G. Remuzzi. 2020. Covid-19 and Italy: What next? *Lancet* 395: 1225–1228.
16. Park, S.W., B.M. Bolker, D. Champredon, et al. 2020. Reconciling early outbreak estimates of the basic reproductive number and its uncertainty: Framework and applications to the novel coronavirus (SARS-CoV2) outbreak. *Journal of the Royal Society Interface* 17: 20200144. Disease models use mathematics to predict what might happen in the future and to evaluate control measures. They present 'quantitative authority' although they are not based on direct observation but they cannot eliminate uncertainty, for example because data is limited or unavailable, and assumptions about human behavior prove to be wrong. Mansnerus, E. 2013. Using model-based evidence in the governance of pandemics. In *Pandemics and emerging infectious diseases: The sociological agenda*, edited by R. Dingwall, L. M. Hoffman, and K. Staniland. Chichester: Wiley, 110–121; Kucharski, A. 2020. *The rules of contagion. Why things spread – and why they stop*. London: Profile Books.
17. "Lives were transformed into mathematical summaries." Horton, R. 2020. *The Covid-19 catastrophe*. Cambridge: Polity, viii.
18. Islam, T., A.H. Pitafi, V. Arya, et al. 2020. Panic buying in the Covid-19 pandemic: A multi-country examination. *Journal of Retailing and Consumer Services*, October 23.
19. Cori, L., F. Bianchi, E. Cadum, and C. Anthonj. 2020. Risk perception and Covid-19. *International Journal of Environmental Research and Public Health* 17 (9): 3114.
20. Dryhurst, S., C.R. Schneider, J. Kerr, et al. 2020. Risk perception of Covid1-9 around the world. *Journal of Risk Research* 23 (7–8): 994–1006.
21. Seale, H., A.E. Heywood, J. Leask, et al. 2020. Covid-19 is rapidly changing: Examining public perceptions and behaviors in response to this pandemic. *PLoS One* 15 (6): e0235112.
22. Horii, M. 2014. Why do the Japanese wear masks? A short historical review. *Journal of Contemporary Japanese Studies* 14 (2).
23. Moore, E.H., and A. Burgess. 2020. Risk rituals? *Journal of Risk Research* 14 (1): 111–124.
24. Lipsitch, M. 2020. We know enough now to act decisively against Covid-19. Social distancing is a good place to start. *STAT*, March 18.
25. COMEST. 2005. *The precautionary principle*, 14. Paris: United Nations Educational, Scientific and Cultural Organization.
26. Peeples, L. 2020. What data say about wearing face masks. *Nature* 586: 186–189.
27. "…search for perfect evidence may be the enemy of good policy." Greenhalgh, T., M.B. Schmid, T. Czypionka, et al. 2020. Face masks for the public during the Covid-19 crisis. *British Medical Journal* 369; m1435; Cheng, K.K., T.H. Lam, and C.C. Leung. 2020. Wearing face masks in the community during the Covid-19 pandemic: altruism and solidarity. *Lancet*, April 16.

28. Boin, A., W. Overdijk, C. van der Han, J. Hendriks, and D. Sloof. 2020. *Covid-19. Een analyse van de nationale crisisresponse*, 84 ff. Leiden: The Crisis University Press.
29. Wright, L. 2021. *The plague year. America in the time of Covid*. London: Allen Lane.
30. Baldwin, P. 2021. *Fighting the first wave. Why the coronavirus was tackled so differently across the globe*. Cambridge: Cambridge University Press, 3, 9 ff, 25, 168–169.
31. Rutter, H., M. Wolpert, and T. Greenhalgh. 2020. Managing uncertainty in the Covid-19 era. *British Medical Journal* 320: m3349.
32. Blastland, M., A.L.J. Freeman, S. van der Linden, et al. 2020. Five rules for evidence communication. *Nature* 587: 362–364.
33. Wong, A., S. Ho, O. Olunsanya, M.V. Antonini, and D. Lyness. 2021. The use of social media and online communications in times of pandemic Covid-19. *Journal of the Intensive Care Society* 22 (3): 255–260.
34. Zhang, L. 2021. *The origins of Covid-19. China and global capitalism*. Stanford: Stanford Briefs.
35. Tsao, S.-F., H. Chen, T. Tisseverasinghe, Y. Yang, L. Li, and Z.A. Butt. 2021. What social media told us in the time of Covid-19: A scoping review. *Lancet Digit Health* 3: e175–e194.
36. Buchanan, K., L.B. Aknin, S. Lotun, and G.M. Sandstrom. 2021. Brief exposure to social media during the Covid-19 pandemic: Doom-scrolling has negative emotional consequences, but kindness-scrolling does not. *PLoS One* 16 (10): e0257728.
37. Harding, L. 2020. 'Weird as hell': The Covid-19 patients who have symptoms for months. *The Guardian*, May 15.
38. World Health Organization. 2020. *Coronavirus disease (Covid-19): Similarities and difference with influenza*. March 17.
39. Keulemans, M., E. de Visser, and T. N. Jansen. 2020. Het virus tot zover. *De Volkskrant*, 17 October, 4–11.
40. Kolata, G. 2005. *Flu. The story of the great influenza pandemic of 1918 and the search for the virus that caused it*, 297. New York: Atria paperback.
41. McKeever, A. 2021. We still don't know the origins of the coronavirus. Here are 4 scenarios. *National Geographic*, April 2.
42. Mallapaty, S. 2020. Where did COVID come from? WHO investigation begins but faces challenges. *Nature* 587: 341–342; Mallapaty, S. 2020. Meet the scientists investigating the origins of the Covid pandemic. *Nature* 588: 208; Lei, R., R. Qiu, and P. Jia. 2021. WHO-China report on Covid: Important step forward, more to be done. *The Hastings Center*, April 9.
43. World Health Organization. 2021. *WHO-convened global study of origins of SARS-CoV-2: China part*. Geneva: WHO.
44. Beaumont, P. 2021. UK and US criticize WHO's Covid report and accuse China of withholding data. *The Guardian*, March 30; Mallapaty, S. 2021. After the WHO report. What's next in the search for Covid's origins. *Nature* 592: 337–338.
45. Agence France-Press. 2021. 'Last chance': WHO reveals new team to investigate Covid origins. *The Guardian*, October 14.
46. Beaumont, P. 2021. Did Covid come from a Wuhan lab? What we know so far. *The Guardian*, May 27; Agence France-Presse. 2021. US joins calls for transparent, science-based investigation into Covid origins. *The Guardian*, May 26; Greve, J. E. 2021. Joe Biden orders US intelligence to intensify efforts to study Covid's origins. *The Guardian*, May 27; Maxmen, A. and S. Mallapaty. 2021. The Covid lab-leak hypothesis: What scientists do and don't know. *Nature* 594: 313–315.
47. Fenn, E.A. 2001. *Pox Americana. The great smallpox epidemic of 1775–82*, 5. New York: Hill and Wang.
48. De Waal, A. 2021. *New pandemics, old politics. Two hundred years of war on disease and its alternatives*, 180 ff. Cambridge: Polity Press.
49. Maxmen, A. 2021. US Covid origins report: Researchers pleased with scientific approach. *Nature* 597: 159–160.

50. Mercatelli, D., and F.M. Giorgi. 2020. Geographic and genomic distribution of SARS-CoV-2 mutations. *Frontiers of Microbiology* 11:1800; Callaway, E. 2020. Making sense of coronavirus mutations. *Nature* 585: 174–177.
51. McKie, R. 2020. What is the new Covid strain – and will vaccines work against it? *The Guardian*, December 19; McCarthy, M., and A. Caplan. 2020. Coronavirus mutation panic. *The Hastings Center*, December 22; Le Page, M. 2021. Threats from new variants. *New Scientist*, January 9: 8–9.
52. Reardon, S. 2021. The most worrying mutations in five emerging coronavirus variants. *Scientific American*, January 29; Burki, T. 2021. Understanding variants of SARS-CoV-2. *Lancet* 397: 462.
53. Wei-Haas, M. 2021. Why some coronavirus variants are more contagious – And how we can stop them. *National Geographic*, January 27.
54. Wise, J. 2021. Covid-19: The E484K mutation and the risks it poses. *British Medical Journal* 372: n359.
55. Callaway, E. 2021. Multitude of coronavirus variants found in the US – But the threat is unclear. *Nature* 591: 190; Reuters, 2021. New coronavirus variant, described as 'double mutant', reported in India. *The Guardian*, March 25.
56. Geddes, L. 2021. *Five things we know about the Delta variant (and two things we don't.* GAVI, June 15; Glenza, J. 2021. The Delta variant is spreading. What does it mean for the US? *The Guardian*, June 16.
57. For example, in mid-February 2021 more than half of infections in France are presumable caused by variants. Haim-Boukobza, S., B. Roquebert, S. Trombert-Paolantoni, et al. 2021. Detecting rapid spread of SARS-CoV-2 variants, France, January 26 – February 16, 2021. *Emerging Infectious Diseases* 27 (5): 1496–1499.
58. Kissler, S.M., C. Tedijante, E. Goldstein, Y.H. Grad, and M. Lipsitch. 2020. Projecting the transmission dynamics of SARS-VCoV-2 through the postpandemic period. *Science* 368: 860–868.
59. Callaway, E., and H. Ledford. 2021. How to redesign Covid vaccines so they protect against variants. *Nature* 590: 15–16.
60. Harding, 'Weird as hell.'
61. Shi, Z., H.J. de Vries, A.P.J. Vlaar, et al. 2021. Diaphragm pathology in critically ill patients with Covid-19 and postmortem findings from 3 medical centers. *JAMA Internal Medicine* 181 (1): 122–124.
62. Frontera, J.A., S. Sabadia, R. Lalchan, et al. 2020. A prospective study of neurological disorders in hospitalized Covid-19 patients in New York City. *Neurology*, October 5.
63. Centers for Disease Control and Prevention. 2020. *Covid-19 in children and teens*. Centers for Disease Control and Prevention, September 17; Parshley, L. 2020. 'Super antigens' tied to mysterious Covid-19 syndrome in children. *National Geographic*, October 16.
64. Huang, C., L. Huang, Y. Wang, et al. 2021. 6-month consequences of Covid-19 in patients discharged from hospital: A cohort study. *Lancet* 397: 220–232.
65. Centers for Disease Control and Prevention. 2021. *Post-Covid conditions*, April 8.
66. Keulemans, M., E. de Visser, T.N. Jansen, Het virus tot zover, C.H. Sudre, B. Murray, T. Varsavsky, et al. 2021. Attributes and predictors of long Covid. *Nature Medicine* 27: 626–631.
67. Nogrady, B. 2020. What the data say about asymptomatic Covid infections. *Nature* 587: 534–535; Oran, D.P., and E.J. Topol. 2020. Prevalence of asymptomatic SARS-CVoV-2 infection. A narrative review. *Annals of Internal Medicine* 173 (5): 362–367.
68. Oran, and Topol. 2020. Prevalence of asymptomatic SARS-CVoV-2 infection.
69. Cevik, M., M. Tate, O. Lloyd, et al. 2021. SARS-Co-V-2, SARS-CoV, and MERS-CoV viral load dynamics, duration of viral shedding, and infectiousness: A systematic review and meta-analysis. *The Lancet Microbe* 2: e13–e22.
70. Eng, K.F. 2020. 5 things that scientists now know about Covid-19 – and 5 things they're still figuring out. Ideas.Ted.com, September 16.

71. Schuetz, A.N., P. Hemarajata, N. Mehta, et al. 2020. When should asymptomatic persons be tested for Covid-19? *Journal of Clinical Microbiology*, 6 October; Rajan, S., J. Cylus, and M. McKee. 2020. Successful find-test-trace-isolate-support systems. *Eurohealth* 26 (2).
72. Moghadas, S.M., M.C. Fitzpatrick, P. Sah, et al. 2020. The implications of silent transmission for the control of Covid-19 outbreaks. *Proceedings of the National Academy of Sciences of the United States of America* 117 (30): 17513–17515.
73. Műller, M., P.M. Derlet, C. Mudry, and A. Aeppli. 2020. Testing of asymptomatic individuals for fast feedback-control of Covid-19 pandemic. *Physical Biology* 17 (6): 065007.
74. Holt, E. 2020. Slovakia to test all adults for SARS-CoV-2. *Lancet* 396: 1386-1387; Lewis, T. 2021. Slovakia offers a lesson in how rapid testing van fight Covid. *Scientific American*, April 8; Pavelka, M., K. Van-Zandvoort, S. Abbott, et al. 2021. The impact of population-wide rapid antigen testing on SARS-CoV-2 prevalence in Slovakia. *Science*: eabf9648.
75. Peto, J. 2020. Covid-19 mass testing facilities could end the epidemic rapidly. *British Medical Journal* 368:m2263; Peto, J., N.A. Alwan, K.M. Godfrey, et al. 2020. Universal testing as the UK Covid-19 lockdown exit strategy. *Lancet* 395 (10234): 1420–1421.
76. Dinnes, J., J.J. Deeks, S. Berhane, et al. 2021. Cochrane COVID-19 diagnostic test accuracy group. Rapid, point-of-care antigen and molecular-based tests for diagnosis of SARS-CoV-2 infection. *Cochrane Database of Systematic Reviews* 2021, issue 3. Art. No.: CD013705.
77. European Center for Disease Prevention and Control. 2021. *Considerations on the use of self-tests for Covid-19 in the EU/EEA*. Stockholm: ECDC.
78. See, for example, Audureau, W., G. Dagom, A. Maad, and J. Parienté. 2020. Covid-19: 54 scientifiques évaluent la stratégie sanitaire. *Le Monde*, September 30.
79. Peto, Alwan, Godfrey, et al. Universal testing as the UK Covid-19 lockdown exit strategy, 1420.
80. Kretzschmar, M.E., G. Rozhnova, M.C.J. Bootsma, et al. 2020. Impact of delays on effectiveness of contact tracing strategies for Covid-19: A modelling study. *The Lancet Public Health* 5: e452–e459.
81. Keulemans, M. 2020. Steeds meer besmettingen, maar waar zit het lek? *De Volkskrant* 9 September: 4.
82. Cox, C., and M. Dixon-Woods. 2020. Need for ethical framework to guide mass testing for asymptomatic covid-19. *British Medical Journal* 371: m4567.
83. Raffle, A.E., A.M. Pollock, and L. Harding-Edgar. 2020. Covid-19 mass testing programmes. *British Medical Journal* 370: m3263.
84. Jung, J., H. Jang, H.K. Kim, et al. 2020. The importance of mandatory Covid-19 diagnostic testing prior to release from quarantine. *Journal of Korean Medical Science* 35 (34): e314.
85. Peto, Alwan, Godfrey, et al. Universal testing as the UK Covid-19 lockdown exit strategy, 1420–1421.
86. Van Doremalen, N., T. Bushmaker, D.H. Morris, et al. 2020. Aerosol and surface stability of SARS-CoV-2 as compared to SARS-CoV-1. *New England Journal of Medicine* 382 (16): 1564–1567; Goldie, S., A. Hill, D. Eagles, and T.W. Drew. 2020. The effect of temperature on persistence of SARS-CoV-2 on common surfaces. *Virology Journal* 17: 145.
87. Goldman, E. 2020. Exaggerated risk of transmission of Covid-19 by fomites. *Lancet Infectious Diseases* 20: 892–893; Mondelli, M.U., M. Colaneri, E. M. Seminari, et al. 2021. Low risks of SARS-Co-V-2 transmission by fomites in real-life conditions. *Lancet Infectious Diseases* 21: e112; Goldman, E. 2021. SARS wars: The fomites strike back. *Applied and Environmental Microbiology* 87 (13): e00653–e00621.
88. Lewis, D. 2021. Covid-19 rarely infects through surfaces. So why are we still deep cleaning? *Nature* 590: 26–28.
89. Lednicky, J.A., M. Lauzardo, Z.H. Fan, et al. 2020. Viable SARS-CoV-2 in the air of a hospital room with Covid-19 patients. *International Journal of Infectious Diseases* 100: 476–482; see also: Santarpia, J. L., D. N. Rivera, V. L. Herrera, et al. 2020. Aerosol and surface contamination of SARS-CoV-2 observed in quarantine and isolation care. *Scientific Reports* 10, 12732.

90. See, for example, Cyranoski, D. 2020. How to stop restaurants seeding Covid infections. *Nature* 587: 344.
91. Lebrecht, N. 2020. Concertgebouw chorus is devastated after pre-Covid Bach passion.
92. Morawska, L., and D.K. Milton. 2020. It is time to address airborne transmission of coronavirus disease 2019 (Covid-19). *Clinical Infectious Diseases* ciaa939.
93. Chu, D.K., E.A. Akl, S. Duda, et al. 2020. Physical distancing, face masks, and eye protection to prevent person-to-person transmission of SARS-CoV-2 and Covid-19: A systematic review and meta-analysis. *Lancet* 395: 1973–1987.
94. Jones, N.R., Z.U. Qureshi, R.J. Temple, et al. 2020. Two metres or one: What is the evidence for physical distancing in Covid-19? *British Medical Journal* 370: m3223.
95. McLaws, M-L. 2020. What is the Covid 'bubble' concept, and could it work in Australia? *The Conversation*, August 31; Rankin, J. 2020. Belgium experiments with 'corona bubbles' to ease social restrictions. *The Guardian*, May 10.
96. Long, N.J. 2020. From social distancing to social containment. Reimagining sociality for the coronavirus pandemic. *Medicine Anthropology Theory* 7 (2): 247–260.
97. Whang, O., and K. Elliott. 2020. Poll finds more Americans than ever think we should wear masks. *National Geographic*, October 5.
98. WHO. 2020. Advice on the use of masks in the context of Covid-19. *Interim Guidance* (June 5).
99. See, for example: Chu, Akl, and Duda, et al. Physical distancing, face masks, and eye protection to prevent person-to-person transmission of SARS-CoV-2 and Covid-19; Chan, T.K. 2020. Universal masking for Covid-19: Evidence, ethics and recommendations. *BMJ Global Health* 5: e002819.
100. Gandhi, M., C. Beyrer, and E. Goosby. 2020. Masks do more than protect others during Covid-19: Reducing the inoculum of SARS-CoV-2 to protect the wearer. *Journal of General Internal Medicine* 35 (10): 3063–3066.
101. Bundgaard, H., J.S. Bundgaard, D.E.T. Raaschau-Pedersen, et al. 2021. Effectiveness of adding a mask recommendation to other public health measures to prevent SARS-CoV-2 infection in Danish mask wearers. *Annals of Internal Medicine* 174 (93): 335–343.
102. Greenhalgh, T. 2020. Face coverings for the public; laying straw men to rest. *Journal of Evaluation in Clinical Practice* 26: 1070–1077.
103. Ji, D., X. Li, and S. Ramakrishna. 2020. Addressing the worldwide shortage of face masks. *BMC Materials* 2: 9.
104. Spooner, J.L. 1967. History of surgical face masks. *AORN Journal* 5 (1): 76–80.
105. Jefferson, T., C.B. Del Mar, L. Dooley, et al. 2011. Physical interventions to interrupt or reduce the spread of respiratory viruses. *Cochrane Database of Systematic Reviews* 2011, Issue 7. Art. No.: CD006207.
106. Wu, J., F. Xu, W. Zhou, et al. 2004. Risk factors for SARS among persons without known contact with SARS patients, Beijing. China. *Emerging Infectious Diseases* 10 (2): 210–216.
107. Van der Vliet, N., K. van der Swaluw, M. Zonneveld, et al. 2020. Gedragswetenschappelijke literatuur rond mondkapjesgebruik. *Een rapid review van de literatuur*. Bilthoven: Gedragsexpertiseteam RIVM.
108. Macer, D. 2020. Wearing masks in Covid-19 pandemic, the precautionary principle, and the relationships between individual responsibility and group solidarity. *Eubios Journal of Asian and International Bioethics* 30 (4): 129–132.
109. IHME COVID-19 Forecasting Team, R.C. Reiner, R.M. Barber, et al. 2021. Modeling COVID-19 scenarios for the United States. *Nature Medicine* 27: 94–105.
110. Lynteris, C. 2018. Plague masks: The visual emergence of anti-epidemic personal protection equipment. *Medical Anthropology* 37 (6): 442–457.
111. Horii, Why do the Japanese wear masks? Burgess, A., and M. Horii. 2012. Risk, ritual and health responsibilisation: Japan's 'safety blanket' of surgical face mask-wearing. *Sociology of Health & Illness* 34 (8): 1184–1198; Sand, J. 2020. We share what we exhale. *Times Literary Supplement*, May 1: 22–23.

112. Earnest, M. 2020. On becoming a plague doctor. *New England Journal of Medicine* 383: e64.
113. Howard, J., A. Huang, Z. Li, et al. 2020. Face masks against Covid-19: An evidence review. *Preprint* 2020040203.
114. Sunstein, C.R. 2020. The meaning of masks. *Journal of Behavioral Economics for Policy* 4 (S): 5–8.
115. Burgess, A., and M. Horii. 2012. Risk, ritual and health responsibilisation: Japan's 'safety blanket' of surgical face mask-wearing. *Sociology of Health & Illness* 34 (8): 1194.
116. Syed, Q., W. Sopwith, M. Regan, and M.A. Bellis. 2003. Behind the mask. Journey through an epidemic: Some observations of contrasting health responses to SARS. *Journal of Epidemiology and Community Health* 57: 855–856.
117. Edridge, A.W.D., J. Kaczorowska, A.C.R. Hoste, et al. 2020. Seasonal coronavirus protective immunity is short-lasting. *Nature Medicine* 26: 1691–1693.
118. Abbasi, J. 2021. Study suggest lasting immunity after Covid-19, with a big boost from vaccination. *JAMA* 326 (5): 376–377.
119. Ibarrondo, F.J., J.A. Fulcher, D. Goodman-Meza, et al. 2020. Rapid decay of anti-SARS-CoV-2 antibodies in persons with mild Covid-19. *New England Journal of Medicine* 383 (11): 1085–1087; Ward, H., G. Cooke, C. Atchison, et al. 2020. Declining prevalence of antibody positivity to SARS-CoV-2: A community study of 365,000 adults. *Preprint*.
120. Hansen, C.H., D. Michlmayr, S.M. Gubbels, et al. 2021. Assessment of protection against reinfection with SARS-CoV-2 among 4 million PCR-tested individuals in Denmark in 2020: A population-level observational study. *Lancet* 397: 1204–1212.
121. Aschwanden, C. 2020. The false promise of herd immunity. *Nature* 587: 26–28.
122. Bramstedt, K.A. 2020. Antibodies as currency: COVID-19's golden passport. *Bioethical Inquiry* 17: 687–689; Voo, T. C., H. Clapham, and C.C. Tam. 2020. Ethical implication of immunity passports during the Covid-19 pandemic. *The Journal of Infectious Diseases* 222: 715–718.
123. Jabbari, P., and N. Rezaei. 2020. With the risk of reinfection, is Covid-19 here to stay? *Disaster Medicine and Public Health Preparedness* 14 (4): e33.
124. Milne, G., T. Hames, C. Scotton, N. Gent, A. Johnsen, R.M. Anderson, and T. Ward. 2021. Does infection with or vaccination against SARS-CoV-2 lead to lasting infection? *The Lancet Respiratory Medicine*, October 21.
125. Dolgin, E. 2021. Covid vaccine immunity is waning – How much does it matter? *Nature* 597: 606–607.
126. Hawkins, D. 2020. Differential occupational risk for COVID-19 and other infection exposure according to race and ethnicity. *American Journal of Industrial Medicine* 63 (9): 817–820.
127. Kiaghadi, A., H.S. Rifai, and W. Liaw. 2020. Assessing COVID-19 risk, vulnerability and infection prevalence in communities. *PLoS One* 15 (10): e0241166.
128. Rubin, R. 2020. Investigating whether blood type is linked to Covid-19 risk. *JAMA* 324 (13): 1273.
129. Zeberg, H., and S. Pääbo. 2020. The major genetic risk factor for severe COVID-19 is inherited from Neanderthals. *Nature* 587: 610–612.
130. Centers for Disease Control and Prevention. 2020. *Health equity considerations and racial and ethnic minority groups*. July 24.
131. Chaudhry, R., G. Dranitsaris, T. Mubashir, J. Bartoszko, and S. Riazi. 2020. A country level analysis measuring the impact of government actions, country preparedness and socioeconomic factors on Covid-19 mortality and related health outcomes. *EClinicalMedicine* 25: 100464; Changeux, J. P., Z. Amoura, F. A. Rey, and M. Miyara. 2020. A nicotine hypothesis for Covid-10 with preventive and therapeutic implications *Comtes Rendu Biologies* 343 (1): 33–39.
132. World Health Organization. 2020. *Coronavirus disease (Covid-19): Tobacco.* May 27; see also: Vardavas, C., and K. Nikitara. 2020. COVID-19 and smoking: A systematic review of the evidence. *Tobacco Induced Diseases*, March 20.

133. Van Westen-Lagerweij, N.A., E. Meijer, E.G. Meeuwsen, et al. 2021. Are smokers protected against SARS-CoV-2 infection (Covid-19)? The origins of the myth. *npj Primary Care Respiratory Medicine* 31, 10; Davey, M. 2021. Scientific paper claiming smokers less likely to acquire Covid retracted over tobacco industry links. *The Guardian*, April 22.
134. Centers for Disease Control and Prevention. 2020. *Covid-19 and animals*., December 4.
135. Oude Munnink, B.B., R.S. Sikkema, D.F. Nieuwenhuijse, et al. 2020. Transmission of SARS-CoV-2 on mink farms between humans and mink and back to humans. *Science* (10 November).
136. Centers of Disease Control and Prevention. 2020. *If you have pets*. September 9.
137. Kiros, M., H. Anudalem, R. Kiros, et al. 2020. Covid-19 pandemic: Current knowledge about the role of pets and other animals in disease transmission. *Virology Journal* 17: 143.

Chapter 6
Linking Experience and Reflection

Abstract How Covid-19 is imagined in public discourse is examined in this chapter. Policies are commonly presented as a war against an invisible enemy. The problematic aspects of such militarized language are pointed out, and an alternative framing is offered, emphasizing connectedness and balance as appropriate metaphors to interpret the experience of the pandemic. Finally, the notion of common home is introduced. This has the advantage to articulate the shared rather than individualistic dimension of pandemic experience, and to relate it to the global threat of climate change.

Keywords Balance · Common good · Common home · Connectedness · Ecological perspective · Liminality · Military metaphor

6.1 Introduction

This is the first time in human history that most of the world population has been required to stay at home for an extended period of time. In some countries it is indeed confinement, giving people no choice. In other countries it is 'sheltering at home', a voluntary choice that is made by the overwhelming majority of citizens, feeling responsibility for each other. At the same time, in many countries numerous people do not have a home or only a small apartment, while in some circumstances home is not really a shelter but a place of abuse and violence. This experience of isolation will deeply influence humans. It triggers loneliness and desperation, especially in older and ill people in care institutions who cannot receive visitors and see friends and family. At the same time, it encourages actions of solidarity and togetherness. The term 'social distancing' is wrong; what is requested is 'physical distancing' while in many cases emotional connectedness is growing. Nonetheless, human interactions are interrupted and intimate connections such as touching, hugging, shaking hands and personal meetings are disrupted and even regarded as threatening. There is a need to reflect on how these pandemic experiences impact on human living reality. Today, we luckily have communication technologies that facilitate

H. ten Have, *The Covid-19 Pandemic and Global Bioethics*, Advancing Global Bioethics 18, https://doi.org/10.1007/978-3-030-91491-2_6

interpersonal contacts even if distanced, so that some effects of isolation and distancing can be mitigated, at least for those who have access to these technologies. Staying at home brings us back to the essence of human existence. It helps to escape from the world of abstraction that bombards us with numbers (deaths, infected, and hospitalized) and to reflect on the significance of relationality.

One major ethical implication of the pandemic is the rehabilitation of the notion of common good. The overwhelming majority of populations support measures to slow the spread of the virus even when they temporarily restrict individual freedom and hurt the economy. The pandemic reinforces the awareness that already existed for some time especially in European bioethics, that individual autonomy is a limited and basically a relational notion. It can only flourish in a context that nourishes and sustains it. In various circumstances there are important ethical considerations that override personal autonomy, and public health is one of them. The common good is not in opposition to individual liberty. Restricting freedom of movement not merely protects oneself but also shows social responsibility and solidarity since it does not expose fellow citizens to a deadly virus.

The pandemic experience is furthermore an opportunity to rehabilitate the idea of 'common home.' The imagery of 'sheltering at home' creates a new awareness of the significance of 'home' as the safe haven and stable base of ordinary life. At the same time the pandemic highlights that Covid-19 is not a crisis on its own but connected to other major crisis: in the economy, in migration policies, discrimination and racism, democracy and ecology. Many people become more aware that the pandemic is particularly linked to environmental degradation and global warming as the other major threat of our time. While Covid-19 and its impact will be temporary, the impact of climate change is already there and will continue. From the perspective of global bioethics, the focus on 'home' in the pandemic experience can therefore be extended to focus on the planet as our common home.

The current pandemic has reactivated ancient metaphors (especially military ones) but also initiated a new vocabulary: flattening the curve, social distancing, lockdown, self-isolation, and sheltering in place. Some words (quarantines, containment zones, and home confinement) are less often used as the virus moved from East to West. Terminology is not ethically neutral but reflects prevailing value systems. What is missing in current ways of talking about pandemic experiences is an ecological vocabulary. As indicated in Chap. 3, it is known for a long time that emerging infectious diseases are associated with the destruction of functioning ecosystems and biodiversity. As the planet is our common home, the major metaphor to explore is sheltering at this home.

This chapter will first analyze the language in which experiences with the pandemic are commonly expressed. The attractive as well as problematic aspects of the war metaphor will be highlighted. It is argued that this metaphor is inadequate since it does not take into account the social and environmental context in which a viral threat emerges and disseminates. Since the enemy is invisible, hostility is often projected on people who could be contaminated, and thus carry the disease. Furthermore, military language transforms global relations in competitive rivalry and nationalist discourse. The next paragraph of this chapter points out that metaphorical language not only influences public conversations but also reflects an interpretation of reality in order to make sense of experiences. Framing experiences and reality in terms of fight, conflict and violence presents the ethical question whether this is the best way to understand the meaning of so-called 'limit situations.' Arguably, the ecological perspective may provide a more appropriate frame of interpretation and understanding. It articulates the metaphors of connectedness and balance which underline coexistence with nature, community solidarity and the importance of care, support, and cooperation. Furthermore, it proposes the notion of common home as metaphor that is more inclusive than the individualistic language of sheltering at home, and that connects the pandemic with the global discourse of climate change.

6.2 Pandemic Language

Many new words have entered popular discourse. Previously used in epidemiology, they are now disseminated to advance images of control and containment. They visualize the spread of the virus, what we can do to avoid infection and to become aware of the effects of our behavior. These images are associated with the war metaphor dominating current policies. This metaphor is connected with the older discourse of bio-invasion and biosecurity. Since the 1990s, many countries have developed policies to protect native biodiversity against invasive species. Non-native species are considered as aliens, enemies; we need to act to protect nature. The assumption is that nature and humans are separated. Nature, and wildlife in particular is regarded as a source of danger. However, bio-invasion is impossible to prevent. When mobility and interconnectedness are the hallmarks of globalization, bio-invasion is a typically global phenomenon since borders are irrelevant for most species. The same applies to new, 'emerging' viruses. Viral diseases have been with us since the beginning of humanity. They are now more easily disseminated through global traffic. The point is that these 'new' viruses do not simply 'emerge' as natural events. Their impact is the result of human activities.

6.2.1 Fighting Covid-19

In March 2020, French President Macron addresses the nation and declares: "we are at war." In the United States, Covid-19 is considered as "our Pearl Harbor." Other leaders use the same terminology [1]. The tone was actually set at the beginning of the pandemic when China mobilized its army and 90 million members of the Communist Party to enforce strict lockdowns and patrol the streets, and imposing a new security system based on QR health codes. President Xi called Wuhan the 'hero city.' Victory was declared in April 2020 with the reopening of the city [2].

In most countries, public discourse and the media have popularized the idea that there is an invisible enemy, attacking and invading innocent victims [3]. Politicians are in command, strategically supported by scientists as generals, hospitals are war-zones, and health care professionals are soldiers and heroes fighting in the front-lines. The body is the battlefield. Although there are no magic bullets yet, the ammunition in this fight are masks, gloves, sanitizers, and ventilators. But victory is in sight because all resources are mobilized: capacities to resist are surging (more tests, protective equipment, and beds) and hopefully soon, vaccines will be available. The New York governor compares ventilators to bombs in the Second World War, and a navy hospital ship is deployed in New York harbor. Many countries call on the army to support medical services, arrange the logistics of testing or survey compliance with lockdowns [4].

In the past, epidemic diseases have been associated with war. Thucydides described how escalating war with the Spartans incited people to withdraw inside the city walls of Athens, leading to overcrowding conditions worsening the pestilence. The plague was associated with the Mongol conquest, cholera with British military operations in Bengal, while the Spanish flu emerged during the last stage of the First World War. Microbes have facilitated conquest, illustrated with the famous example of the subjugation of the Aztec and Inca empires by a limited number of Spaniards because of the introduction of smallpox within a population that was not immune [5].

The use of the war metaphor to combat infectious disease with its language of threat, attack, enemy, weapons, urgency, sacrifices, tactics, strategic thinking, and stringent action, introduces a specific perspective. When there is an unknown, airborne threat of a deadly and infectious pathogen, and the horror specter of the Spanish flu looms, narratives are needed that provide hope and perspectives on the future. Military language suggests that we can fight against the disease, and win the war against the silent killer. When we all mobilize our resources, we can defeat our common enemy. The war metaphor is simple and clear, and thus attractive. Like all metaphors, it gives sense to the world and what is happening, now that ordinary life is seriously disturbed. Remarkably, it is often forgotten how horrible experiences wars actually are.

6.2.2 Implications of the War Metaphor

Interpreting the Covid-19 pandemic in military terms has consequences that might be problematic and confusing [6].

- Massive mobilization of resources and collective action is necessary because there is a threat to everyone. All people, in all regions of the world, are at risk. Especially with the more contagious Delta variant, everybody, – young and old, rich and poor,- is under threat. Wars are communal affairs affecting all citizens. But this assumption should be qualified. Epidemic diseases do not strike equally.

In the Spanish flu, young people were particularly affected while SARS-CoV-2 is more dangerous for older persons. The virus is also more harmful for racial and ethnic minorities and for people in institutions such as elderly homes and prisons.

- Another assumption is that effective strategies are available to combat the virus. In war, the focus is very much on technical approaches and solutions: better and more weaponry, massive and continuous assaults, and annihilating the enemy's infrastructure. But with Covid-19 there are no 'magic bullets.' Responses to the viral threat depend on human behavior rather than medical technology. The war metaphor reinforces the image that humans are powerful and can control nature and other forms of life because they have technical instruments. But the microbe cannot survive on its own; it circulates and disseminates through humans. Responses should therefore not merely address the virus but the social, cultural and ecological context in which it is transmitted. The war metaphor tends to suppress other perspectives, notably in connection to ecology and inequity [7].
- Military language prioritizes action to overcome the threat and limit the loss of lives and economic damage. In the heat of battle there is little time to discuss the effectiveness of actions, only after they have taken place. The imperative of action implies that measures cannot be discussed; they need to be imposed, often after a state of exception is declared. The collective effort to win the war requires obedience and compliance. Dissent cannot be tolerated because critical discourse undermines the public resolve and is regarded as treason. Physicians and nurses who spoke up against shortages of protective equipment have been silenced and punished [8]. Furthermore, urgency demands that normal mechanisms of evaluation and control are bypassed or abridged, so that new medication and vaccines are fast-tracked without the ethical appraisal that is usual. Military language moreover evokes a strong style of leadership with sometimes censorship and authoritarian interventions.
- Strategies are aimed at victory. The ultimate aim of war is defeat of the enemy. All efforts are concentrated and prioritized to accomplish this aim. Hardship and sacrifices for the greater good will be required, and collateral damage should be accepted. Instead of protecting vulnerable groups, essential workers are identified. Similar patterns are recognized in the responses to Covid-19. Everybody should be in isolation except key workers; first of all, healthcare providers, police, but also people who supply grocery stores, deliver food, generate electricity, and collect garbage. Doctors and nurses are lauded as heroes. Sometimes, it is argued that older people may be sacrificed to save or restart the economy [9]. Because regular care has been scaled down, patients with kidney disease who need regular dialysis, with diabetes and heart disease, or with cancer needing surgery are regarded as 'collateral damage' in the war against Covid-19. The irony is that while war is aimed at taking lives or at least at accepting that lives are lost, the policy measures against the pandemic are primarily guided by the wish to save as many lives as possible [10]. Another irony is that martial language is used to overcome fear and to demonstrate that the threat can be over-

come, while at the same time it evokes fear because the threat is so seriously articulated and the danger magnified.

These implications raise questions about the appropriateness of the war metaphor. Countering a pandemic is different from warfare since in a real war essential infrastructure is destroyed, communications are interrupted, and hospitals bombarded. The virus is unlike a human enemy since it is not intimidated by martial language. Fighting opponents requires courage while combating the virus requires caution. Attacking SARS-CoV-2 means retreating and staying at home. In a real war there are at least some safe areas while in a pandemic the entire world is affected [11]. Besides, it is curious that the war metaphor is not taken seriously in many countries. Militant narratives are frequently politicized. Self-declared wartime leaders downplay the dangerousness of the virus, debilitate effective countermeasures, and issue inconsistent policies [12].

6.2.3 The Difficulty with Military Language

War metaphors are often used in the history of medicine (e.g., the 'war against cancer') but also to address social problems ('war on drugs' and 'war on terror') [13]. Militarized responses to all kinds of threats are problematic. Metaphors promote a particular framing and interpretation of the world in order to make sense of events and processes. Viewing the virus as an aggressive enemy and foreign invader directs all efforts towards the microbe that needs to be eradicated. Examples of previous pandemics show that a contagious disease is a social phenomenon, and not merely a medical one [14]. To mitigate the effects of the infectious agent changes in the behavior of citizens are necessary, and to prevent the emergence of the agent attention should be given to the conditions that facilitate the origin and spread of the virus. Both concerns look wider than the viral actor and focus on the context, specifically the determinants of the infection. Such broader view follows from the understanding of the notion of 'virosphere' discussed in Chap. 3. Microbes have always been with us and will stay with us; life will not continue without them. The coronavirus cannot be eradicated, and war against it will be perpetual. Human beings have to learn to accommodate to the microbial world. The 'new normal' means living with microbes [15]. Other metaphors are better equipped to emphasize the interdependence of humans and non-humans, and the need for balance rather than domination, aggression, and violence.

The inadequacy of the war metaphor is furthermore displayed in the idea that there is a clear enemy. However, the real enemy is invisible and often not manifested by symptoms. It is a threat nonetheless through the people it has invaded, so that every adult is a potential source of infection. In practice, some social groups are regarded as more contaminating and riskier for others because they show a lack of personal responsibility. Military language encourages a distinction between good and bad citizens, patriots and others, which can lead to discrimination and stigmatization. In the coronavirus pandemic, many have witnessed examples of xenophobia, and discrimination of Asian people in Western countries [16]. Reports from India mention that healthcare providers have been attacked and evicted from their accommodation, fearing that they might be carriers of disease [17]. Conspiracy theories flourish. Imagining the pandemic in terms of threat, conflict and violence is risking to alienate people from each other and to instigate stigmatization and exclusion. Other metaphors are needed to express the importance of care and community, solidarity and compassion in the face of widespread disease [18].

Military language has an impact on global relations. It is associated with national pride and rivalry between countries, resulting in political competition to show who is the best in handling the pandemic and who is the first to develop remedies. In the early stage of Covid-19 protective materials are the object of diplomacy and propaganda. Since face masks and personal protective equipment are scarce in European countries, China exported these with much public pomp to countries such as Italy. Russia sent 15 airplanes with emergence aid as well as medical teams to Italy. Most of the aid was not very useful but it highlighted the feeling that wealthy Western countries are not successful in combating the pandemic, and that European solidarity is absent. But in fact, more materials were shipped from Germany and France to other countries than from China and Russia. Similar competitive mechanisms are noticeable in the development of vaccines: who is the first to discover and develop a vaccine, to test it, and to bring it on the market? While the majority of countries in the world is committed to a global alliance to secure equitable access to Covid-19 vaccines in developing nations, large countries such as China and Russia, and initially the United States do not participate. While the funds provided by the alliance are insufficient to cover entire populations, China sees a commercial and political opportunity to promote its own vaccines, donating them to neighboring countries, and building production facilities in Brazil, Morocco and Indonesia, while almost nothing is known or published about the safety and efficacy of the Chinese vaccines [19]. Covid-19 has exacerbated existing inequalities within and between countries, and the militarization of policies will not resolve but often exploit them.

6.3 Changing Language and Experience

War rhetoric has a profound influence on everyday language. Many militarized terms have entered ordinary discourse: lockdown, shut down, quarantine, sheltering at home, stay at home orders, all referring to the restriction of freedom of movement. Because treatments and vaccines are the most powerful weapons, public health measures are regarded as secondary option, as 'nonpharmaceutical interventions' [20]. The new language dominating public conversation is influencing human experiences. Military discourse is often experienced as invasive and impersonal. Terms as 'lockdown' and 'mandates' are perceived as more negative than 'stay at home' and 'protocols.' The emphasis on infection and hospitalization rates feels distant and impersonal; they bring us into a world of abstraction [21]. The speed of vaccine development neglects that people must be assured of safety and effectiveness [22]. Policies against Covid-19 impact on all human experiences. Funeral and marriage ceremonies for example have drastically changed. The quality of social support and human interactions is transformed because we cannot touch the people we love. The essential role of physical intimacy in human existence is lost. People feel anxious and lonely. In public, we try to avoid other humans, and adapt our lives, behaviors and walking patterns. We cover the lower part of our faces, hiding our identity and making it more difficult to communicate. We become irritated and sometimes rude when people do not practice rules such as social distancing [23]. Consequently, the framing power of the war metaphor has ethical implications: it makes humans more afraid, more vulnerable, more obedient.

When coping with Covid-19 is framed as a moral struggle against evil, significantly transforming human experiences, an ethical question is whether this is the best way to approach a disease phenomenon. Are other metaphors not more appropriate and sensible?

Pandemics are transformative experiences. As discussed in Chap. 2, they signify a disruption of normal ways of living, acting and thinking, and as liminal events they call for a transition to a new phase [24]. Old patterns are dissolving and continuities of life disturbed but new ones are not established, producing experiences of ambiguity, fear, and disorientation. The concept of liminality expresses that we are in a borderland, and that it is impossible to return to the pre-existing reality of life. Existing frames of references become especially uncertain in what the psychiatrist and philosopher Karl Jaspers has called 'limit situations.' Human beings are always situated, i.e., they find themselves in situations of specific and concrete circumstances, dependent on gender, age, race, character, education and particular circumstances. Some situations, however, are inescapable: nobody can avoid death, serious illness and suffering. They enunciate experiences of loss, accompanied by feelings of anxiety, dread and guilt. The security of existence disappears and humans are confronted with their vulnerability. In such situations, humans become aware not merely of the limitations but also of the possibilities of existence, and they can go beyond them in communication with others. Such limit situations provide the possibility of a transition to a new orientation of life [25]. The war metaphor denies liminality. It is an effort to manage and to dismiss the transience of human life. It mobilizes human power to dispel the threat of disease, to neutralize the possibility of death, and to return to the situation that existed before the pandemic. The new normal will not be fundamentally different from the situation before Covid-19 and the new vocabulary will become outmoded. Other narratives are perhaps better able to overstep the calculating and numerical models as well as the lists of casualties that confront us for months, to overcome anxiety, to outline opportunities for change, and to appeal to the imagination to provide perspectives on the future.

6.4 The Ecological Perspective

Infectious diseases emerge because of the intricate relationship between humans and the environment. The global threat of pandemics is not a natural event but the product of human behavior, exploiting the planet and destroying biodiversity. Pandemic management inspired by the war metaphor, does not pay much attention to the conditions which promote the emergence of diseases [26]. Their emphasis is on controlling the effects of the virus (with medical treatment, self-isolation, and social distancing) rather than exploring the underlying causes of the pandemic so that future pestilences can be avoided. Environmental degradation is associated with an economic global order that assumes a separation of humans and nature and that regards nature as a resource to be exploited and commodified. Policies as military exercises against the virus are specifically aimed at preserving this global order. The focus is on domination, discipline, control, continuity, and return to normality. There is no motivation to regard the disaster as an opportunity for change, to make a fresh start, and to question the ways of life and economic dogmas that have led to the pandemic in the first place [27].

The ecological perspective offers a different view. It provides a framing in terms of peaceful coexistence with nature, solidarity at the community level, and the importance of care, support and cooperation [28]. In this perspective, two metaphors are advanced: connectedness and balance.

6.4.1 Connectedness

The metaphor of connectedness articulates the connections between humans themselves and between humans and nature. First, it emphasizes that human persons are characterized by relationality. They are embedded in communities. Rather than merely interacting with each other, they belong together and are mutually dependent, engaging, taking responsibility and shaping their lives in sharing the world with each other. Relationships with other persons define who a person is. In bioethical discourse, the significance of relationality has redefined the concept of individual autonomy as 'relational autonomy.' One implication of this redefinition is the important role of encounter, meeting and dialogue. The person becomes himself in meeting other persons. Authentic being is being-together. Personalist philosophers such as Gabriel Marcel argue that intersubjective connections are constitutive for humanity. The primordial experience of human beings is that they share the world with others. Marcel uses the notion of 'presence' to explain that we are not simply beings that are there and then relate to world. Encounter involves being present to each other. Consciousness of being in the world implies that we are not locked into our individuality. An essential characteristic of the person, in Marcel's view, is availability (*disponibilité*) [29]. Being available and receptive to others is not simply openness. It means the opportunity to undergo new experiences and to enhance one's life. The difficulty is, according to Marcel, that the world is broken, i.e. the essential connection between human beings and the world is fractured as the result of processes of abstraction and objectification. The human person himself is often treated as an object like other things in the world, so that personal relationships with other persons and the world have become problematic. The ideology of the market has reduced interpersonal relationships to contractual and impersonal interactions. Societies constantly move toward depersonalization, for example by categorizing persons in abstract terms as 'immigrants' or 'muslims.' However, in exceptional circumstances such as in confrontation with disease, disability and death, the lived experience of being together returns [30]. During the Covid-19 pandemic one of the most commonly used sayings refer to togetherness. All people, regardless where they are, are in this scourge together, and even when they cannot meet, physical distancing makes them more aware of the importance of relationality, the experience of 'we'. It is curious that the notion of person is derived from the Etruscan term for mask. In Antiquity, actors are wearing a mask to present the character they are playing. The 'persona' or the mask refers to the individual but at the same time to the social role in society. Wearing a mask during the pandemic is therefore not hiding individuality but shows a social appearance as concerned and engaged citizen.

Being situated in a web of connections is a precarious experience. Relationality is not an option that we can select or not. Humans are necessarily related. Because the body is our positioning in the world, we are exposed to the world and other persons. That implies fundamental vulnerability. Authentic relationships are both a blessing and a wound. The other person that we encounter is a blessing since we otherwise cannot become a person ourselves and flourish as a human being. But the other may also wound us. Life in community with others ('*communitas*') is ambivalent; it entails pain, harm and injury. Social life is thus related to suffering. Neoliberal ideology is focused on the possible wounds rather than blessings: others are potential negative factors, and the emphasis should be on independency. In this ideology, a specific model of cooperation is promoted with core notions such as self-interest and contractual relations with the purpose to avoid being wounded by others. Such model attempts to provide '*immunitas*' instead of accepting '*communitas*' [31].

The metaphor of connectedness has implications for bioethical discourse. As a social enterprise based on reciprocity and solidarity, it articulates the idea of human community, togetherness across borders, with interdependency rather than interactions between separate individuals. As social beings, humans are evaluative beings; they do not have a neutral or indifferent stance towards the world. Particularly since they are vulnerable beings, they want to know how important things are because they may produce suffering or enhance well-being. In other words, they have relations of concern to the world [32]. It matters to them what happens and what is done. That explains why they are sometimes outraged and feel that their dignity is not respected. Ethical theory cannot be abstracted from the social context in which human beings live and act. This conclusion also relates to the purpose of ethical discourse: avoiding harm and promoting well-being and human flourishing. Ethics cannot be reduced to principles and norms but is pertinent to our sense of flourishing and harm, taking into account our capabilities as well as vulnerabilities.

The consequence is that the language of connectedness directs attention to moral notions that are important in global bioethics (which will be elaborated in Chap. 9).

An example is vulnerability. In the dominant perspective, voices of vulnerable groups are often not heard. Justice and equality are significant concerns. 'Togetherness', highlighted frequently in the Covid pandemic, does not apply to everyone alike. Not all are in the same boat, even if they are in the same tempest. Another important notion is the common good, notably public health. Such good is desirable for society as a whole, and advantageous to every citizen. It cannot be achieved by individual efforts but requires collective action and the engagement of all citizens, and therefore solidarity and social responsibility. All members of society benefit from this common good although it can only be attained by mutual and communal endeavors. The common good is not just a description of what is important for societies but it is a normative concept. It refers to values, virtues and ideals that express what it means to have a good life together within society [33].

The second dimension of connectedness underlines the embeddedness of human beings in the natural world. The starting point for bioethical discourse is therefore not individual autonomy but the broader context of which individuals are part. Self-isolation in this perspective is not merely protecting oneself but first of all protecting fellow citizens. But the experience of togetherness is not restricted to humans but involves all forms of life, even viruses. Regarding humans as part of an interconnected web of life means moving from an anthropo-centric to a bio-centric approach with a relational concept of the self, dependent on biodiversity. This shift has been advocated by many environmental ethicists as well as by indigenous worldviews [34]. The ecological perspective implies that the military language of the pandemic is distorting the view of human embeddedness in the natural world. Emphasizing the antagonism between humans and the virus as enemy, blames nature for diseases ignoring that diseases emerge due to environmental degradation as a result of human exploitation of biodiversity. It also disregards that microbes are inhabitants of our world; they have been and will be always there. Rather than eradication, cohabitation will be required.

One consequence of the ecological perspective is a broader notion of health. Human health, animal health, and healthy ecosystems are linked [35]. Human beings cannot be healthy when the planet is not healthy. The implication is that governance of health problems should be global, focusing on health as one, as a planetary concern. Applying the vision of One Health means monitoring and surveilling human connections with animals, specifically in the bioindustry. Another implication is that attention should not merely focus on morbidity and mortality but on prevention. Future pandemics will continue to emerge if environmental destruction and loss of biodiversity are not addressed.

6.4.2 Balance

Ecology has the same etymological root as economy (οίκος: family house or household), referring to the proper administration and management of resources. It focuses on the relations between living organisms and their environment, presupposing the ideas of balance, harmony and stability of nature. Concepts such as natural order, equilibrium, self-regulation, and homeostasis are often used, for example by Potter [36]. Traditionally, the concept of health is understood as balance between the human body and the cosmos. Living in harmony with nature is regarded as the pinnacle of health. Being healthy is not merely an individual quality but a reflection of the way one lives and works within environing conditions. Health is a relational concept, the result of a balance between humans and nature. Currently there is increasing awareness of the importance of social and economic determinants of health. These determinants are not the same for everyone, and they lead to significant inequalities in health. Poor people die earlier than wealthy people because they lack access to healthcare; some people die from starvation; poor and marginalized populations are disproportionately exposed to environmental harm. Unemployment, unsafe workplaces, urban slums, gender and racial discrimination are all related to deteriorated health. It is evident that Covid-19 unmasks and often reinforces existing inequalities within countries and among countries. Capacities to cope with the virus outbreak are also very different in various countries. For example, simple hygienic measures such as regular hand washing are impossible when safe water is not available. Policy measures inspired by the war metaphor usually do not pay much attention to these imbalances in health and healthcare. The metaphor of balance takes inequality, and thus justice, as a major ethical concern [37].

Another aspect is that the concern with balance can resist the tendency of militant policies to advance a technical and biomedical approach to Covid-19. Not just the virus should be the center of attention but people are ill, some are severely suffering and dying, many are afraid and mentally disturbed, and the health of others is jeopardized because regular care is scaled down. Balance means a broader notion of health and illness, paying attention to the psychological, spiritual, and social dimensions of being ill, rather than chiefly focusing on physical and biological aspects [38]. Ethics requires a comprehensive and relational approach since human beings are affected.

The military vocabulary furthermore assumes that although human beings are interacting with nature, only humans have value. The notion of balance transcends the dualism between humans and nature. When the fate of nature and human beings cannot be disconnected, value should be attributed to the relationship, and not uniquely to humans or nature. This view implies that we are not locked into a perpetual conflict with microbes. The virus is not an enemy to be eradicated but it is 'a long-term life fact' [39]. Acknowledging the interdependency of humans and non-humans, we have to learn to live with microbes [40].

6.5 Common Home

After the above considerations we propose that from the perspective of global bioethics, there is a need for less violent metaphors which are inclusive, caring, supportive, and emphasize worldwide shared experiences [41]. One common experience for numerous people is staying at home for prolonged periods of time. The notion used in official recommendations is 'sheltering in place.' This is reminiscent of the Cold War. In many Western countries civil exercises taught people to shelter when there is an imminent threat, signaled by public alarm sounds in all cities. The advice is that people take refuge in a small, interior room, lock all windows and remain indoors, or move to special shelters nearby. This self-isolation usually takes a few hours, not days or weeks [42]. The concept of sheltering in place is also used for biological and chemical threats, as well as for extreme weather events. The basic idea is to wait until the worst is over; then go outside and resume normal activities. A specific view of disasters is presupposed: they are sudden events with immediate but often localized impact such as tornadoes or bioterrorist attacks. Pandemics, however, are gradual disasters; they have slow and incremental impact, and move across the globe, potentially affecting everyone. The difference is important for the degree of preparedness. Since pandemics come in waves, not all regions and countries are simultaneously affected. This leaves time for preparation. It also means that sheltering will be extended for weeks perhaps months, not knowing how long it will provide security.

During a pandemic, 'sheltering at home' seems a more acceptable term. It avoids military connotations of mandatory quarantine or isolation. It is restricting freedom of movement but appeals to the responsibility of individuals. Contrary to shelter-in-place, sheltering at home is more lenient since it allows to go out for essential business and walking, if physical distance to other people is observed. The notion of 'home' generally has positive connotations. In distinction to 'place' it is not a neutral location. It is where people live together in a space of intimacy and privacy, often regarded as a haven or refuge, a secure site to retreat and feel comfortable, a setting for caring relationships and conviviality [43]. At home means more than residing in a specific place. Homes are related to activity; dwelling places are transformed into homes, making them into *our* places. Being-at-home is the result of shaping and being shaped by the world [44]. A home is located in a specific space, so that the concept refers to territory where people are rooted. It is furthermore a center of self-identity, a place where values are imbued, where people can be themselves, where they belong. Finally, a home is a social and cultural unit, providing an emotional environment in which relationships and activities can grow and flourish [45]. For philosopher Gaston Bachelard 'home' refers to everyday experience [46]. Of course, not all homes are the same (e.g., nursing homes), not all homes are safe (e.g., domestic abuse), and not everybody has a home. Now that more than half of the world population is confined to their homes, Bachelard reminds us that being-at-home is more fundamental than activities such as working. Nevertheless, the emphasis on sheltering and isolation reflects an individualistic perspective, assuming that people can easily withdraw from social interactions. In many cases, connections with environing conditions cannot be severed. For numerous people, especially in low- and middle-income settings, this is not an option; they cannot survive if they don't earn an income. Also, poor people in affluent countries cannot shelter but have to expose themselves in order to subsist.

Nowadays, the home metaphor is used to refer to planet Earth. References to the planet as the home of humankind have been used by scientists, for example in describing the various qualities of the planet as a geophysical body [47]. The metaphor is employed in policy documents such as the *Rio Declaration on Environment and Development* (1992). Its Preamble states that the earth is our home [48]. The *Earth Charter*, a civil society initiative declaring principles and values for a sustainable global future, launched in 2000, repeats the same metaphor [49]. Having found its way in scientific, policy and popular discourses, the home metaphor is now frequently applied in connection to the global threats of climate change, biodiversity loss, and environmental degradation. A major impetus was given in the Encyclical letter *Laudato Si'* of Pope Francis in 2015 with its subtitle 'On care for our common home' [50]. Home is not a specialist concept. It is familiar to everyone as the place where human beings inhabit the world, where they used to be born and die. The metaphor is therefore often associated with positive emotions [51]. Applying the metaphor of home to the planet reformulates the connotations of the concept. The earth is the physical structure where humans live but it is also the place where they belong and should feel safe and secure. Like a house is divided in different rooms, the world is divided but the planet as common home brings humans together as inhabitants of the same dwelling. If home is an accomplishment since it depends on our action transforming this dwelling place into home, it implies the responsibility to take care and make ourselves at home in the world [52]. Being-at-home is the experience of locality but now the local is global since human beings are part of the environmental world.

Imagining the planet as common home has two advantages [53]. Firstly, it presents a perspective broader than the individualistic one associated with sheltering at home, where everybody is shielding in his or her own private sphere. This perspective clarifies that human beings are integrated in the world. Planet Earth is not merely a globe, an external object that they inhabit, a territory that can be explored and occupied. As argued in Chap. 1, it is a sphere in which humans are embedded and that defines their lifeworld. The planet is the location on which people are at home but it is not the world within which they live. Humans cannot be separated from the context and physical environment in which they live and that they share with other beings. The second advantage of the metaphor of common home is that it brings together the experience of the pandemic and today's main challenge of climate change. This is not a long-term problem threatening the survival of future generations but it is a crisis happening now. Climate disasters occur almost every week. Tsunamis, earthquakes, wildfires, cyclones, droughts, and nuclear accidents cause immense destruction. Climate change makes the unimaginable a reality for many people across the world. Predicted destructive effects such as rising sea levels, violent weather, loss of biodiversity, extinction of species, desertification, oceanic dead zones, and climate refugees are no longer hypothetical but now observable everywhere. The unprecedented scale of disastrous change creates a sense of calamity. In 2018, the Intergovernmental Panel on Climate Change warned that we only have twelve years left to avoid global catastrophe [54]. This decade is therefore crucial to mitigate the effects of climate change. Although global warming is known for at least a century (the greenhouse effect was identified in the midst of the nineteenth century), global policies have not been effective. While the impact of

Covid-19 will be temporary, the effect of climate change is already there and will request sustained action for a very long time. Both global threats are usually interpreted from a similar perspective: humans are not embedded in nature. The earth is a resource that can be commodified and exploited; it is not regarded as a living organism that creates and nourishes life (humans, animals and micro-organisms). The fact that many of us are now confined to our homes is hopefully an incentive to realize that we all share a common home. It demonstrates not merely interconnectedness of human beings manifested in society and culture but furthermore their embeddedness in the natural world. What affects the health of the planet will unquestionably deteriorate human health. Concepts such as 'one health' and 'planetary health' articulate the connection between health care and earth care. They make clear that the focus of bioethics should go beyond individual health, and that effective policies should be based on collective action.

6.6 Conclusion

In Chap. 2, we maintain that pandemic experiences used to be explained in two different ways. The contamination frame views disease as the result of contagion, i.e., person-to-person transmission. Disease is an invading enemy, attacking the body from the outside. The configuration frame regards disease as the result of a specific constellation of circumstances and as the consequence of certain stimuli. Cholera for example is the result of social organization and change, and thus the symptom of underlying social pathology. Dirt, crowding and poverty make people more vulnerable to infection. This last frame stresses the role of context and interconnection. It clarifies that a pandemic disease is a biological as well as social phenomenon, a form of drama, unfolding in time with a particular pattern (start, slow increase, accelerated spread, crisis, and ebbing to a closure).

The military metaphor employed today to express experiences and policies in regard to Covid-19 generates an interpretative framework that emphasizes contagion. In this chapter the implications and difficulties of such interpretation are analyzed. The language of this framework refers to danger and casualties, domination and control, discipline and surveillance. There is more interest in the virus and how to cope with it, than in the context in which infections arise. The focus on the virus as enemy, although invisible, may induce stigmatization and discrimination of carriers of the virus. Military language is also associated with nationalistic sentiments and competition between countries. This chapter presents an alternative framework, based on the perspective of ecology. The metaphor of connectedness asserts the connections between humans and between humans and nature. Relationality is a main characteristic of human beings. They are at the same time embedded in the natural world. The metaphor of balance affirms that health and disease are not simply a matter of biological constitution but dependent on psychological, social and economic determinants; it promotes a broad and holistic view of illness. Another appropriate metaphor is common home. Sheltering at home has a double meaning:

retreating within one's personal sphere (home) and protecting oneself by shielding the planet that is our fundamental dwelling place (common home). Considering Earth as common home transcends the individualistic connotations of home by emphasizing the shared experience of living together on the same and only planet. This metaphor connects the two major global threats of this moment: pandemic infectious disease and climate change.

The diverse language to explain Covid-19 is not morally neutral. Metaphors conceptualize the world and shape experiences. They help to make sense of what is happening and they give meaning to what we endure. At the same time, they open up or close possibilities for interpretation and understanding, and guide actions in specific directions. Confronted with lethal threats and serious illness, the metaphors used are important to navigate the ethical challenges that arise in the context of treatment, care, and prevention.

References

1. See, for example, Rajandran, K. 2020. 'A long battle ahead': Malaysian and Singaporean prime ministers employ metaphors for Covid-19. *GEMA Online Journal of Language Studies* 17 (2): 163–176; United Nations. 2020. *COVID-19 Response. This war needs a war-time plan to fight it*; Euronews. 2020. We need a 'war economy' to deal with COVID-19 crisis, UN chief Antonio Guterres tells Euronews. *Euronews Online*.
2. Zhang, L. 2021. *The origins of Covid-19. China and global capitalism*. Stanford: Stanford Briefs.
3. See, for example, Adam, M. 2020. An enemy to fight or someone to live with, how Covid-19 is metaphorically described in Indonesian media discourse. *Conference paper*, July, 2nd National Webinar on English Linguistics and Literature; Wicke, P., and N.M. Bolognesi. 2020. Framing COVID-19: How we conceptualize and discuss the pandemic on Twitter. *PLoS ONE* 15 (9): e0240010.
4. Pasquier, P., A. Luft, J. Gillard, et al. 2020. How do we fight Covid-19? Military medical actions in the war against the Covid-19 pandemic in France. *BMJ Military Health*; Opillard, F., A. Palle, and L. Michelis. 2020. Discourse and strategic use of the military in France and Europe in the Covid-19 crisis. *Tijdschrift voor Economische en Sociale Geografie* 111 (3): 239–259.
5. McNeill, W.H. 1998. *Plagues and people*. New York: Anchor Books.
6. Walker, I.F. 2020. Beyond the military metaphor. *Medicine Anthropology Theory* 7 (2): 261–271; Cipolletta, S., and M.C. Ortu. 2020. Covid-19: Common constructions of the pandemic and their implications. *Journal of Constructivist Psychology*; Bates, B.R. 2020. The (in) appropriateness of the WAR metaphor in response to SARS-CoV-2: A rapid analysis of Donald J. Trump's rhetoric. *Frontiers in Communication* 5: 50; De Waal, A. 2021. *New pandemics, old politics. Two hundred years of war on disease and its alternatives*. Cambridge: Polity Press.
7. "Our fears and efforts were targeted on individual germs, and not on the ecologies that generate them or the society and economy that enable them to spread." De Waal, *New pandemics, old politics*, 13.
8. Rohela, P., A. Bhan, D. Ravindranath, et al. 2020. Must there be a 'war' against coronavirus? *Indian Journal of Medical Ethics* 5 (3).
9. Buruma, I. 2020. Virus as metaphor. *New York Times*, March 28.
10. Isaacs, D., and A. Priesz. 2021. Covid-19 and the metaphor of war. *Journal of Paediatrics and Child Health* 57: 6–8.

11. Fairbanks, E. 2020. A pandemic is not a war. *HuffPost*, April 14.
12. Bell, A., and T. Gift. 2020. 200,00 losses later. Some lessons for the 'wartime president'. *Newsweek*, September 22.
13. Chapman, C.M., and D.S. Miller. 2020. From metaphor to militarized response: The social implications of 'we are at war with Covid-19' – Crisis, disaster, and pandemics yet to come. *International Journal of Sociology and Social Policy* 40 (9/10): 1107–1124.
14. This is the reason that De Waal uses the term 'pandemy.' It more adequately refers to a societal crisis, including ecological and societal pathologies. De Waal, *New pandemics, old politics*, 15.
15. "Life is not only a bloody confrontation: it also depends on myriad living interactions..." Ryan, F. 2020. *Virussphere*. London: William Collins, 178; WHO. 2020. *Coronavirus Outbreak (COVID-19): WHO Update* (13 May 2020). United Nations. 13 May.
16. Craig, D. 2020. Pandemic and its metaphors: Sontag revisited in the Covid-19 era. *European Journal of Cultural Studies* 23 (6): 1025–1032.
17. Rohela, Bhan, Ravindranath, et al. 2020 Must there be a 'war' against coronavirus? *Indian Journal of Medical Ethics* 3: 222–226.
18. Sabucedo, J-M., M. Alzate, and D. Hur. 2020. Covid-19 and the metaphor of war. *Revista de Psicologia Social/International Journal of Social Psychology* 35 (3).
19. Hassoun, N. 2020. What is COVAX and why does it matter for getting vaccines to developing nations? *The Conversation*, October 2.
20. Miller, F.G. 2020. Pandemic language. *Hastings Center*, July 2020.
21. Rose, J. 2020. Pointing the finger. *London Review of Books* 42 (9): 3, 6–8, 10.
22. De Beaumont Foundation. 2020. *Poll: New national conversation about Covid-19 urgently needed to overcome partisan divide and save lives*. November 30.
23. Shaw, D. 2020. The many meanings of 'stay safe' in a pandemic: Sympathy, duty, and threat. *Journal of Medical Ethics Blog*, May 13.
24. Bedyński, W. 2020. Liminality: Black Death 700 years later. What lessons are for us from the medieval pandemic? *Society Register* 4 (3): 129–144.
25. Thornhill, C., and R. Miron. 2020. Karl Jaspers. In *The Stanford encyclopedia of philosophy*, ed. E.N. Zalta.
26. Carrington, D. 2021. World leaders 'ignoring' role of destruction of nature in causing pandemics. *The Guardian*, June 4.
27. George, D.R., E.R. Whitehouse, and P.J. Whitehouse. 2016. Asking more of our metaphors: Narrative strategies to end the 'war on Alzheimer's' and humanize cognitive aging. *American Journal of Bioethics* 16 (10): 22–24.
28. Walker, Beyond the military metaphor, 265, 268; Nie, J-B., A. Gilbertson, M. de Roubaux, et al. 2016. Healing without waging war: Beyond military metaphors in medicine and HIV cure research. *American Journal of Bioethics* 16 (10): 3–11.
29. Marcel, G. 1962. *Homo viator. Introduction to a metaphysic of hope*, 23. New York: Harper & Row.
30. ———. 1984. Reply to John E. Smith. In *The philosophy of Gabriel Marcel*, ed. P.A. Schilpp and L.E. Hahn, 350–353. La Salle: Open Court.
31. Bruni, L. 2012. *The wound and the blessing. Economics, relationships, and happiness*. Hyde Park/New York: New City Press.
32. Sayer, A. 2011. *Why things matter to people. Social science, values and ethical life*. Cambridge: Cambridge University Press.
33. In the words of Michael Sandel: "The common good is about how we live together in community. It's about the ethical ideals we strive for together, the benefits and burdens we share, the sacrifices we make for one another. It's about the lessons we learn from one another about how to live a good and decent life." Quoted in Friedman, T. 2020. Finding the 'common good' in a pandemic. *New York Times*, March 24.
34. Ten Have, H. 2019. *Wounded planet. How declining biodiversity endangers health and how bioethics can help*. Baltimore: Johns Hopkins University Press.

35. Hinchliffe, S. 2015. More than one world, more than one health: Re-configuring interspecies health. *Social Science & Medicine* 129: 28–35.
36. Potter, V.R. 1971. *Bioethics. Bridge to the future*. Englewood Cliffs: Prentice-Hall; Potter, V.R. 1988. *Global bioethics. Building on the Leopold legacy*. East Lansing: Michigan State University Press.
37. Ten Have, *Wounded Planet*, 44ff.
38. Nie, Gilbertson, de Roubaux, et al. *Healing without waging war.*
39. World Health Organization. 2020. *Coronavirus outbreak (COVID-19), Update* (13 May).
40. Benezra, A., J. DeStefano, and J.I. Gordon. 2012. Anthropology of microbes. *Proceedings of the National Academy of Sciences of the United States of America* 109 (17): 6378–6381.
41. Oswick, C., D. Grant, and R. Oswick. 2020. Categories, crossroads, control, connectedness, and change: A metaphorical exploration of Covid-19. *The Journal of Applied Behavioral Science* 56 (3): 284–288.
42. American Red Cross. 2003. *Fact sheet on shelter-in-place.*
43. Mallet, S. 2004. Understanding home. A critical review of the literature. *Sociological Review* 52 (1): 62–89.
44. "Home is lived in the tension between the given and the chosen, then and now, here and there." Mallet, Understanding home, 80.
45. Smith, S.G. 1994. The essential qualities of a home. *Journal of Environmental Psychology* 14: 31–46; Moore, J. 2000. Placing *home* in context. *Journal of Environmental Psychology* 20: 207–217.
46. Bachelard, G. 2014. *The poetics of space*. New York: Penguin Books, 28. According to Levinas: "To exist…means to dwell." Levinas, E. 1991. *Totality and infinity: An essay on exteriority*. Dordrecht: Kluwer Academic Publishers, 156. See also: Dekkers, W. 2011. Dwelling, house and home: Towards a home-led perspective on dementia care. *Medicine Health Care and Philosophy* 14: 291–300.
47. Selâl Şengör, A.M. 1991. Our home, the planet earth. *Diogenes* 39 (155): 25–51.
48. *Rio Declaration on Environment and Development.*
49. *The Earth Charter.*
50. Francis, Pope. 2005. *Encyclical letter Laudato Si' – On care for our common home*. Rome: Vatican.
51. Thibodeau, P.H., C. McPherson Frantz, and M. Berretta. 2017. The earth is our home: Systemic metaphors to redefine our relationship with nature. *Climatic Change* 142 (1): 287–300.
52. Jacobson, K. 2009. A developed nature: A phenomenological account of the experience of home. *Continental Philosophy Review* 42: 355–373.
53. Ten Have, H. 2020. Sheltering at our common home. *Journal of Bioethical Inquiry* 17: 525–529.
54. Intergovernmental Panel on Climate Change. 2018. *Global warming of 1.5°C*. Cham: IPCC.

Chapter 7
Treatment, Care, and Ethics

Abstract This chapter examines the ethical challenges of treating and caring for Covid patients. Healthcare professionals are at risks to be infected by SARS-CoV-2, and in all countries illness and death has affected them as well as their families. The ethical discussion focuses on whether professionals have a duty to treat and to care, when there is substantial personal risk, particularly when sufficient protective equipment is not available. Ethical issues also exist for patients. They can experience various kinds of harm as a result of having contracted an infectious disease, and as the result of being in isolation during treatment in the healthcare facility. Patients with other diseases are harmed because modalities of treatment and care are cancelled or postponed since priority is given to Covid patients. A further ethical concern relates to the difficulty to maintain ordinary standards of care in conditions of emergency. Specific attention is subsequently given to ethical questions of research. The only way to improve the treatment of Covid patients is sustained research to test and develop medication. Intensive public debate has emerged on the subject of triage. If resources, especially in intensive care, are limited, which patients will be selected for treatment, and which criteria are ethically justified? The last paragraph of this chapter will focus on end-of-life care, and the need to provide palliative care to seriously ill Covid patients.

Keywords Care ethics · Death · Duty to care · Intensive care · Non-Covid patients · Publication ethics · Research · Standards of care · Triage · Utilitarianism

7.1 Introduction

In the first wave of the infectious disease many healthcare systems have been overwhelmed by the rapidly growing numbers of seriously ill patients who needed hospital treatment and frequently intensive care. The ethical debate was therefore focused on the challenge of saving the maximum number of lives. Shortages of intensive care beds and equipment were the main concern of policymakers, while professional organizations developed guidelines for prioritization. In this stage,

H. ten Have, *The Covid-19 Pandemic and Global Bioethics*, Advancing Global Bioethics 18, https://doi.org/10.1007/978-3-030-91491-2_7

several proposals for criteria of triage have been advanced, often based on utilitarian ethics. After a few months, more scientific insight has emerged into potential remedies and treatments. However, the challenge has remained how much evidence will be needed to implement policies and treatment guidelines. Furthermore, victory is declared after a wave is over, leaving societies no longer prepared for the next wave which again puts healthcare systems under pressure.

In the field of healthcare ethics, public health is commonly distinguished from clinical medicine. While the first is focused on the population and common interest, the later is primarily concerned with the interest of the individual patient. Policy measures against the coronavirus aim at slowing the spread of the pathogen among the population so that the healthcare system can cope with the rising numbers of infected and seriously ill patients. Specific ethical considerations at the level of public health refer to the notions of solidarity and social responsibility. Restrictive interventions such as isolation, quarantine, lockdown, and travel advisories request the balancing of individual freedom and the collective good. The ethical issues of public health will be discussed in Chap. 8. Clinical medicine aims at providing the best possible treatment and care for each individual in need, respecting personal autonomy, informed consent, and voluntary cooperation. SARS-CoV-2 particularly affects vulnerable populations (the elderly, the immunocompromised, people with multiple morbidities, and socially disadvantaged persons). Because effective medication is not available in the early stages, caring for Covid patients is especially challenging, even more because protective equipment is not sufficiently provided to elderly and nursing homes.

Medical treatment and care during a pandemic with a massive influx of infected and ill patients raise specific ethical challenges that will be discussed in this chapter. In a major catastrophe such as Covid-19 the distinction between public health interventions and patient-centered care is compromised. Public health measures aim at the collective good, often appeal to personal responsibility but this insistence on an individual perspective is frequently insufficient to control the spread of the virus. At the same time, a focus on the population will only work when individuals change their behavior. Patient-centered care, on the other hand, is confronted with the fact that the usual focus on ethical principles at the individual level is insufficient. Respect for autonomy is difficult to maintain in emergency conditions, and when there are serious risks for other people, especially patients, visitors, healthcare providers, and support staff. Scarcity of resources introduces the perspective of rationing with the question how setting priorities can be as fair as possible. The priority of treating Covid patients leads to scaling down or cancelling other medical services and treatments, with serious impacts on the health of patients with other diseases. The interest of a particular individual patient that usually has primacy in medical treatment is problematic in a massive disaster where many are victims. What is done for one person affects what can be done for other potential patients.

The question is raised whether under the exceptional circumstances of the pandemic, the usual ethical principles of clinical ethics still apply. Are moral responsibilities and duties modified in catastrophic situations? It is argued that these principles do not provide sufficient guidance in an unprecedented crisis [1].

Moreover, in a pandemic, ethical standards are shifting from maximizing the benefit for each patient to maximizing the benefit for the majority [2]. Furthermore, there is a tendency to shift from deontological to teleological moral reasoning in times of a disaster [3]. These arguments illustrate the point made in Chap. 1 that the principles of mainstream bioethics are too narrow for contemporary challenges. Respect for autonomy, beneficence, non-maleficence, and justice are established principles in clinical medicine but they do not offer adequate assistance to clarify and address global challenges. A broader ethical perspective is needed that transcends the focus on individual moral agents. The emerging discipline of global bioethics provides such broader ethical view, emphasizing not merely the established principles but as a social ethics also the concerns with vulnerability, solidarity, cooperation, social responsibility, and common good. Rather than underlining that in a pandemic ethical principles have shifted or are applied in a relaxed or altered way, the experiences with Covid-19 demonstrate what has been argued for a long time since the coining of the term 'bioethics': faced with the challenges of globalization, a broader framework is necessary for understanding and coping with contemporary ethical problems.

This chapter will discuss areas where specific ethical questions emerge in clinical medicine and care. First, treating patients with a serious and sometimes lethal infectious disease calls into question the duty to care. Second, the pandemic influences and often transforms the experiences of patients who are not in control of their own destiny. Third, in disaster conditions traditional standards of care can no longer be maintained, so that questions arise about the quality of treatment provided. Fourth, since there is a desperate need for effective remedies, Covid patients may be subjected to experimental treatments, while at the same time, popular discourse and politicians advocate unproven drugs. Research is the only way to make progress in treatment, especially when physicians are confronted with a novel virus and an unknown disease. But how should research be conducted in disaster circumstances? Fifth, major ethical concerns regard intensive care interventions. A substantial number of patients require ventilatory support often for a prolonged time. ICU beds and equipment are scarce so that the possibility of triage is haunting medical and popular debate. Finally, the pandemic challenges the ethics of care since it demonstrates and amplifies the vulnerability of people in care and nursing homes, confronting many with experiences of death, suffering, and dying, necessitating extra efforts in palliative care.

7.2 Healthcare Professionals

When Covid-19 starts to disseminate across the globe, diverse images capture the public mind. Images from China articulate control and massive efforts of the government, rapidly building a gigantic new hospital, and imposing draconian rules for quarantine and masking. The images from Italy are different. They show an overwhelmed healthcare system with hospitals in chaos, overburdened healthcare workers, patients in hospital corridors, and deceased people who could not be buried. In

April 2020, it is reported that 16,991 healthcare workers test positive for the coronavirus (more than 10% of the total number of positive cases in Italy). The death toll is considerable: 119 medical doctors and 51 nurses and nursing aids, a higher number than in any other country at that time. Of all medical disciplines, general practitioners are most affected (with 66 deaths), mostly due to lack of protective equipment in general practice [4]. That healthcare providers are a special vulnerable group in all countries becomes clear in the following months. In November 2020, it is reported that almost 300,000 care workers in 37 countries had been infected with Covid-19, and more than 2500 have died by August 15th [5]. One year into the pandemic, more than 3600 have reportedly died from Covid-19 in the United States alone, mostly nurses and support staff, while two-thirds have been people of color [6]. Almost 2 years after the beginning of the pandemic, the WHO estimates that between 80.000 and 180.000 health and care workers could have died from Covid [7].

A common response to pandemic diseases has been to run away and escape to safer places, as exemplified in the classic work of Boccaccio. Stories are known of physicians who abandoned affected cities and areas during the Black Death and cholera. The famous physician Galen left Rome when the Antonine Plague broke out in 166 and fled to Pergamon. Nowadays, a huge majority of healthcare workers are taking care for infected patients, although in some countries doctors and nurses have resigned or not reported to work [8]. Popular media have published numerous stories of healthcare professionals who sacrificed themselves caring for Covid patients [9]. This is impressive given the risk of infection. Reflecting the war metaphor, healthcare professionals are celebrated as heroes: they are battling in the frontlines to save as many lives as possible. There are numerous manifestations of appreciation for these warriors (clapping and singing, banners, flowers), especially in the early stages of the pandemic. The language of heroism and sacrifice is problematic, however [10]. Healthcare workers have expressed that they feel uncomfortable with this language [11]. They just do their job and continue to do what they have always done, giving priority to the interests of individual patients. The image of the hero also diverts attention from the context that contributes to the risks, notably the lack of personal protective equipment, insufficient testing capacity, and poor compliance of the population with public health restrictions. Applauding care providers therefore shifts responsibility and minimizes the role of other actors in managing the pandemic [12]. Furthermore, the image suppresses critical debate. Heroes are not supposed to complain and protest. That they are vulnerable human beings with anxiety, physical stress and moral distress cannot be recognized.

The ethical debate is often focused on the duty to treat and the duty to care. The common assumption is that healthcare professionals have a moral obligation to provide treatment and care during a pandemic. Fear of contagion cannot be a reason to refuse care. This obligation is incorporated in professional codes of ethics. Doctors and nurses should give priority to the interest of patients over their personal interests [13]. The duty is based on the ethical principle of beneficence. Doctors have unique and special expertise to treat disease, and patients engage in a relationship with them, trusting that their interest will prevail. Health professionals have freely chosen their profession, knowing that they may encounter more risks than the general population. Moreover, there is a social contract between healthcare professionals and society; they have been given a special status (e.g., self-regulation, and development of practice standards) in the expectation that they will respond to threats to health, especially during infectious diseases [14].

The duty to treat and to care is criticized for three reasons. One argument is that the duty is too broad: healthcare workers may have conflicting obligations. They not only have the duty to care but also the duty to protect themselves and the duty to protect others. Particularly when the risks are high and serious, as with Covid-19, concerns may arise about other people, such as colleagues, loved ones, family members and other patients. Physicians are not just soldiers who follow commands but autonomous persons who make moral judgments about what they should do. During a public health emergence, the context of care has changed, and ethical considerations not only apply to individuals but also to the community. All things considered, the duty to care is shorthand for a range of duties [15]. The language of heroism neglects that the duty to care is limited by other duties. The level of personal risk might be unacceptable in case of healthcare workers with medical conditions that make them more susceptible to Covid-19. It may also be not acceptable when protective equipment is not available, and the health and safety of care workers is not protected. Professional guidelines on this point are sometimes ambiguous. The General Medical Council in the United Kingdom states that all available steps should be taken to minimize the risk, practitioners should raise concerns with their employer and record their decisions [16]. The ethical guidelines of the British Medical Association are clearer: there are limits to the risks doctors are expected to expose themselves to: "You are under no obligation to provide high-risk services without appropriate safety and protection" [17].

The second argument is that the duty to treat and to care is too restricted. The duty is often focused on physicians and nurses, but it can only be exercised in a context where many other professionals and nonprofessionals make important contributions. Without housekeepers and kitchen staff for example care institutions cannot function. To cope with the threat of a pandemic, solidarity among all workers within a health facility is necessary. The labelling of some professionals as heroes ignores that a duty to care must be accepted by all workers, and that organizations and institutions must facilitate and nurture the implementation of this duty [18]. Treating Covid patients is a common effort. While the focus on duties directs attention to specific individual professionals, a broader ethical perspective articulates cooperation and solidarity.

The third argument emphasizes reciprocity. If the duty to care is grounded on a social contract, healthcare workers exposing themselves to personal risks, may expect in return social obligations [19]. Healthcare institutions should provide adequate protection and create a safe environment so that risk of illness is as minimal as possible. Governmental authorities have similar responsibilities. It is important that the general population takes its responsibility by adhering to public health measures, so that the burden on the care system will diminish. Taken together, healthcare workers are risking their well-being and lives within a wider context of social responsibilities of multiple actors. Applauding them as heroes is an easy way to deflect from policy inertia and to excuse lax implementation of measures [20]. An additional consideration is that the anxiety and stress associated with working in contagious conditions is underestimated. The pandemic has a major psychological impact on healthcare workers [21]. They suffer from mental and physical exhaustion due to excessive workload; they face difficult decisions, and see patients and colleagues succumbing. They may be scared to go to work, knowing that they can be the next victim. Fearing that they might infect loved ones and relatives, they take elaborate arrangements of disinfection and disrobing when they return home or isolate themselves from family members. Workers who complain about workload or lack of protective materials have been silenced, sanctioned, or suspended from work [22]. Healthcare workers furthermore experience moral distress since they cannot adhere to the usual standards of care. Routine activities are cancelled so that they are not able to provide regular care for non-Covid patients. The point is that contrary to other emergencies, the pandemic is continuing. Even if PPE shortages have been addressed, care workers have not recovered from the first wave when confronted with the second wave. Rather than appeals to the duty to treat and to care, healthcare institutions and policymakers should provide support for people who assume disproportional burdens to help patients [23].

From the above considerations it can be concluded that the focus on the duty to treat or care presents a narrow view. Providing treatment and care is a common endeavor requiring the cooperation of many people, – professionals and nonprofessionals. Moreover, it implies reciprocity: employers and the government should engage in efforts to protect those involved in care. Instead of the ethical approach of mainstream bioethics, a broader perspective is appropriate, asserting global bioethics principles such as solidarity, social responsibility, and vulnerability [24]. This perspective operates with a context of connectedness that informs moral experiences. Rather than accentuating abstract rules and duties, it highlights virtues such as empathy, courage and resilience that motivate healthcare workers to continue caring in stressful and precarious circumstances. But virtues are not simply individual characteristics. They are also expressed and sustained at the level of organizations and the community.

7.3 Patients

In the early stages of the pandemic, it is unclear what the symptoms of Covid-19 precisely are and how the disease can be diagnosed, particularly since tests are scarce and only available in the hospital setting. These uncertainties influence patient experiences, especially in the home situation when also treatment and prognosis are unsure. Anxiety and stress are higher among certain groups of people such as the elderly and persons with underlying disorders for whom SARS-CoV-2 is particularly threatening. But healthy individuals can be affected too, while access to hospitals is not always possible. People are quarantining at home where they must cope with the disease often without adequate professional support. Many patients share their stories in journals and social media [25]. In hospitals, public health strategies are adopted with isolation and admission in special wards with restrictions or prohibitions of visits, and with doctors and nurses in full protective equipment, when available. Managing individual patients in such circumstances makes it difficult to maintain the usual relationship between healthcare providers and patients as well as a holistic approach to patient care. Doctors complain that they cannot fully communicate with patients because they are covered in protective equipment. Patients cannot see their faces, ordinary human connection through handshakes and touching is impossible, and physical examinations and visits are as short as possible. [26] Patients point out that personal interaction and communication with healthcare staff is limited and difficult, and that their isolation is stressful and depressing.

Although studies of the experiences of Covid-19 patients are limited, previous practices of infection control demonstrate two types of effect: the psychological effects of isolation, and the effects of having contracted an infectious, and potentially lethal novel disease. Isolation is associated with feelings of anxiety, uncertainty, loneliness, depression, but also anger and stigma. Patients may receive less care since staff is entering isolation rooms less frequently, and contact time with healthcare professionals is reduced. Some feel confined in a narrow physical space with a closed door and without windows, completely disconnected from the outside world [27]. Since infection control measures are not primarily aimed at the interest of the individual patient, he or she can feel that care is unsatisfactory and that individual needs are not sufficiently respected. At the same time, many patients are concerned that they may have infected others, especially family and friends, and may experience guilt when others fall ill. They blame themselves for the misery of other afflicted people [28].

These findings indicate that sufficient attention should be given to the experiences of patients in Covid wards. At least two challenges need to be addressed. In ordinary circumstances, healthcare is holistic and personal. Confronted with a serious disease such as Covid-19, hospital care tends to focus on severe cases requiring ICU intervention and ventilator support while patients with moderate symptoms are carefully observed for any deterioration of their condition. Concentrating on physical parameters such as saturation, temperature, and blood values should be complemented with paying attention to possible emotional distress and psychological problems. Even in emergency conditions, holistic care should be provided as much as possible with concerns for mental health, social support, and good communication. The ethical challenge is how the experience of the patient in isolation can be enhanced. Quality of care may be reduced since attention is exclusively focused on the physical condition of the patient while the psychological effects of isolation are not recognized. Creative approaches attempt for example to overcome barriers to communication, and to humanize the experience with the use of tablets and other digital technology to restore the ability to communicate [29]. Caring for Covid-19 patients has initiated a major shift to virtual care now that in-person care is more difficult. The other challenge is that personal care for the individual patient is subordinated to the public health perspective. Individual autonomy is restricted, and the focus on the traditional principles of beneficence and non-maleficence in patient care is overridden by the prior concern to protect the health of other persons, – first of all, healthcare workers, but also other patients, and family members. The main challenge is how to minimize the harm to individual patients. It requires that the concern with physical health is complemented with psychosocial care and mental health support. Even when the conditions are more difficult, patients will have the same need for empathy as in other circumstances [30]. The commitment to relate to patients places the clinician in an ethical dilemma: how to control the possible spread of the infection while providing the best possible care for individual patients? In a public health emergency, it is argued that the responsibilities of physicians to the common good have priority over responsibilities to individual patients. But altered responsibilities do not imply that the physician-patient relationship should not be maintained, and that physicians should not act in the best interests of patients. Within the framework of public health measures such as isolation, patients should trust that treatment and care is provided in their interest to the extent possible [31]. The language of war and heroism adds another dimension: it reinforces the asymmetry of the doctor-patient relationship, regarding the patient as a defenseless subject not able to actively participate in treatment and care [32].

While the above ethical challenges to patient care are visible and increasingly addressed, another challenge is less noticeable and explored. The surge of the pandemic is limiting the services for patients with other illnesses than Covid-19. Care for numerous people is disrupted, not only hospital treatment but also primary care. In hospitals, patient visits to emergency departments substantially diminished. Planned hospital treatment decreased because fewer patients were referred and because non-urgent care was cancelled or postponed. In primary care the number of appointments significantly dropped, while many appointments were changed to distance contact, e.g., by telephone. For example, in England, the number of patients seen by a specialist following an urgent referral by a general practitioner for suspected cancer dropped with 60% in April 2020. Although there was a substantial effort to reopen services and resume care, in October 2020 still 8% fewer patients were seen by a specialist than in October 2019 [33]. In many countries the focus on Covid not only reduced treatment and care for non-Covid patients, but also other healthcare activities such as vaccination of children, preventive cancer screening, and in some cases even hemodialysis. This overall contraction of healthcare is usually explained by two factors. One is care avoidance: people are postponing or foregoing services since they feel unsafe to visit a hospital or doctor's office, fearing that they might be infected in these corona hotspots. This is the consequence of stay-at-home orders and social distancing measures, making people cautious to engage in social contacts. The other explanation is that healthcare services are deliberately restricted, with the hope to release capacity and to prevent that critical care is overwhelmed.

The decision to cancel or postpone non-Covid care is often presented as a factual necessity. The argument is that Covid patients need quick intervention. If oxygen levels are low and persons experience dyspnea, they need acute admission to the hospital. The level of care for such patients is intensive. Each Covid patient in the

ICU requires 5–6 care workers per 24 h. Patients also stay for a long period in the hospital (up to 3–4 weeks). Adequate care requires shifts in manpower, beds, and equipment. Surgical areas are turned into ICUs and human resources are reallocated within the healthcare facility. Therefore, there is no choice rather than postpone care for other patients. Using the war metaphor, the resulting harm is 'collateral damage' that cannot be avoided [34].

However, giving priority to Covid patients is an ethical choice. The assumption is that avoiding harm and saving the life of these patients is preeminent and overrides the harms of delaying care for other patients. It becomes increasingly clear that many patients will be seriously harmed if interventions and treatments are postponed. For example, when acute cardiovascular care is compromised, the prognosis of patients will be worse if they see a specialist later, with the result that they require a longer stay in the hospital or the ICU [35]. This is particularly worrying when urgent operations are not performed, for example surgery for cancer or appendicitis [36]. Also, the medical condition of people with chronic illnesses may worsen. Postponing treatment and care will furthermore impact the quality of life of many persons who need for example hip replacement and other interventions when they experience disability. Delays may moreover increase risks for other people, for instance when treatment of patients with tuberculosis is postponed. Concentration on Covid patients is thus associated with a host of negative effects for other patients. The daily reporting of Covid infections, hospitalizations, and deaths reflects only part of the story of the pandemic because significant indirect effects produce excess morbidity and mortality that is not directly attributed to SARS-CoV-2. These indirect consequences may outweigh the direct effects of the pandemic [37].

From an ethics perspective, scaling down healthcare for non-Covid patients is a form of rationing resulting from policies and patient behavior. Although in some countries postponing or cancelling care is recommended by policy agencies or professional organizations, in other countries individual healthcare facilities decide without any coordinated efforts so that specific types of care are available in some areas while inaccessible in other. The decision to cancel or postpone treatment and care is often broadly justified with the argument that the functioning of healthcare systems should be sustained. The implicit assumption is that prevention of harm to Covid patients is more compelling than the harm inflicted on non-Covid patients. An explicit weighing of various kinds of harm is not provided so that it is unclear why priority is given to certain kinds of harm. Why is treating tuberculosis, stroke or myocardial infarction regarded as less important than treating Covid-19? In rationing practices, a distinction is made between essential and non-essential services. In the emergency of the pandemic, it is taken for granted that Covid treatment is essential. The problem is that other services can be equally essential because they also aim to save lives. Even if the practical circumstances are dire, with shortages of beds, healthcare personnel and protective equipment, other arrangements are possible. For example, in a major hospital in South Korea, treatment areas for Covid and non-Covid patients are strictly separated, so that essential services for patients with acute myocardial infarction, stroke, severe trauma, and acute appendicitis do not decrease [38]. Similar strategies of designating health centers for Covid and non-Covid are applied in other countries, not only reducing the risk of contamination but also continuing care for some categories of patients [39]. Focusing hospital resources on Covid patients does not need to leave other seriously ill patients untreated [40]. The problem remains however that the distinction between essential and non-essential, or urgent and non-urgent care is not clear-cut. While some interventions are considered as non-urgent, the condition of some patients (for example with advanced heart failure) may rapidly deteriorate so that in hindsight they should have received urgent treatment [41]. The decision about what is essential is a value judgment involving the likelihood, seriousness, and evolution of significant harm.

Additionally, implicit rationing follows from patient behavior to avoid healthcare. It is doubtful whether such avoidance is the outcome of deliberate choices of patients. Many fear to be exposed to the virus, particularly the elderly and those with underlying conditions. They do not want to put themselves and their families at risk, while the horror stories in the media provide further lack of motivation to go to medical centers where infection is rampant. Hospitals ask patients not to come to the emergency department with mild symptoms. During the first and second wave of the pandemic government policies discourage or even prohibit people to leave their homes. Public health policies therefore constrain the options of patients. Other constraints are the social and economic effects of lockdowns. Millions of people lost their jobs and income, as well as their health insurance, at least in countries without national health insurance systems. Many of them simply cannot access healthcare even if they would. Rather than accepting care avoidance as a source of rationing, policies should include clear messaging that people should not be afraid to seek treatment and care.

Since rationing concerns the life and well-being of numerous citizens, it should not be an implicit process but based on explicit criteria that are consistent, transparent, and open to public discussion. Allocation of resources for patient care is not only a matter of balancing various kinds of harms (suggesting that health professionals are the experts who can decide) but moreover an issue of fairness. Healthcare should be provided equitable to all who are needing it, infected or not. Nonetheless, ethical discussions, and national ethical frameworks for treatment prioritization are generally absent and primarily focused on triage for ICU intervention [42].

7.4 Standards of Care

Isolated patients in hospitals may receive lower quality of care. Patients at home may not be easily seen by general practitioners but have remote or virtual contacts. Non-Covid patients are confronted with postponed consultations and interventions, while many people with chronic disorders experience lack of continuity of care. Because protective equipment is in short supply, the use of N95 masks in healthcare settings is extended, and masks are disinfected and reused. This is against the safety standards that recommend change after each patient contact. While these practices are scaring healthcare workers since it is not known what protection is offered, they are regarded as better than no protection at all [43]. In these cases, optimal treatment and care, requested by the standards that usually determine the provision of healthcare, are not offered. Such lower standards of care are justified with the argument that in the context of a catastrophic disaster usual standards cannot be maintained, and that other standards apply in crisis situations. Only in such exceptional conditions, medical decisions are allowed that would otherwise not be acceptable under ordinary circumstances. That patients do not receive the best possible care according to the current state of medical knowledge is not a voluntary choice but the outcome of *force majeure* [44]. It is not merely due to overwhelming demand but also the consequence of poor planning and preparation that could have avoided the need to change standards. In the current clinical situation, there are shortages of materials, and even more, insufficient numbers of staff while many patients need treatment and care. If standards are not adapted, only a limited number of patients receive optimal treatment and care. The ethical implication is that inequalities will result because some patients will be treated and others not. It will then be necessary to select patients who qualify for the best possible treatment, which will engender complicated ethical debates on the criteria used. Another implication is that given the potential lethality of Covid, only a limited number of lives will be saved. When standards are adapted, more lives could be saved, and all patients have an equal chance to benefit from treatment, although the potential benefit may be lower than in the case of maintaining ordinary standards of care. The rationale for applying crisis rather than conventional standards of care is that in a massive health emergency the goal of healthcare shifts from promoting the wellbeing of every individual to preserving population benefit. Saving as many lives as possible, and providing the best possible care to most patients takes precedence over the wishes and needs of individual patients [45].

Crisis standards of care are defined as the best possible care that can be delivered in disastrous situations. The Institute of Medicine in the United States has emphasized that the provision of care in such circumstances should be based on seven ethical considerations: fairness, duty to care, duty to steward resources, transparency, proportionality, and accountability [46]. The IOM guidelines have been developed before the Covid-19 pandemic but are essentially re-endorsed during the outbreak [47].

Implementation of crisis standards is not a signal that ethical norms during disasters are changed. Because the context of application has been transformed, they receive a different weight. This refers particularly to the ethical principle of respect for autonomy which in ordinary circumstances is regarded as one of the most, perhaps the most important principle to guide clinical decisions. In pandemic conditions, healthcare workers continue to be obliged to provide the best possible care but in disastrous circumstances they do no longer focus only on the individual patient under their care. This change in moral perspective is justified if certain conditions are fulfilled. One is that maximum efforts are made to avoid the situation in which healthcare systems are overwhelmed. That means proactive planning, so that everything possible is done to prevent or delay the need to enact crisis standards of care [48]. Particularly when ethical challenges are foreseen in a public health emergency, healthcare institutions have a duty to plan in order to manage potential ethical quandaries associated with triage, withdrawal of life-sustaining treatment, and palliative care [49]. It furthermore means that crisis standards are not implemented abruptly but along a continuum. When conventional care is no longer possible, the next phase is contingency care. Demand starts to exceed the available resources of medical staff, equipment or medication, and in this phase functionally equivalent care must be provided. Only if this type of care is impossible due to depleted resources, crisis care will begin [50]. Another condition is that the enactment of crisis standards of care is a policy decision. It should not be left to healthcare professionals or institutions in their particular setting; rather the shift to population-based healthcare requires governmental guidance, as well as public debate, engaging the community and the providers. It means that governments should formally declare an emergency and provide a legal and ethical framework to authorize the provision of care in these circumstances. This ensures that policies are consistent so that there is a uniform approach. A third condition requires that implementation is coordinated at a higher level. When a healthcare facility faces increasing pressure, and threatens to be overwhelmed, resources should be pooled and used within an entire area, state or country. Covid patients have indeed been moved to other regions of a country, and even to other countries. Germany for example has provided ICU treatment of patients from the Netherlands. A further condition is that procedures are clear and transparent. Public trust can be maintained if there is honest communication about the state of the healthcare system and its facilities. Finally, as soon as possible a return to the normal situation with the usual standards of care should occur.

The above conditions as such are not problematic but in practice they are often not met. This is particularly true for the first one. Proactive planning in many countries has clearly been deficient as demonstrated in the shortages of protective equipment and testing materials, necessitating health workers to provide care in a less safe manner. Only after the pandemic has emerged, strenuous efforts are made to address shortages. The condition of consistent policies is frequently not fulfilled. Guidelines for crisis standards of care have not been developed at the global level so that each country has its own mechanism to determine when these standards are implemented. In the United States, all states established policies since federal guidance was almost non-existent with the result that there are 50 different policies. For example, most states give lower priority to some patients with cancer while such patients are categorically excluded in 8 states [51]. Criteria for allocating resources vary widely. Some states exclude certain categories of patients from access to critical care, such as those with neurological conditions like severe dementia, or patients with comorbid conditions such as chronic lung disease and end-stage renal disease. Other states include societal value as a criterion, giving priority to healthcare workers [52]. This variety imposes significant emotional and psychological stress on caregivers who have to make an individualized assessment of the patient's condition. The experience that they provide substandard care to their patients may lead to moral distress since they are unable to do what is morally right [53]. Criteria such as comorbid conditions exacerbate already existing inequities in healthcare. They may lead to maximizing the number of lives saved whereas it is known that these underlying conditions are unequally distributed between racial and ethnic groups. People with disabilities are also disadvantaged because they have a greater incidence of comorbid health conditions [54]. The requirement that the enactment of crisis standards of care is a policy decision demands the involvement of the public. In many countries, however, citizens are hardly included in the debate; decisions are usually left to the medical profession. Compared to the United States, public debate about the need of governmental policies for crisis standards of care in Europe is minimal.

While the aim of planning for crisis standards of care is to manage resources in such a way that situations are avoided that necessitate to select patients for treatment and care, the increasing burden of Covid-19 may lead to cases where selection is unavoidable. At that moment, the daunting ethical problem of triage arises, particularly in regard to intensive care (see below).

7.5 Medication and Research

When Covid-19 spreads around the world, no effective treatment is known. In the early phase, the virus is novel and the symptomatology uncertain. Most patients have mild to moderate symptoms, a substantial number of patients has severe symptoms (with dyspnea, hypoxia and lung involvement), while a small number (5–15%) has critical symptoms (respiratory failure, septic shock, or multiorgan system dysfunction). All deaths occur in this last group. In the absence of viral pneumonia and hypoxia, patients with mild symptoms will stay at home, and do not require hospitalization. Patients with severe and critical symptoms require supportive management of the most common complications in the hospital (e.g., thrombosis and respiratory failure) [55]. It becomes clear that Covid has a range of clinical manifestations, and that as pulmonary disease it can lead to multiple complications (cardiac, hematological, hepatic, neurological, renal, and dermatological). That implies that a range of interventions must be necessary, particularly for critically ill patients. It also becomes clear that in the early stages of the disease replication of the virus is the main factor so that antiviral therapies will be most effective. In later stages, the disease can lead to an exaggerated immune response that is damaging tissues. Immunosuppressive therapies will be more beneficial in these stages [56]. While effective remedies are absent in the beginning of the pandemic, two medications come into use in the course of 2020. Remdesivir is an antiviral agent, approved for emergency use for the treatment of hospitalized patients who require supplemental oxygen. The drug was initially provided by the producing company for compassionate use but a few months later withdrawn since priority was given to use the limited supply of the drug for clinical testing. Remdesivir needs to be injected, has a range of adverse effects, and is expensive. As the first approved treatment for Covid patients, there was a scramble among countries to secure sufficient doses, especially after the United States has bought virtually all stocks of the medicine [57]. Continuing clinical research, however, delivers mixed results. An international study concludes that the evidence is weak; the drug may reduce recovery time but may have no effect on the length of hospital stay. The World Health Organization recommends in November 2020 against the use of remdesivir because at the time there is insufficient evidence that it is beneficial. It points out that use of this novel therapy may divert resources from the best supportive care for severely ill Covid patients. Because of intravenous administration and its price, global application of remdesivir will be limited anyway [58]. In June 2020, randomized clinical trials show that systemic corticosteroids reduce mortality in seriously ill patients in need of mechanical ventilation with 20% [59]. The WHO now recommends steroids for patients with severe or critical Covid infection. Corticosteroids are widely and globally available, and are cheap. Physicians have vast expertise with this type of medication. Later, it was found that combining dexamethasone therapy with monoclonal antibodies (e.g., tocilizumab) improves the survival of very ill patients in the ICU. Modulating the immune response to Covid-19 is the purpose of other interventions, for example convalescent plasma (i.e., from patients who have recovered from infection). Further studies have indicated that its use should only be recommended in clinical trials [60]. During 2021, several new medications are recommended for patients with severe or critical Covid-19, such as immunomodulators (tocilizumab and sarilumab) and anti-SARS-CoV-2 monoclonal antibodies (casirivimab and imdevimab). An interesting new development is molnupiravit as the first oral antiviral since it does not require intravenous application, is less expensive than monoclonal antibodies, and can be taken at home [61].

Progress in treatment and care for Covid patients requires undoubtedly a huge effort in medical research. The rapid increase of cases, the lethality of the virus, the fear and anxiety of populations, the threat of collapse of healthcare systems, and the social and economic impact of the pandemic create enormous pressure on researchers and research oversight to find remedies and make them available as quickly as possible. In this context of urgency, specific ethical problems have emerged. Amidst uncertainties, despair, and the search for a panacea, false hopes may be created, especially within a climate of misinformation and conspiracy theories [62]. The case of hydroxychloroquine is an example. Approved for the treatment of arthritis and malaria, the drug is promoted by some political leaders as a potential cure and prevention for Covid-19. Although data on efficacy and safety are lacking, hydroxychloroquine is allowed to be used in hospitals, after an Emergency Use Authorization by the FDA in March 2020. A large number of clinical trials with the drug are initiated. When randomized trials do not demonstrate benefits but report serious adverse effects, the authorization is revoked in June 2020. The WHO advises against the use of this medication for Covid patients, arguing that "almost all well-informed patients would not want to receive hydroxychloroquine given the evidence…" [63]. This case shows that political and public pressures influence decision processes about permitting the clinical use of medication which is ordinarily assumed to be based on factual evidence and thorough review processes. It furthermore shows that the public eagerness for some remedies may impede enrollment in clinical trials [64]. Many patients are requesting to be treated with the drug, and doctors are prescribing it off-label, so that usually no evidence is gathered. For example, in many African countries, hydroxychloroquine is widely used, leading to sometimes fatal consequences, shortages of the drug for malaria treatment as well as distribution of counterfeited forms of the drug [65]. Without research studies it is impossible to know whether patients have benefitted from the use of medication or whether they have been harmed [66]. The story of false claims is repeated in 2021 when some doctors and vaccine skeptics promote the use of ivermectin as a 'miracle drug' to treat or prevent Covid. This medication is on the market against parasitic diseases in humans and animals, and increasingly prescribed off-label and advocated through social media while its benefits and harms are untested in Covid cases [67].

A related ethical challenge is associated with Emergency Use Authorizations that may be issued by regulatory agencies during a public health emergency. Authorizations have been issued for hydroxychloroquine, remdesivir and convalescent plasma. These are all criticized as premature because they are based on limited evidence, few studies of low quality, and unreliable data [68]. Emergency authorizations are intended to make treatment quickly available without the normal regulatory process. The risk is that they allow treatments that are not sufficiently tested. Previous examples of authorized medical interventions have harmed patients because they were not safe or not effective according to post-approval studies [69]. In the context of a pandemic emergency authorization may lead to absorbing hospital resources, to confusion and undermining of trust in the regulatory process, especially when the authorization has to be revoked later. It can also lead to exacerbation of existing inequities in the healthcare system, especially when patients have no (or no longer) adequate health insurance [70].

Furthermore, priorities in medical research change. With the surge of Covid-19, many ongoing clinical trials are halted or modified. Staff has no longer time for research, and the safety of research subjects cannot be guaranteed, while laboratories are forced to close. Research capacity and funding are primarily focused on Covid [71]. Moreover, the public promotion of certain drugs against Covid initiated a large number of clinical trials with these drugs. For example, many studies are started with hydroxychloroquine, duplicating resources and complicating enrolment [72]. These changes in research are ethically problematic since there is no balanced consideration of the consequences. Whereas all efforts are directed at Covid, research that might benefit patients with chronic and underlying conditions who are most vulnerable to SARS-CoV-2 should continue with special measures to protect their safety. Setting priorities should take into account that research not only targets medication and treatment but also addresses the psychosocial impact of the pandemic, e.g., the social effects of lockdowns, mental health problems, and moral distress [73].

The above issues illustrate that research ethics may be compromised in the search for novel therapies. Confronted with an unknown virus, enormous health needs, social and economic disruption, public health measures and policies are necessary as experiences with past pandemics learn. In distinction to the past, contemporary societies have a powerful means to cope with infectious diseases: science will be able to provide reliable knowledge about the characteristics of the virus and its impact on human beings; it will also be able to develop treatments and vaccines that will eventually reduce and possibly eliminate the viral disease. Given the potential benefits of this means, there is arguably a moral duty to expand and apply scientific research in the face of the harms wrought by the virus. In normal circumstances, science has its regular procedures and mechanisms to safeguard that it is reliable and objective. In extraordinary circumstances, acceleration of research processes and expedited use of research findings in clinical practice is advocated [74]. Expedited research, however, is risky. The case of hydrochloroquine illustrates why. When in March 2020 a study on the use of this drug in Covid patients is disseminated via YouTube, and a few days later as an unreviewed preprint, its positive findings for the healing and contagiousness of infected patients are immediately picked up by the global media. The drug becomes widely used, and is promoted by some political leaders, while the evidence of this one small, nonrandomized study is weak. One week later, the FDA issues an authorization to use the drug as treatment for Covid, while numerous trials are started to investigate its application, demonstrating a few months later its lack of efficacy. The case illustrates that ordinary scientific processes of reflection, testing and review have a purpose and are based on previous experiences with errors, hypes, misinterpretations, and false conclusions. A single small study should not lead to widespread use of a drug but in the current context of desperate patients, overburdened healthcare systems, and helpless policymakers every indication of hope and promising 'cures' will instigate direct application [75].

Research may be expedited through streamlining of administrative processes [76]. Research ethics committees receive a rapidly growing number of protocols, either new or modified. Their main concern is to protect research participants. Pressure to expedite may impact the quality of reviews, and therefore insufficiently protect individual rights. There are however ways to simplify the paperwork without affecting the ethical assessment. These administrative measures may not be sufficient to address the challenges of research, and then the question arises whether the ethical framework of research and research review should be adapted. Like in clinical care when crisis standards are applied in emergency conditions, it is advocated that research in similar conditions should modify its ordinary standards, not only methodological (e.g., randomization, blinding, and controls) but also ethical ones. It is argued, for example, that ethics committees should be 'more flexible' in the explanation and application of ethical principles, for example the requirement of informed consent [77]. The justification is not simply that extraordinary circumstances demand exceptional approaches because the urgency and need for treatments is unprecedented but also that the objective of research during a pandemic is different: not merely enlarging the pool of scientific knowledge, but providing practical solutions to relieve the burden of the virus.

In spite of the urgent circumstances, the argument of exceptionalism in research should be rejected [78]. First of all, it is artificial to identify a special goal of research during a pandemic. All medical research is aimed at reducing uncertainty, providing new knowledge and better ways to enhance health. Second, when the basic methodological requirements for research are not implemented, decisions in healthcare will be based on insufficient or incomplete evidence, with doubtful effects and potential harms. This furthermore may waste resources, as exemplified in the concentration of efforts to test hydroxychloroquine. The urgent need to produce new knowledge in the face of Covid-19 should therefore not lead to abandoning the core ethical principles that guide research. The discussion on exceptionalism is especially intensive when it concerns the development and testing of vaccines (as will be discussed in Chap. 8).

The debate on exceptionalism reflects the tension that is always existing in healthcare between the normative frameworks of deontology and consequentialism. Within the setting of emergency and urgency, it is argued that consequentialist (particularly utilitarian) reasoning should have priority. In disasters, the ethical perspective, it is claimed, shifts from a deontological to a consequentialist one [79]. In my opinion, this view is too simplistic. Medicine always has a strong consequentialist tendency since medical intervention is first of all directed at results: cure, prevention, and palliation. Good healthcare is not just the implementation of duties but the accomplishment of effects that are in the interest of the patient. This tendency is even stronger in medical research: it is valuable if it produces new knowledge that improves healthcare. At the same time, when these tendencies are not embedded within a deontological framework, things go wrong. If a patient is treated against his or her will because the physician is convinced that it delivers a good effect, this medical act is considered as wrong. It is precisely the achievement of modern bioethics that it has set certain limits to the decisional power of professionals, asking respect for human dignity and human rights. Even more disturbing is medical research carried out with the argument that it is important for scientific progress or has high social value because it may produce benefits for a large population, so that it is justified to harm or sacrifice a few individuals. Following a range of research scandals in the 1950s and 1960s, in the wake of experiments in Nazi Germany and Japan, stringent guidelines for medical research have been formulated. They motivated the rise of bioethics, emphasizing that the interests of the individual have always priority over the interests of science and society [80]. Even in disaster conditions some deontological requirements need to be observed. For example, it is often pointed out that resource allocation decisions driven by utilitarian considerations (maximizing the good for the largest number of people) should not use categorical exclusion criteria (such as age or disability); that would infringe the principle of respect for persons, as well as the principle of non-discrimination. Rationing decisions should be based on an individualized assessment of each patient [81].

The tension between normative frameworks is again highlighted in publication ethics. The number of Covid related publications has increased sharply; according to a PubMed search from 301 in 2019 to 92,943 in 2020 and 112,154 in 2021 (October 26), covering a wider range of research areas. Publishing research findings has been expedited in two ways: fast-track review and online uploading on preprint servers. The effort of journals to speed up the review process will make results of studies more quickly available but it is risky. Care should be taken that scientific quality and integrity are not compromised. That risks are present, is shown in the fact that tens of publications and preprints have been retracted and withdrawn for a variety of reasons: falsification of data, methodological concerns, conflict of interest, or performance beyond the scope of research ethics committee approval [82]. An example, mentioned in a previous chapter, is a publication on the protective effect of smoking against coronavirus disease. It created a hype with increased consumption of tobacco and nicotine products. The article is retracted because the authors have been funded by the tobacco industry in the past [83]. Papers on preprint servers have not been peer-reviewed, and the readers are warned not to use the findings for clinical decision-making. Nonetheless, in many cases research results are announced through press releases so that they find their way in public media, and thus be translated into clinical actions. As with the other ethical challenges discussed above, speed and urgency should not compromise quality and integrity of research. Here again, the opposition between consequentialist and deontologist frameworks is false. Admittedly, there is a need for fast scientific information, but this information is useless and even harmful if it is not reliable, and not accommodating ethical principles of integrity and honesty. Speed, though important, is not the only consideration; time for critical reflection will enhance quality of information, and reduce errors and risks.

The ethics of Covid research finally implies considerations that are especially important from the perspective of global bioethics. One consideration is that vulnerable groups should receive special attention. Global power differences inhibit that decisions about the type and place of research include the voices of people who are most affected by the pandemic. The impact of Covid is differentiated across the world and across populations. Research needs differ in situations of poverty and lack of access to healthcare. The research response should therefore include multiple actors and agencies from many different countries. This highlights the significance of international cooperation. Important as it is for setting priorities in research, collaboration expresses that humankind is confronted with a global challenge that requires solidarity. A coordinated approach moreover avoids duplication and waste of resources. One such international partnership is now operational: the Solidarity clinical trial, led by the WHO and including patients from more than 30 countries across the world. Another global consideration is sharing of benefits. Many major publishing companies provide open access to Covid related research articles. Relevant information is therefore available free of charge to everybody in every country. Individuals and communities that participate in research, should also have access to the benefits that result from their participation, but it is less clear how that requirement is met [84]. This refers to another global consideration: fairness. It is one of the values of the Ethical Compass proposed by the Nuffield Council on Bioethics [85]. The research effort should be inclusive and equitably distribute benefits and burdens. Not all populations and regions of the world are affected similarly by the pandemic. Special attention should be paid to the needs and perspectives of these groups. Study protocols should be finalized after the input of local communities and researchers. Selection of participants in clinical trials should be fair and equitable, so that the burdens and benefits of research are equitably distributed, and the results are applicable to the entire population [86].

7.6 Intensive Care

In responses to the pandemic, the number of corona patients in intensive care units is one of the most significant parameters for the severity of the disease, and an indicator for the effectiveness of policy measures. Every night, the public is confronted in television news reports with the amount of people hospitalized, admitted to ICUs, and deceased. Casualties in the war against the virus indicate whether we are winning or not. Dying means defeat, and all efforts are directed at saving the maximum number of lives. Approximately 5–15% of Covid patients require ventilatory support, often for a prolonged time [87]. In the first stage of the pandemic, in most countries the rapidly increasing number of infected patients puts growing pressure on available ICU capacity. Efforts are made to increase the number of beds. Regular care is reduced to make sure that all hospital beds can be used for Covid patients. National platforms are created to coordinate care so that patients can be distributed across the country, state or region. There is not only a shortage of beds but also of medication, oxygen, equipment and staff (for example, intensive care nurses and anaesthetists), aggravated by the fact that many care providers become infected. New units and beds are established, and thousands of respirators are ordered by governments [88]. The problem is that the pandemic disturbs the global supply chain, and that only very few companies manufacture ventilators so that alternative solutions to production have to be explored [89]. These experiences with real or imminent deficient ICU capacity come back during subsequent waves of the pandemic when in the final months of 2020, and in the Spring and Autumn of 2021 again the number of critically ill patients escalates. At that time, there is a growing shortage of personnel, exhausted by the continuing workload [90].

In March 2020, when Italy is the epicenter of the pandemic, stories emerge how doctors are forced to decide which patients should have ventilatory support. They watch people die because this support could not be offered to all patients. Doctors are agonized by these decisions and are hesitant to explain how they are made. In practice it seems that age is often the major consideration. Patients older than 75 years are not offered mechanical ventilation [91]. These stories raise the specter of triage as one of the most horrifying ethical challenges in the pandemic. This challenge is frightening because the consensus is that it should not happen. Everything must be done to avoid or delay the need to ration; this is "the most ethical course of action" [92]. Scarcity of resources can never be solved by triage. It is imperative to develop alternative strategies to avoid the need for triage, for example, relocating patients to other regions or countries or transporting ventilators and staff to facilities in need as a matter of solidarity [93].

That rationing becomes unavoidable is not merely the result of the surge of critically ill patients. The stage is set by previous health policies to restrict intensive care capacity as much as possible since surpluses are deemed inefficient. The number of hospital beds in countries is very different, with the highest number (beds per 1000 inhabitants) in Japan (13) and Germany (8). Other Western countries have much lower hospital capacity: the Netherland 3.2, Italy 3.1, the United Kingdom 2.5, and the United States 2.9 [94]. The number of intensive care beds shows similar disparities. Among OECD countries, Germany has the highest capacity (33,9 beds per 100.000 population) while the US also is well provided (25.8). Much lower capacity is maintained in Italy (8.6), the Netherlands (6.7) and Japan (5.2) [95]. In developing countries, intensive care capacity is extremely limited. Uganda, for example has 0.1 beds per 100.000 population, and Sri Lanka 1.6 [96]. Some countries have no ventilators at all.

The notion of 'triage' is used in emergencies when healthcare resources are too limited to treat everyone. At the policy level, decisions are taken how available resources are distributed, if not all needs can be addressed, for example when protective equipment is scarce. The notion is also used in the planning of care, for example in determining how many intensive care beds will be sufficient in a country or region. At this level, 'macro-triage' occurs since at the level of the healthcare system specific dimensions of a disaster are selected for immediate relief. In the global context, macro-triage leads to decisions to provide various forms of humanitarian aid to countries affected by earthquakes, inundations or other disasters [97]. At the local level and within the clinical context, it is decided what kind of resources and how many are available for specific healthcare needs. It is here that triage is most disturbing since it directly determines who will live and who will die.

Although triage has become one of the most hotly debated ethical issues during the Covid-19 pandemic, it is not a new topic. It is used in transplantation medicine to determine who has priority in receiving an organ transplant. After the global spread of the avian influenza virus in 2008, many countries developed guidelines for allocating ventilators and other life sustaining treatments [98]. The term 'triage' is most often applied at the clinical (micro)level when patients have to be selected for treatment when not all can be treated. In ordinary circumstances, when there is an enormous number of casualties, resources are ultimately sufficient to treat everybody, but it is just a matter of time before all can be treated. The wounded will be classified so that those with the most urgent needs will receive assistance immediately. Less urgent cases will be treated on the basis of first-come, first-served. In extraordinary circumstances, this conventional triage is no longer adequate. The number of people in need is higher than the available healthcare services so that not all can receive lifesaving care. In this situation of nonconventional triage, selection will focus on saving as many lives as possible, with the consequence that patients will die who in ordinary circumstances would have survived [99].

In the context of the Covid-19 pandemic, triage has emerged as one of the most daunting and complex ethical challenges. The dominant war metaphor in the fight against SARS-CoV-2 reactivated triage as a practice that was first used in military medicine, notably by the French military surgeon Dominique-Jean Larry. Confronted with mass casualties in the Napoleonic wars, he sorted the wounded in three categories: dying even when treated, living without treatment, and dying unless treated. Those in the last category should receive care first. This practice has defined the characteristics of triage: when resources are limited, the medical condition of the patients has to be assessed, and some sorting system must be applied to determine who has treatment priority. Triage is nowadays used in different contexts: in emergency departments, in inpatient care, especially intensive care, in major incidents, battlefields, and natural or manmade disasters [100].

The moral problem with triage is that it is an explicit decision about life and death that is not guided by the best interest of a particular patient. Determining who is the first to receive potentially life-saving treatment implies that somebody is expected to benefit most, and that other patients may die. As said, such decisions are not new to medicine. In the 1960s, the invention of the arteriovenous shunt made repeated dialysis treatment possible for patients with end-stage renal disease but the shortage of hemodialysis machines raised the question how to select patients for this lifesaving intervention. It is argued that this query has been the origin of bioethics [101]. In order to determine who would live or die, a special committee was established of mostly non-medical members in the United States, indicating that such decisions are no longer in the remit of traditional medical ethics but require the involvement of a broad set of stakeholders. The criteria used for decisions are reflecting values that are beyond the scope of medical expertise, and that should be subjected to public moral reflection and analysis.

Faced with the shortages of ICU beds, equipment and staff, many guidelines have been developed for triage but with a wide variety of approaches and criteria. Again, a common assumption is that in a pandemic the moral perspective should shift from a deontological to a consequentialist one [102]. Especially a utilitarian perspective is advocated: the primary goal of healthcare is to save as many lives as possible. A basic value in this perspective is efficiency: scarce resources should be used in ways that maximizes their output, i.e., beneficial effects for as many persons as possible. Patients should be selected for treatment who are most likely to benefit from intervention, and are expected to recover. Patients who are likely to survive may not need ICU intervention, and should not be prioritized, just as patients unlikely to survive regardless of the treatment provided. Determination of the prioritized patients requires a medical estimation of the severity of acute illness, the prognosis, the urgency of intervention, and the chances to respond to treatment. This criterion of the best immediate-term survival is endorsed in most guidelines since it will save most lives. However, it is not uncontroversial. The acuteness and severity of Covid-19 is not equally distributed among populations but associated with racial and ethnic groups, and socio-economic status. The criterion will therefore be less favorable towards certain categories of patients [103]. That rationing decisions are inherently moral is illustrated by the decision of the Turkish government to airlift hospital patients from Europe. For example in the Netherlands, when doctors decided to discontinue ICU treatment for patients because their quality of life in the longer run was not considered acceptable, and the chances of recovery low, they are transferred to Turkey at the request of family members. The country has a very high ICU bed capacity (47 beds per 100,000 inhabitants) [104].

When patients with a low likelihood of immediate survival are excluded from mechanical ventilation, it might still be the case that demand is higher than available resources so that other selection criteria must be used. One is the long-term prognosis of patients. The argument is that in addition to maximize the number of lives saved it is important to maximize the number of life-years saved. Since people with pre-existing health conditions have a worse long-term prognosis, this criterion is unfavorable for patients with major comorbidities such as cancer and dementia. In some guidelines they are categorically excluded from priority treatment. The criterion also de-prioritizes older patients. The moral problem is that in this way the ethical principle of justice is violated, especially when comorbidity and age are used as explicit and categorical criteria. Most guidelines agree that de-prioritization of categories of patients should be rejected as unfair and discriminatory [105]. But even when the assessment is individualized, the application of the criterion of long-term survival is problematic since it de-prioritizes vulnerable populations that are severely affected by Covid-19 while they also have more health problems because of race, ethnicity, and socio-economic status. Comorbidities are often the result of lack of access to healthcare, poor and unsafe living conditions, lack of employment, and environmental degradation. Minority groups (e.g., people of color, or indigenous populations) have a higher risk of getting infected and dying from Covid-19. Already disadvantaged in normal circumstances, they are now doubly affected, not only by the virus but also by criteria that deny them priority in treatment [106]. Other patient groups are equally impacted. People with disabilities, chronic illnesses, and older people all have a condition that reduces long-term life expectancy. At the same time, there is wide variety in prognosis and life expectancy among these groups of persons. Scoring systems to determine disease severity and prognosis are usually developed for use in the study of populations, and are less accurate for predicting the mortality in individuals [107]. They suggest to be objective means of evaluation while they do not exclude subjective assessments [108]. The risk is that the execution of triage is influenced by bias and thus may be unfair. An example is the assumption that people with disabilities have compromised health and are less likely to benefit from treatment. Similar bias exists when the quality of life after survival is taken into account; this is usually estimated to be much lower by non-disabled people than by disabled people themselves. These biases are the expression of 'ableism,' i.e., the idea that society normally values and promotes a certain kind of body. This idea engenders social prejudice, stigma, and discrimination against people with physical, intellectual or psychiatric disabilities who do not comply with this standard of normality [109]. It violates the principle of human dignity that demands respect for every human being, regardless of any disabilities. The principle requires that attention should be given to each particular individual and his or her medical history, rather than inserting them in a general category that is inferior [110]. The risk of stigmatization is particularly strong when elderly persons with dementias are concerned. Nursing and elderly homes are severely hit by Covid-19, and a significant number of residents died while public health measures were not prioritized in these institutions with minimal provision of protective equipment and testing. In some countries, older patients were not even transported to a hospital or ICU [111]. If they are also excluded in triage systems, it seems that their lives are expendable [112].

Some triage schemes apply additional criteria. One is societal value or utility. The argument is that certain groups of people should receive preferential treatment because they play a crucial role in society and provide critical services to protect everyone in times of social and economic disruption. A related argument emphasizes instrumental value. When the healthcare system threatens to collapse giving healthcare workers priority treatment enables the system to keep up under pressure. This will have a multiplier effect since saving the life of a healthcare worker who can return to clinical work will possibly save the lives of many others. It will also boost morale among a workforce that is under stress for such a long time [113]. Moreover, it is regarded as compensation for the risks that health professionals take in caring for infected patients [114]. Another related argument highlights reciprocity. People who take risks to help others, such as research participants and healthcare workers, should have extra protection in return [115]. These arguments, though reasonable, are complicated in practice. Social value has been proposed as a criterion in the allocation of dialysis machines in the 1960s. The established committee recommended that caregivers, professionals, and heads of family should receive priority. This led to huge public protest and to the policy decision to make the treatment available to everyone. Social value as a selection criterion is problematic because it is difficult to determine who is worthy of saving more than others. Why would only healthcare workers be regarded as a privileged class while many others are necessary for the proper functioning of healthcare services (providing for example security, transportation, housekeeping, and food)? Others should also be regarded as key workers during a pandemic, for example researchers testing new medication that can be lifesaving, or vaccine developers. Outside of the hospital setting many are responsible for maintaining basic services such as electricity, telecommunications, water and sewerage systems [116]. A list of categories of essential workers will rapidly expand, and if a limited choice is made, it seems rather arbitrary, and thus unfair. The arguments are furthermore problematic from an ethical point of view. Prioritization of healthcare workers is justified because they have instrumental value in saving others. Key workers should be saved not because they have more value as individuals but because of the value of their work [117]. This justification is based on a utilitarian view of human beings: they are valued because they provide benefits to others; they are not recognized as valuable in themselves as persons with dignity. This view is opposed to the position that human dignity cannot be reduced to social contributions. In this perspective everyone has an equal moral claim to treatment.

While consensus is lacking in regard to the ethical criteria for triage, the emphasis in numerous guidelines, policy documents, and publications is on the procedural ethics of the process of decision-making [118]. One requirement is transparency. The criteria used in triage should be clear, explicit and known. They should be the result of public discussion and engagement [119]. The same criteria should apply to all patients who need intensive care, and Covid patients should not be prioritized over patients with other diseases. It should be explained that all individuals have equal value, and that triage is not a judgment that some persons have more value and are more worth saving than others. It is often underlined that treating physicians should be shielded from triage decisions [120]. Their duty is to consider the best interest of their patients, even in extraordinary circumstances, and they should not be brought into the situation to decide which patients will not qualify for lifesaving treatment, and probably will die, for the benefit of other patients. Such decisions are agonizing and create moral distress among clinical staff. Separate triage teams or special triage officers should make these decisions, so that the responsibility for allocating scarce resources is removed from care providers who are responsible for direct patient care.

However, the problem is that it is often not clear whether these requirements of procedural justice are met or can be met. Rationing in healthcare is a social taboo because it is highly controversial in society, and often regarded as a death sentence and a valuation of the worth of persons [121]. Views of appropriate criteria widely differ, and between and within countries guidelines can vary significantly [122]. Comparison of triage guidelines in several European countries, for example, shows substantial discrepancies in terminology, goals, criteria, and procedures [123]. There also might be divergence of the assessment of allocation criteria between medical professionals and the general public, although empirical research on preferences in the context of Covid-19 is scarce [124]. If rationing happens, the practice is frequently not made public. For example, doctors in Italy who watched patients die because of insufficient resources, were mostly silent about rationing or prohibited to talk to the press [125]. In countries like France, triage decisions are made by experts, particularly intensivists, and triage criteria are exclusively discussed within the medical community without involvement of the public or ethicists [126]. In general, the connection of triage and military context, especially that of an urgent war with SARS-CoV-2, facilitates a paternalistic discourse, restraining individual freedom and human rights for the sake of the public good, with experts and 'authorities' taking control [127]. Patients are affected by priority decisions but usually not consulted. Ethicists are involved in drafting guidelines, but contrary to the United States, in Europe, ethics consultation in the execution of triage, is rarely recommended [128].

The predominance of utilitarian reasoning during the pandemic is understandable from the point of view of public health which emphasizes the common good and collective interest. From the perspective of ethics, it clarifies that triage decisions are not scientific appraisals but value judgments. The value of maximizing lives saved, as articulated in a utilitarian framework, is not the only relevant value, but should be considered in relation to at least two other values: human dignity and justice [129]. The value of human dignity requires equal treatment and respect for personal autonomy. Human dignity is one of the fundamental principles of global bioethics as well as the foundation of human rights discourse. It affirms that every human being has intrinsic value and all humans have the same dignity. The moral principle of respect for autonomy is based on human dignity. It is argued that in a public health emergency the focus of healthcare moves from an individual to a community perspective, from individual autonomy to the common good of public health. Interventions are assessed not for their effectiveness for the individual patient but for the largest group of patients, and the wishes of the patient are subordinated to the interest of the entire population of patients. The key decision-maker should be the triage officer or team, not the patient, his or her surrogates or family, and not the caregiver. In ordinary circumstances, there are two reasons for not providing treatment, i.e., futility, and patient refusal, but in extraordinary conditions the reason is different. If lifesaving treatment cannot be provided to all patients who need it, the condition of one patient is weighted against that of other patients. If the expectation is that someone else will benefit more from available treatment, lifesaving treatment is denied or withdrawn from this individual patient. It seems that the focus is on the utility of the ventilator: how can it be used most efficiently, maximizing its output for the largest number of patients. This balancing is sometimes also the case in ordinary circumstances, notably in the allocation of organs for transplantation when the best recipient is identified in order to maximize the utility of the organ. It is argued that this is not a value judgment about patients. It does not necessarily conflict with human dignity as long as only the medical condition is taken into account (the urgency of need, and the probable outcome of the intervention), and non-medical characteristics of patients such as age, gender, disability, race or ethnicity are disregarded. Nonetheless, the prevalence of utilitarian considerations does not imply that respect for individual autonomy should no longer be a concern. First of all, decisions need to be discussed with patients. They cannot request treatment but they may still refuse interventions. If the treatment has disproportionate burdens and is not or minimally beneficial, patients can decide not to receive it. Respect for human dignity requires that patients receive adequate information, that informed consent is obtained, that care and treatment are discussed [130]. Lack of engagement with the patient or his proxies is particularly problematic with decisions to withdraw life support. Such decisions are not based on the clinical judgment that the intervention is futile or disproportionate; the Covid patient clearly benefits from it. However, the intervention is terminated (and most probably the patient's life) in favor of another patient who is expected to benefit more from treatment. Physicians decide that the life of a patient will be sacrificed to enable another patient to live. This decision is regarded by some as an act of killing, and therefore problematic for clinicians [131]. It defies the ethical responsibilities of care, and the relationship between doctor and patient which assumes that medical efforts are focused on the individual patient's wellbeing [132]. Moreover, the autonomy (and thus dignity) of the patient is not respected when the decision is taken without his or her involvement. When they are only "made aware of this possibility at admission," they are denied the opportunity to choose themselves for discontinuation of treatment [133].

Similar concerns refer to the principle of justice. Utilitarian triage criteria have a negative impact on vulnerable patients with pre-existing health conditions, the elderly and socially disadvantaged groups. The emphasis on long-term prognosis is unfavorable towards persons with disabilities and chronic ailments. Sometimes categories of patients (e.g., persons with cancer) are excluded from access to treatment. Such criteria infringe the principle that all people have the right to the same treatment if they have equal medical needs. Applying these criteria will be unfair to some groups of patients who will have no equal opportunity to benefit from intervention, while it makes socially disadvantaged groups victims of further injustices, added to the ones they suffer from social and economic determinants. An egalitarian approach based on equality, demands that random assignment is used when the same resource is needed by several patients, and their clinical prognosis and duration of need are similar [134]. When patients are equally ill and have a similar prognosis, a lottery system is unbiased and thus fair [135]. This is preferable to a system of first-come first-served since this favors people who have ready access to care, live close to health facilities, have high health literacy, and are able to leave their jobs. Existing inequalities will not be eliminated in this last system. The same principle of justice rejects categorical exclusions, i.e., systematic de-prioritization of categories of patients. They are not only unfair but may furthermore lead to stigmatization and discrimination [136].

Because the utilitarian approach conflicts with the principles of human dignity and justice, and sometimes overrides them, it is uncomfortable and problematic. Some important values are sacrificed on the altar of efficiency and the quest to save as many lives as possible [137]. The irony is that the utilitarian aim of triage schedules lacks scientific evidence. The assumption is that they save more lives than random assignment but they have not been tested to observe whether they indeed save lives [138]. In the previous epidemic of avian influenza, 70% of patients survived with continued ventilation while according to the rationing plan they should have been removed from ventilators [139]. Lastly, the utilitarian approach is paradoxical: in order to save lives some lives have to be sacrificed. Its basic normative presupposition is that saving lives is unquestionably morally good. The point in this approach is that the focus is not on individual life as a good, but on the total number of lives. There is always a balance between lives to be saved and lives to be risked. Not all lives are equal, as demonstrated in the argument that healthcare workers should have priority treatment. Emergency ethics celebrates the value of human life but is simultaneous associated with inequality [140]. This paradox illustrates that the value of human life continuously competes with values like human dignity and justice. Human life is good but not an absolute good. In humanitarian discourse during disasters the perspective of the recipients of assistance is often absent and silent; they are regarded as victims, persons in need who require immediate help, compassion, and treatment. They are not supposed to participate in determining their own destiny. With the focus on ICU capacity and the possibilities for treatment, it can be forgotten that some patients prefer care over mechanical ventilation.

7.7 Ethics of Care

During 2020, the Covid-19 pandemic killed 1.82 million people (on a total of 83.54 million confirmed cases), while the numbers are still rising in 2021. While in the earlier pandemic of the Spanish flu, children and young adults were primarily affected, the SARS-CoV-2 virus is most lethal for the elderly. For every 1000 infected people older than 75 years, approximately 116 will die, while for people younger than 50 years, almost none will perish [141]. Early in the pandemic, the incidence of Covid-19 infections was highest among older adults, but during the second part of the year a larger proportion of younger people became infected.

Assessment of the response to the pandemic is in most countries based on fatalities in hospitals and particularly ICUs, with the major concern that the healthcare system might be overwhelmed. It has become clear that this focus is narrow, and overlooks many other components of the system. Elderly and nursing homes are disproportionally affected since the virus spreads rapidly and widely through such homes with an increasing death toll, making these homes one of the most harmed areas of society. In several countries, a significant proportion of SARS-CoV-2 related deaths occurred in long-term care facilities: in Canada 78.2%, in Spain 68.8%, and in Belgium 49.6% [142]. In England and Wales more than a third of the care homes reported a confirmed outbreak by May 2020. Larger and crowded homes were more severely affected [143]. There are several reasons why care homes are disproportionally affected. Residents are older, frail and more vulnerable than those living outside. In one European study, the mean age is 83.4 years; most of them suffer from various degrees of disability as well as cognitive impairment, while comorbidities are frequent [144]. Many people living in nursing homes are not able to feed themselves, have speech impairment, and need help with walking. Caring for them requires physical interactions with care workers, often for a prolonged time. The condition of many residents is commonly associated with limited understanding of the need for restrictive measures. Furthermore, residents live together in a community (a 'home') where they interact and share the same facilities, enhancing the dissemination of the virus. Frequently the number of care workers is limited; they are often low-paid with minimal or no sick pay, as a result of long-term underinvestment in the care home sector [145]. Such working conditions make it difficult for staff to self-isolate or quarantine at their homes. Covid-19 revealed that long-term care in many countries has been neglected for decades in health policies [146]. A significant factor contributing to the care homes tragedy are the policy measures that, at least in the first wave of the pandemic, do not pay much attention to care homes. For a relatively long time, facilities are confronted with shortages of protective equipment and testing materials. In the early phase of the pandemic, patients who need hospital care are discharged to nursing homes, in order to relieve the pressure on beds in the hospitals. It was clear that care homes were facing many asymptomatic infections among staff and residents, with atypical presentation of symptoms which were initially not recognized and did not qualify for testing [147]. Later, the situation could be better addressed since more protective materials became available, and designated units for confirmed cases were established. Another consequence of the focus on intensive treatment is that substantial excess mortality in care home residents is also caused by non-Covid diseases because regular care has been downsized [148].

The dire situation in care and nursing homes has become a significant moral problem. Horrific stories emerged about care homes where residents have been abandoned because care workers left [149]. On the other hand, in some cases care workers locked themselves in with the residents in order to keep infections outside the care home [150]. In some instances, critically ill residents were not transported to the hospital since ambulances refused to bring them or because authorities had ordered physicians not to refer them. These experiences show that the focus of policies on saving lives is almost exclusively concentrated on high-tech medicine with resources diverted to the acute hospital sector. The broader dimension of care has not received much attention, at least in the earlier stages of the pandemic.

Ethics of care is also compromised in the responses to the Covid crisis [151]. Most homes are closed for visitors, and residents are isolated in their rooms as much as possible. Most activities are cancelled. Isolating and restricting movement not only reduce patient autonomy but also increase the possibility of psychological and physical harms. Lack of supervision may enhance the risk of falls and injury, aggravated by the absence of staff due to self-isolation because of illness or positive tests [152]. The psychological impact of Covid-19 is considerable. Care workers are faced with high mortality rates among residents which whom they often have longtime relationships. Lack of contacts with family, relatives, and friends has a negative influence on many residents who feel lonely and depressed. Many will not understand the necessity of the measures, and feel unhappy. From the perspective of ethics of care, a better balance between infection control and quality of life in care homes should therefore be established [153]. Particularly in the final stages of life, social connections and relationships are critical. More flexible approaches to social distancing are therefore implemented in a later stage of the pandemic. With adapted schedules and safety measures, family and friends can offer support and comfort. Alternative ways of interacting have been developed. Although virtual visits are a poor alternative for in-person visits, they are better than complete isolation.

Significant concerns are related to the mortality of SARS-CoV-2. Many patients, in care homes as well as intensive care units, are dying alone, without the presence of loved ones. This has drastically changed end-of-life care. It is argued that such dying experiences are unacceptable in patient-centered care, and that infection control should not exclude family presence at the bedside of critically ill patients [154]. Care of the dying requests the specific expertise of palliative care. The ethical discussion of Covid-19 predominantly concentrates on treatment, vaccination, and resource allocation, while a broader perspective than saving the maximum number of lives is easily lost in the turbulence of the pandemic. Growing awareness of the impact on care and nursing homes reactivates the moral imperative that healthcare should be a personal relationship and founded on respect for the individual, and should therefore respond to the needs of the dying patient. Instead of a utilitarian perspective, palliative care emphasizes that patients cannot be completely isolated and abandoned for whatever purpose. It rejects the idea that when treatment is no longer effective, nothing more can be done. When patients are not selected for mechanical ventilation, they still need medical attention. Alternative and second-best treatments must be explored such as non-invasive ventilatory support. Regardless how poor the condition of the patient is, palliative care should be provided and suffering addressed. Facing the final stage of life is such an exceptional situation for patient and family that maximal application of prevention measures seems cruel. Palliative care furthermore stresses that care should be personal. Patients should not be classified in abstract categories that are excluded from healthcare services but attention should go to each individual person who needs compassion and understanding, with respect for his or her dignity and autonomy [155].

While prior to the pandemic it was highlighted that even in disaster conditions when resources are scarce, palliative care is "a moral imperative of a humane society," this type of care is not prioritized in the early phase of the pandemic [156]. Triage is even less acceptable when patients who do not qualify for life-sustaining treatment, do not receive comfort care [157]. Guidance is almost absent; for example, the WHO guidelines on maintaining essential services in June 2020 do not mention palliative care [158]. Hospices cannot accept Covid patients because of a lack of protective equipment [159]. The provision of palliative care is also impeded by shortages of medication. Drugs to sedate patients or to mitigate pain are in short supply. Particularly in low and middle-income countries, palliative care, already insufficient before the pandemic, is unable to provide adequate relief of suffering. For example, inexpensive opioids to alleviate breathlessness were hardly available in these countries, and are now even more difficult to access [160]. Furthermore, shortage of time, rapid progression of illness, and infection control measures challenge one of the basic requirements of end-of-life care, and in fact all care, i.e., communication with patients and family members. Even if patients do not have choices, it is important to explore patient's values [161]. Triage schedules commonly suppose that there are two reasons not to admit patients to the ICU: patients who will likely recover without intervention, and patients who are seriously ill and will probably die. They forget a third reason: patients who choose not to be admitted and opt for palliative care.

Ethics of care requires the involvement of patients. Facing the end of life, it is important to discuss the goals of treatment and care. That means identifying patient's wishes and values. Some patients want every possible intervention but others prefer care that is focused on symptom relief and comfort above intensive treatment. Usually, such preferences are clarified in a process of shared decision-making. The wishes of patients or their surrogates can be expressed in dialogue with physicians, or directly in advance directives. With the general impact of the pandemic, it can be assumed that many people, especially older people and care home residents and their relatives have reflected on what they want to do in case of serious illness. Do they want to be transferred to the hospital, admitted to the ICU or do they want supportive care and palliative care? [162]. These questions underline the urgency of proactive (and if necessary, virtual) advance care planning, especially for older patients with preexisting conditions [163]. Although clarification of patient values is more difficult with acutely ill patients in the emergency department, serious efforts can be successfully made, particularly by embedded palliative care staff, as experiences in New York City show [164]. There is no reason why the voice of patients and their relatives should be silenced in the pandemic, not only in policy-making but also in clinical care [165]. Although little is known about patient preferences during Covid-19 so far, results from previous studies indicate that many hospitalized frail and older patients with comorbidities do not wish life-prolonging measures such as mechanical ventilation. They as well as their family members do not want to be kept alive with little hope for recovery. Rather, they prefer honest communication about their disease, and the opportunity to complete their life in a meaningful way and prepare for the end of it [166]. The outcomes for Covid patients requiring mechanical ventilation are poor. One study in November 2020 reports that 84.3% of patients older than 70 years died in the hospital, while 67.4% of patients younger than 70 years survived to hospital discharge [167]. Those who survive ICU intervention often have a low long-term quality of life; full recovery will not be attained in many cases. These preliminary findings indicate that the appropriateness of treatment cannot be assumed for all patients, and that the focus on saving lives and aggressive intervention which is a rational approach for healthcare providers should be discussed with patients or their surrogates. Recently, it is reported that a growing number of patients prefer to stay at home, and die amidst family members, rather than be transferred to a hospital or nursing home [168]. The problem is that palliative care itself is a limited resource in most countries. The number of experts is usually small so that it is argued that every care provider should offer compassionate, palliative care [169].

Finally, the challenge of caring for Covid-19 patients has broader implications for how societies imagine ageing and finiteness as fundamental human experiences. The disparities of the effects of SARS-CoV-2 between younger and older people may engender intergenerational tensions. Young people are the least affected by Covid but are required to stay at home, cannot attend school, participate in events and parties, or meet with friends. Sacrifices are demanded in order to show solidarity with older generations with much higher morbidity and mortality. In some cases, negative attitudes have been expressed towards the elderly, arguing that the pandemic will weed out 'dry wood,' eliminating the weak, disabled, and already diseased persons. Others argue that older people had their 'fair innings' so that it is less tragic when they die rather than young people. The debate on herd immunity and the use of age as a criterion for triage has implications for the social image of the elderly. On the other hand, seniors may blame younger persons for not following public health measures while they themselves strictly adhere to social distancing guidelines because they have more risks. Such tendencies complicate the notions of vulnerability, marginalization and stigmatization of certain groups in society. These issues will be discussed in Chap. 9.

Like previous pandemics, the extensive confrontation with death has implications for how modern people imagine the finiteness of life. Patients may be dying alone without opportunity to see or touch family or loved ones; they have limited interaction with doctors and nurses; they have no access to pastoral workers. At the same time, funerals may be rushed, and rituals cancelled [170]. Funeral companies have difficulties in coping with the number of deceased persons. Shocking images circulated around the world of bodies piling up in corridors, improvised mortuaries, or mass graves [171]. These experiences have a deep impact on relatives who face a difficult mourning process. It also influences thinking about the value of preserving physical life at all costs, making more people aware that there may be things worse than death or reducing the risk of dying. Since the last few decades, there is the general assumption that, at least in developed countries, life expectation will continuously grow. Against this backdrop, it is unacceptable that a virus is killing an increasing number of people. All imaginable measures are taken to prevent fatalities, even if the health of society (as well as economy, education, culture) itself is negatively affected. Death is regarded as a failure, and it cannot be acknowledged that humans are vulnerable beings with a limited prospect of control.

7.8 Conclusion

In disastrous circumstances the normal way of life is disrupted and upset. When a pandemic occurs, the normality that dominates everyday life is suddenly lost. One of the main threads in this chapter is that a similar disruption affects the ethical principles and norms that used to guide medical practice. It is pointed out that exceptional circumstances are associated not only with exceptional measures but also with a transformation of the ethical framework that normally applies. The general assumption that health professionals should not abandon their patients because they have a duty to treat and care, is questioned. Conventional standards of care, personal and holistic care, equality of treatment, informed consent, and research integrity are compromised, while the life of one patient can be sacrificed to save another person. Similar arguments of exceptionality are used at the policy level, as argued in previous chapters. The justification is that in the face of a massive event such as a pandemic the ethical perspective should shift toward consequentialism, and especially utilitarianism. It is not possible to provide all patients with the treatment and care that they would receive when there is no pandemic. This chapter claims that these arguments are problematic in the context of clinical medicine. The emphasis on utility and efficiency disregards the needs and values of ill people, as is evidenced in the deficient care for the elderly, residents in nursing homes, and dying patients. It also disregards the context in which clinical decisions are made, particularly the pre-existing situation of inequality and social injustice which determines that some patients have been more vulnerable to diseases, and are therefore more disposed to Covid-19 than others, reducing their chances to receive ventilatory support if triage decisions are based on long-term survival and quality of life. Disadvantaged patient populations will often be more disadvantaged when clinical medicine is only guided by utilitarian considerations [172]. These considerations do not cancel the importance of respect for human dignity, human rights, equality and justice. Principles of utility and efficiency are furthermore associated with a certain degree of paternalism. The balancing of various ethical principles is ultimately left to clinicians (or triage officers). Accepted practices as shared decision-making are abandoned, and the involvement of patients and the public in determining the goals of treatment and care is sidelined. These tendencies bring the discourse of bioethics back to where it started: returning the subject of the patient in the practice of care and treatment against the dominance of medical and technical perspectives on health and disease [173]. Paternalism is reinforced by the omnipresence of the war metaphor, framing treatment and care as a relentless fight against the virus. Death is regarded as defeat, care workers as heroes, lack of care for non-Covid patients as collateral damage, triage as sacrificing older patients, and the urgency to impose crisis standards of care and to expedite research. It is also reflected in the arguments of people opposed to stringent measure that a larger number of deaths is acceptable in order to maintain social and economic life as it used to be.

Many ethical dilemmas in the treatment and care of Covid-19 patients result from lack of anticipatory action, and deficiencies in preparedness and prevention. Vexing challenges such as implementing crisis standards of care or triage are not solving the problem of scarce resources. They are desperate efforts to make the best of a horrible situation that should not have arisen in the first place. This points to the moral significance of prevention.

References

1. Arora, A., and A. Arora. 2020. Ethics in the age of Covid-19. *Internal and Emergency Medicine* 15: 889–890.
2. Dudzinski, D.M., B.Y. Hoisington, and C.E. Brown. 2020. Ethics lessons from Seattle's early experience with Covid-19. *American Journal of Bioethics* 20 (7): 68.
3. Komrad, M.S. 2020. Medical ethics in the time of Covid-19. *Current Psychiatry* 19 (7): 30.
4. Lapolla, P., A. Mingoli, and R. Lee. 2020. Deaths from Covid-19 in healthcare workers in Italy – What can we learn? *Infection Control & Hospital Epidemiology*: 1–2.
5. Erdema, H., and D.R. Lucey. 2020. Healthcare worker infections and deaths due to Covid-19: A survey from 37 nations and a call for WHO to post national data on their website. *International Journal of Infectious Diseases* 102: 239–241.
6. Spencer, J., and C. Jewett. 2021. Twelve months of trauma: More than 3,600 US health workers died in Covid's first year. *The Guardian*, April 8.
7. World Health Organization. 2021. *Health and care worker deaths during Covid-19*. October 21.
8. Gopichandran, V. 2020. Clinical ethics during the Covid-19 pandemic: Missing the trees for the forest. *Indian Journal of Medical Ethics* 5 (3); Swazo, N.K., M.H. Talukder, and M.K. Ahsan. 2020. A *duty* to treat? A *right* to refrain? Bangladeshi physicians in moral dilemma during Covid-19. *Philosophy, Ethics, and Humanities in Medicine* 15: 7.
9. Tallès, O., et al. 2020. Ils se sont sacrifiés soigner les maladies du Covid. *La Croix*, September 28.
10. Bellieni, C. 2021. Ethical drawbacks of treating doctors as heroes during the COVID pandemic. *Academia Letters*, Article 2700.
11. Higgins, C. 2020. Why we shouldn't be calling our healthcare workers 'heroes'. *The Guardian*, May 27.
12. Pennella, A.R., and A. Ragonese. 2020. Health professionals and Covid-19 pandemic: Heroes in a new war? *Journal of Health and Social Sciences* 5 (2): 169–168.
13. See for example, General Medical Council. 2019. *Good medical practice*.
14. Ruderman, C., C. Shawn Tracy, C.M. Bensimon, et al. 2006. On pandemics and the duty to care: Whose duty? Who cares? *BMC Medical Ethics* 7: 5.
15. Sheahan, L., and S. Lamont. 2020. Understanding ethical and legal obligations in a pandemic: A taxonomy of 'duty' for health practitioners. *Journal of Bioethical Inquiry* 17: 697–701.
16. General Medical Council. 2020. *Working safely*.
17. British Medical Association. 2020. *Covid-19: Refusing to treat where PPE is inadequate*.
18. Brody, H., and E.N. Avery. 2009. Medicine's duty to treat pandemic illness: Solidarity and vulnerability. *Hastings Center Report* 39 (1): 40–48.
19. Lipworth, W. 2020. Beyond duty: Medical 'heroes' and the Covid-19 pandemic. *Journal of Bioethical Inquiry* 17: 723–730.

20. Cox, C.L. 2020. 'Healthcare heroes': Problems with media focus on heroism from healthcare workers during the Covid-19 pandemic. *Journal of Medical Ethics* 46: 510–513.
21. Conti, C., L. Fontanesi, R. Lanzara, I. Rosa, and P. Porcelli. 2020. Fragile heroes. The psychological impact of the Covid-19 pandemic on health-care workers in Italy. *PLoS ONE* 15 (11): e0242538.
22. Turale, S., C. Meechamnan, and W. Kunaviktikul. 2020. Challenging time: Ethics, nursing and the Covid-19 pandemic. *International Nursing Review* 67 (2): 164–167.
23. Adams, J.G., and R.M. Walls. 2020. Supporting the health care workforce during the Covid-19 global epidemic. *JAMA* 323 (15): 1439–1450.
24. Jeffrey, D.I. 2020. Relational ethical approaches to the Covid-19 pandemic. *Journal of Medical Ethics* 46: 495–498.
25. Dishman, L., and V. Schroeder. 2020. A Covid-19 patient's experience: Engagement in disease management, interactions with care teams and implications on health policies and managerial practices. *Patient Experience Journal* 7 (2); Hristova, B. 2020. Recovering COVID-19 patient describes what it was like to have the virus. *CBC News*, 20 March.
26. McNairy, M., B. Bullington, and K. Bloom-Feshbach. 2020. Searching for human connectedness during Covid-19. *Journal of General Internal Medicine* 35 (10): 3043–3044.
27. Barratt, R.L., R. Shaban, and W. Moyle. 2011. Patient experience of source isolation: Lessons for clinical practice. *Contemporary Nurse* 39 (2): 180–193; Pursell, E., D. Gould, and J. Chudleigh. 2020. Impact of isolation on hospitalized patients who are infectious: Systematic review with meta-analysis. *BMJ Open* 10: e030371; Hossain, M., A. Sultana, and N. Purohit. 2020. Mental health outcomes of quarantine and isolation for infection prevention: A systematic umbrella review of the global evidence. *Epidemiology and Health* 42: e2020038.
28. Shaban, R.Z., S. Nahidi, C. Sotomayor-Castillo, et al. 2020. SARS-CoV-2 infection and Covid-19: The lived experience and perceptions of patients in isolation and care in an Australian Healthcare setting. *American Journal of Infection Control* 48: 1445–1450; Sun, N., L. Wei, H. Wang, et al. 2021. Qualitative study of the psychological experience of Covid-19 patients during hospitalization. *Journal of Affective Disorders* 278: 15–22; Sahoo, S., A. Mehra, V. Suri, et al. 2020. Lived experiences of the corona survivors (patients admitted in Covid wards): A narrative real-life documented summaries of internalized guilt, shame, stigma, anger. *Asian Journal of Psychiatry* 53: 102187.
29. McNairy, Bullington, and Bloom-Feshbach, Searching for human connectedness during Covid-19, 3044; Millstein, J.H., and S. Kindt. 2020. Reimagining the patient experience during the Covid-19 pandemic. *NEJM Catalyst/Innovations in Care Delivery*.
30. Halpern, J., and D.J. Opel. 2020. Sustaining clinical empathy during the pandemic. *The Hastings Center Bioethics Forum*.
31. Lo, B., and M.H. Katz. 2005. Clinical decision making during public health emergencies: Ethical considerations. *Annals of Internal Medicine* 143 (7): 493–498.
32. Pennella, and Ragonese, Health professionals and Covid-19 pandemic, 168.
33. Thorlby, R., C. Fraser, and T. Gardner. 2020. Non-Covid-19 NHS care during the pandemic. Activity trends for key NHS services in England. *The Health Foundation*, December 12.
34. Feral-Pierssens, A-L., P-G. Claret, and T. Chouihed. 2020. Collateral damage of the Covid-19 outbreak: Expression of concern. *European Journal of Emergency Medicine* 27 (4): 233-234.
35. Moroni, F., M. Gramegna, S. Ajello, et al. 2020. Collateral damage: medical care avoidance behavior among patients with myocardial infarction during the Covid-19 pandemic. *JACC: Care Reports* 2 (10): 1620–1624.
36. Hűbner, M., T. Zingg, D. Martin, et al. 2020. Surgery for non-Covid patients during the pandemic. *PLoS ONE* 15 (10): e0241331.
37. Bruno, B., and S. Rose. 2020. Patients left behind: Ethical challenges in caring for indirect victims of the Covid-19 pandemic. *Hastings Center Report* 50 (4): 19–23.
38. Lee, K.D., S.B. Lee, J.K. Lim, et al. 2020. Providing essential clinical care for non-COVID-19 patients in a Seoul metropolitan acute care hospital amidst ongoing treatment of COVID-19. *Journal of Hospital Infection* 106 (4): 673–677.

39. Goulabchand, R., P.-G. Claret, and B. Lattuca. 2020. What if the worst consequences of Covid-19 concerned non-Covid patients? *Journal of Infection and Public Health* 13: 1237–1239.
40. Hassan, B., and T. Arawi. 2020. The care of non-Covid-19 patients: A matter of choice or moral obligation? *Frontiers in Medicine* 7: 564038.
41. Rosenbaum, L. 2020. The untold toll – The pandemic's effects on patients without Covid-19. *New England Journal of Medicine* 382 (24): 2368–2371.
42. Huxtable, R. 2020. Covid-19: Where is the national ethical guidance? *BMC Medical Ethics* 21: 32.
43. Dudzinski, D.M., B.Y. Hoisington, and C.E. Brown. 2020. Ethics lessons from Seattle's early experience with Covid-19. *American Journal of Bioethics* 20 (7): 67–74.
44. "… entering a crisis standard of care mode is not optional – it is a forced choice, based on the emerging situation." Institute of Medicine. 2009. *Guidance for establishing crisis standards of care for use in disaster situations: A letter report*. Washington, DC: The National Academies Press, 15.
45. DeBruin, D., and J.P. Leider. 2020. Covid-19: The shift from clinical to public health ethics. *Journal of Public Health Management and Practice* 26 (4): 306–309.
46. Institute of Medicine. *Guidance for establishing crisis standards of care for use in disaster situations*, 27 ff.
47. National Academies of Sciences, Engineering, and Medicine. 2020. *Rapid expert consultation on crisis standards of care for the Covid-19 pandemic (March 28, 2020)*. Washington, DC: The National Academies Press.
48. Hick, J.L., Hanfling, D., M.K. Wynia, and A.T. Pavia. 2020. *Duty to plan: Health care, crisis standards of care, and novel coronavirus SARS-CoV-2. NAM Perspectives*. Discussion paper. Washington, DC: National Academy of Medicine.
49. Berlinger, N., M. Wynia, T. Powell, et al. 2020. Ethical framework for health care institutions responding to novel coronavirus SARS-CoV-2 (Covid-19). *The Hastings Center*, March 16.
50. National Academies of Sciences, Engineering, and Medicine. *Rapid expert consultation on crisis standards of care for the Covid-19 pandemic (March 28, 2020)*, 4.
51. Hantel, A., J.M. Marron, M. Casey, et al. 2020. US state government crisis standards of care guidelines. Implications for patients with cancer. *JAMA Oncology*, December 3.
52. Cleveland Manchanda, E.C., C. Sanky, and J.M. Appel. 2020. Crisis standards of care in the USA: A systematic review and implications for equity amidst Covid-19. *Journal of Racial and Ethnic Health Disparities* 3.
53. Hertelendy, A.J., G.R. Ciottone, C.L. Mitchell, et al. 2020. Crisis standards of care in a pandemic: Navigating the ethical, clinical, psychological and policy-making maelstrom. *International Journal for Quality in Health Care*: 1–4.
54. Cleveland Manchanda, Sanky, and Appel, Crisis standards of care in the USA, 9.
55. Center for Disease Control and Prevention. 2020. *Interim clinical guidance for management of patients with confirmed coronavirus disease (Covid-19)*. December 8.
56. NIH. 2020. *Coronavirus disease 2019 (COVID-19) treatment guidelines*.
57. Boseley, S. 2020. US secures world stock of key Covid-19 drug remdesivir. *The Guardian*, 30 June.
58. Mahase, E. 2020. Covid-19: Remdesivir probably reduces recovery time, but evidence is uncertain, panel finds. *British Medical Journal* 380: m3049; World Health Organization. 2020. *Therapeutics and Covid-19: Living guideline*, 17 December 2020.
59. The WHO Rapid Evidence Appraisal for Covid-19 Therapies (REACT) Working Group. 2020. Association between administration of systemic corticosteroids and mortality among critically ill patients with COVID-19 – A meta-analysis. *JAMA* 324 (13): 1330–1341.
60. NIH. *Coronavirus Disease 2019 (COVID-19) Treatment Guidelines*, 112 ff.
61. Landau, M. D. 2021. How Merck's antiviral pill could change the game for Covid-19. *National Geographic*, October 2.

62. McBride Folkers, K., and A. Caplan. 2020. False hope about coronavirus treatments. *The Hastings Center*, March 20.
63. World Health Organization, *Therapeutics and Covid-19: living guideline*, 12.
64. Bierer, B.E., S.A. White, J.M. Barnes, and L. Gelinas. 2020. Ethical challenges in clinical research during the Covid-19 pandemic. *Journal of Bioethical Inquiry* 17: 717–722.
65. Belayneh, A. 2020. Off-label use of chloroquine and hydroxychloroquine for COVID-19 treatment in Africa against WHO recommendation. *Research and Reports in Tropical Medicine* 11: 61–72.
66. Kalil, A.C. 2020. Treating Covid-19 – Off-label drug use, compassionate use, and randomized clinical trials during pandemics. *JAMA* 323 (19): 1897–1898.
67. Szalinski, C. 2021. Fringe doctors' group promote ivermectin for Covid despite a lack of evidence. *Scientific American*, September 29.
68. Brown, B. 2020. Ethics of emergency use authorization during the pandemic. *The Hastings Center*, October 30.
69. Maschke, K.J., and M.K. Gusmano. 2020. Ethics and evidence in the search for vaccine and treatments for Covid-19. *The Hastings Center*, April 15.
70. Goldstein, R.H., and R.P. Walensky. 2020. The challenges ahead with monoclonal antibodies. From authorization to access. *JAMA* 324 (21): 2151–2152.
71. Ledford, H. 2020. Coronavirus shuts down trials of drugs for multiple other diseases. *Nature* 580 (7801): 15–16.
72. Bierer, White, Barnes, and Gelinas. Ethical challenges in clinical research during the Covid-19 pandemic, 720.
73. Meagher, K.M., N.W. Cummins, A.E. Bharuha, et al. 2020. Covid-19 ethics and research. *Mayo Clinic Proceedings* 95 (6): 1119–1123.
74. Friedrich, A. 2021. Fear of doing too much too soon or too little too late: Research on Covid-19. *The Hastings Center*, August 20.
75. Saag, M.S. 2020. Misguided use of hydroxychloroquine for Covid-19. The infusion of politics into science. *JAMA* 324 (21): 2161–2162.
76. Sisk, B.A., and J. DuBois. 2020. Research ethics during a pandemic: A call for normative and empirical analysis. *American Journal of Bioethics* 20 (7): 82–84.
77. Ma, X., Y. Wang, T. Gao, et al. 2020. Challenges and strategies to research ethics in conducting Covid-19 research. *Journal of Evidence Based Medicine* 13 (2): 173–177.
78. "… the moral mission of research remains the same: to reduce uncertainty and enable caregivers, health systems, and policy-makers to better address individual and public health." London, A.J., and J. Kimmelman. 2020. Against pandemic research exceptionalism. Crises are no excuse for lowering scientific standards. *Science* 368 (6490): 476–477. See also: Yeoh, K-W., and K. Shah. 2020. Research ethics during a pandemic (Covid-19). *International Health*: 1–2.
79. Komrad, M. S. 2020. Medical ethics in the time of Covid-19 *Current Psychiatry* 19 (7): 29–32, 46.
80. Solbakk, J.H., H.B. Bentzen, S. Holm, et al. 2021. Back to WHAT? The role of research ethics in pandemic times. *Medicine, Health Care and Philosophy* 24: 3–18.
81. National Academy of Medicine. 2020. *National organizations call for action to implement crisis standards of care during Covid-19 surge*. December 18, 2020.
82. Bramstedt, K.A. 2020. The carnage of substandard research during the Covid-19 pandemic: A call for quality. *Journal of Medical Ethics* 46: 803–807; Agoramoorthy, G., M.J. Hsu, and P. Shieh. 2020. Queries about on the Covid-19 quick publishing ethics. *Bioethics* 34 (6): 633–634; Lipworth, W., M. Gentgall, I. Kerridge, and C. Stewart. 2020. Science at warp speed: medical research, publication, and translation during the Covid-19 pandemic. *Journal of Bioethical Inquiry* 17: 555–561.
83. Van Westen-Lagerweij, N.A., E. Meijer, E.G. Meeuwsen, et al. 2021. Are smokers protected against SARS-CoV-2 infection (Covid-19)? The origins of the myth. *npj Primary Care Respiratory Medicine* 31: 10.

84. WHO. 2020. *Ethical standards for research during public health emergencies: Distilling existing guidance to support Covid-19 R&D*. March 20; Katib, A.A. 2020. Research ethics challenges during the Covid-19 pandemic: What should and what should not be done. *Journal of Ideas in Health*: 183–187.
85. Nuffield Council on Bioethics. 2020. *Research in global health emergencies – Ethical issues*. January 28.
86. Jansen, M.O., P. Angelos, S.J. Schrantz, et al. 2020. Fair and equitable subject selection in concurrent Covid-19 clinical trials. *Journal of Medical Ethics* 47: 7–11.
87. Möhlenkamp, S., and H. Thiele. 2020. Ventilation of COVID-19 patients in intensive care units. *Herz* 45 (4): 329–331.
88. In the Netherlands for example, the capacity of intensive care units is expanded from 1150 to 1600 beds; as of April 7, 2020 there have been never more than 1424 patients on the ICU. Boin, A., W. Overdijk, C. van der Ham, J. Hendriks, and D. Sloof. 2020. *Covid-19. Een analyse van de nationale crisisresponse*. Leiden: The Crisis University Press, 68 ff.
89. Iyengar, K., S. Bahl, R. Vaishya, and A. Vaish. 2020. Challenges and solutions in meeting up the urgent requirement of ventilators for Covid-19 patients. *Diabetes & Metabolic Syndrome: Clinical Research & Reviews* 14: 499–501.
90. Badshah, N. 2021. British Medical Association says 'time is now' for Covid plan B. *The Guardian*, October 20.
91. Rosenbaum, L. 2020. Facing Covid-19 in Italy – Ethics, logistics, and therapeutics on the epidemic's front line. *New England Journal of Medicine* 382 (20): 1873–1875. See also: Craxi, L., M. Vergano, J. Savulescu, and D. Wilkinson. 2020. Rationing in a pandemic: Lessons from Italy. *Asian Bioethics Review* 12: 325–330.
92. "The most ethical course of action is to do everything possible to delay having to ration." Iserson, K.V. 2020. Healthcare ethics during a pandemic. *Western Journal of Emergency Medicine* 21(3): 481.
93. Wirth, M., L. Rauschenback, B. Hurwitz, et al. 2020. The meaning of care and ethics to mitigate the harshness of triage in second-wave scenario planning during the Covid-19 pandemic. *American Journal of Bioethics* 20 (7): W17–W19.
94. OECD. 2020. *Hospital beds (indicator)*.
95. ———. 2020. Intensive care bed capacity, April 20.
96. Anonymous. 2012. Challenges in critical care in Africa: Perspectives and solutions. *ICU Management & Practice* 12 (4).
97. Ten Have, H. 2014. Macro-triage in disaster planning. In *Disaster bioethics: Normative issues when nothing is normal*, ed. D.P. O'Mathuna, M. Clarke, and B. Gordijn, 13–32. Dordrecht: Springer.
98. White, D.B., M.H. Katz, J.M. Luce, and B. Lo. 2009. Who should receive life support during a public health emergency? Using ethical principles to improve allocation decisions. *Annals of Internal Medicine* 150 (2): 132–138.
99. Barilan, Y.M., and M. Brusa. 2016. Triage. In *Encyclopedia of global bioethics*, ed. H. ten Have, Vol. 3, 2839–2847. Cham: Springer.
100. Iserson, K.V., and J.C. Moskop. 2007. Triage in medicine, part I: Concept, history, and types. *Annals of Emergency Medicine* 49 (3): 275–281.
101. Jonsen, A. 1998. *The birth of bioethics*, 211–217. New York/Oxford: Oxford University Press.
102. See, for example: British Medical Association. 2020. *Covid-19 – Ethical issues. A guidance note*; National Academy of Medicine. 2020. *National organizations call for action to implement crisis standards of care during Covid-19 surge*. See also: Pence, G.E. 2021. *Pandemic bioethics*. Peterborough: Broadview Press.
103. Cleveland Manchanda, Sanky, and Appel, Crisis standards of care in the USA.

104. Ercetin, G., and H. Raba. 2021. Turkije haalt coronapatiënten in Nederland en andere landen op. *NOS Nieuws*, February 11; Daily Sabah, 2021. *Patient left for dead in Netherlands brought to Turkey*, February 11; TRTWorld. 2020. *Why Turkey is better-equipped to tackle coronavirus*, March 16.
105. Hantel, Marron, Casey, et al., US State government crisis standards of care guidelines, 1–13.
106. Cleveland Manchanda, E., C. Couillard, and K. Sivashanker. 2020. Inequity in crisis standards of care. *New England Journal of Medicine* 383 (4): e16.
107. Hick, J.L., L. Rubinson, D.T. O'Laughlin, and J.C. Farmer. 2007. Clinical review: Allocating ventilators during large-scale disasters – problems, planning, and process. *Critical Care* 11: 217; Cleveland Manchanda, Sanky, and Appel. Crisis standards of care in the USA.
108. Solomon, M.Z., M.K. Wynia, and L.O. Gostin. 2020. Covid-19 crisis triage – Optimizing health outcomes and disability rights. *New England Journal of Medicine* 383 (5): e27; Cleveland Manchanda, Couillard, and Sivashanker. Inequity in crisis standards of care.
109. Komrad, Medical ethics in the time of Covid-19, 30.
110. Scully, J.L. 2020. Disability, disablism, and Covid-19 pandemic triage. *Journal of Bioethical Inquiry* 17: 601–605.
111. Orfali, K. 2020. What triage issues reveal: Ethics in the Covid-19 pandemic in Italy and France. *Journal of Bioethical Inquiry* 17: 675–679.
112. Kim, S.Y.H., and C. Grady. 2020. Ethics in the time of Covid. What remains the same and what is different. *Neurology* 94 (23): 1007–1008.
113. Sokol, D., and B. Gray. 2020. Should we give priority care to healthcare workers in the Covid-19 pandemic? *BMJ Opinion*, April 1.
114. Fins, J.J. 2020. Pandemics, protocols, and the plague of Athens: *Insights from Thucydides. Hastings Center Report* 50 (3): 50–53.
115. Emanuel, E.J., G. Persad, R. Upshur, et al. 2020. Fair allocation of scarce medical resources in the time of Covid-19. *New England Journal of Medicine* 382 (21): 2049–2055.
116. British Medical Association. 2020. *Covid-19 – Ethical issues. A guidance note*, 7–9.
117. White, D.B., and B. Lo. 2020. A framework for rationing ventilators and critical care beds during the Covid-19 pandemic. *JAMA*. 323 (18): 1773–1774.
118. Dawson, A., D. Isaacs, M. Jansen, et al. 2020. An ethics framework for making resource allocation decisions with clinical care: Responding to Covid-19. *Journal of Bioethical Inquiry* 17: 749–755.
119. White, D.B., M.H. Katz, J.M. Luce, and B. Lo. 2009. Who should receive life support during a public health emergency? Using ethical principles to improve allocation decisions. *Annals of Internal Medicine* 150 (2): 137.
120. Vincent, J-L., and J. Creteur. 2020. Ethical aspects of the Covid-19 crisis: How to deal with an overwhelming shortage of acute beds. *European Heart Journal. Acute Cardiovascular Care*: 1–5; DeBruin, and Leider, Covid-19: The shift from clinical to public health ethics, 307; Truog, R.D., C. Mitchell, and G.Q. Daley. 2020. The toughest triage – Allocating ventilators in a pandemic. *New England Journal of Medicine* 382 (21): 1973–1975; National Academy of Medicine. *National organizations call for action to implement crisis standards of care during Covid-19 surge*.
121. Bhatia, N. 2020. We need to talk about rationing: The need to normalize discussion about healthcare rationing in a post Covid-19 era. *Journal of Bioethical Inquiry* 17: 731–735.
122. Piscitello, G.M., E.M. Kapania, W.D. Miller, et al. 2020. Variation in ventilator allocation guidelines by US state during the coronavirus disease 2019 pandemic. *JAMA Network Open* 3 (6): e2012606.
123. Ehni, H.-J., U. Wiesing, and R. Ranisch. 2021. Saving the most lives – A comparison of European triage guidelines in the context of the Covid-19 pandemic. *Bioethics* 35: 125–134.

124. Krűtli, P., T. Rosemann, K.Y Törnblom, and T. Smieszek. 2016. How to fairly allocate scarce medical resources: Ethical argumentation under scrutiny by health professionals and lay people. *PLoS ONE* 11 (7): e0159086; Werner, P., and R. Landau. 2020. Laypersons' priority-setting preferences for allocating Covid-19 patients to a ventilator: Does a diagnosis of Alzheimer's disease matter? *Clinical Interventions in Aging* 15: 2407–2414; Asghari, F., A. Parsapour, and E.S. Gooshk. 2020. Priority setting of ventilators in the COVID-19 pandemic from the public's perspective. *medRxiv preprint*; Buckwalter, W., and A. Peterson. 2020. Public attitudes toward allocating scarce resources in the COVID-19 pandemic. *PLoS ONE* 15 (11): e0240651; Wilkinson, D., H. Zohny, A. Kappes, et al. 2020. Which factors should be included in triage? An online survey of the attitudes of the UK general public to pandemic triage dilemmas. *BMJ Open* 10: e045593.
125. Rosenbaum, Facing Covid-19 in Italy, 1873–1875.
126. Orfali, What triage issues reveal, 676, 678. See also: Craxi, Vergano, Savulescu, and Wilkinson, Rationing in a pandemic: lessons from Italy, 328.
127. Ten Have, Macro-triage in disaster planning, 15–16; Lewis, J., and U. Schuklenk. 2021. Bioethics met its Covid-19 Waterloo: The doctor knows best again. *Bioethics* 35: 3–5.
128. Ehni, Wiesing, and Ranisch, Saving the most lives, 132.
129. This point is strongly emphasized in the advice of the French national ethics committee: Comité consultatif national d'éthique. 2020. *Covid-19. Contribution of the French National Consultative Ethics Committee; Ethical issues in the face of a pandemic*. March 13.
130. Kirchhoffer, D.G. 2020. Dignity, autonomy, and allocation of scarce medical resources during Covid-19. *Journal of Bioethical Inquiry* 17: 691–696.
131. "Undoubtedly, withdrawing ventilators or ICU support from patients who arrived earlier to save those with better prognosis will be extremely psychologically traumatic for clinicians…" Emanuel, Persad, Upshur, et al. 2020. Fair allocation of scarce medical resources in the time of Covid-19, 2052. It is regarded as an act of killing by the German Ethics Council. *Solidarity and responsibility during the coronavirus crisis. Ad hoc recommendation*. March 27. See also: Tham, J., L. Melahn, and M. Baggot. 2021. Withdrawing critical care from patients in a triage situation. *Medicine, Health Care and Philosophy* 24: 205–211.
132. Tham, Melahn, and Baggot, withdrawing critical care from patients in a triage situation.
133. Emanuel, Persad, Upshur, et al., Fair allocation of scarce medical resources in the time of Covid-19, 2052–2053.
134. DeBruin, and Leider, Covid-19: The shift from clinical to public health ethics, 307; Daher, M., G. Rouhana, N. Souaiby, et al. 2020. Ethical consideration in response to the Covid-19 pandemic. *Lebanese Medical Journal* 68 (1–2): 101; BMA, *Covid-19 ethical issues*, 5–6; Emanuel, Persad, Upshur et al., Fair allocation of scarce medical resources in the time of Covid-19, 2053.
135. Cleveland Manchanda, Sanky, and Appel, Crisis standards of care in the USA, 9; Silva, D. S. 2020. Ventilators by lottery. The least unjust form of allocation in the coronavirus disease 2019 pandemic. *Chest* 158 (3):890–891.
136. White, and Lo, A framework for rationing ventilators and critical care beds during the Covid-19 pandemic; Hantel, Marron, Casey, et al., US State government crisis standards of care guidelines; National Academy of Medicine. *National organizations call for action to implement crisis standards of care during Covid-19 surge.*
137. "As we adopt an ethics of epidemics to save as many lives as possible, we must be aware that something is lost in our embrace of utilitarianism." Fins, Pandemics, protocols, and the plague of Athens, 53.
138. Barilan, and Brusa, Triage, 2839; Fink, S. 2020. The hardest questions doctors may face: Who will be saved? Who won't? *The New York Times*, March 21.
139. See, Iserson, Healthcare ethics during a pandemic, 481.
140. Ten Have, Macro-triage in disaster planning, 25.
141. Mallapathy, S. 2020. The coronavirus is most deadly if you are old and male. *Nature* 585: 16–17.

142. Thompson, D-C., M-G. Barbu, C. Beiu, et al. 2020. The impact of Covid-19 pandemic on long-term care facilities worldwide: An overview of international issues. *BioMed Research International*, Article ID 8870249.
143. Burton, J.K., G. Bayne, C. Evans, et al. 2020. Evolution and effects of Covid-19 outbreaks in care homes: A population analysis of 189 care homes in one geographical region of the UK. *Lancet Healthy Longevity* 1: e21–e31.
144. Onder, G., I. Carpenter, H. Finne-Soveri, et al. 2012. Assessment of nursing home residents in Europe: The Services and Health for Elderly in Long TERm care (SHELTER) study. *BMC Health Service Research* 12 (5).
145. Hollinghurst, J., J. Lyons, R. Fry, et al. 2021. The impact of Covid-19 on adjusted mortality risk in care homes for older adults in Wales, UK: A retrospective population-based cohort study for mortality in 2016–2020. *Age and Ageing* 50: 25–31.
146. Werner, R.M., A.K. Hoffman, and N.B. Coe. 2020. Long-term care policy after Covid-19 – Solving the nursing home crisis. *New England Journal of Medicine* 383 (10): 903–905.
147. Cousings, E., K. de Vries, and K.H. Dening. 2020. Ethical care during Covid-19 for care home residents. *Nursing Ethics*: 1–12.
148. Hollinghurst, Lyons, Fry, et al., The impact of Covid-19 on adjusted mortality risk in care homes for older adults in Wales, UK, 28 ff.
149. For example, Cook, M. 2020. A ghastly incident in Spain shows what could happen. *BioEdge*, March 29.
150. For example, Bell, V. 2020. French care home where staff locked themselves in with patients for 47 days avoids coronavirus. *Yahoo News* UK, May 4. See also: Murray, J. 2020. Care workers move into Sheffield dementia home to shield residents. *The Guardian*, March 25.
151. See: Cousings, de Vries, and Dening, Ethical care during Covid-19 for care home residents.
152. Gordon, A.L., C. Goodman, W. Achterberg, et al. 2020. Commentary: Covid in care homes – Challenges and dilemmas in healthcare delivery. *Age Ageing* 49 (5): 701–705.
153. Peisag, C., A. Byrnes, I. Doron, et al. 2020. Advocacy for the human rights of older people in the Covid pandemic and beyond: A call to mental health professionals. *International Psychogeriatrics* 32 (10): 1199–1204.
154. Curley, M.A.Q., E.G. Broden, and E.C. Meyer. 2020. Alone, the hardest part. *Intensive Care Medicine* 46: 1974–1976.
155. Sheahan, L., and F. Brennan. 2020. What matters? Palliative care, ethics, and the Covid-19 pandemic. *Journal of Bioethical Inquiry* 17: 793–796.
156. Committee on Guidance for Establishing Crisis Standards of Care for Use in Disaster Situations; Institute of Medicine. 2012. *Crisis standards of care: A systems framework for catastrophic disaster response*, 1–78. Washington, DC: National Academies Press.
157. Downar, J., and D. Seccareccia. 2010. Palliating a pandemic: "All patients must be cared for". *Journal of Pain and Symptom Management* 39 (2): 291–295.
158. WHO. 2020. *Maintaining essential health services: Operational guidance for the Covid-19 context. Interim Guidance*, June 1; see also: Sullivan, D.R., and J.R. Curtis. 2020. A view from the frontline: Palliative and ethical considerations of the Covid-19 pandemic. *Journal of Palliative Medicine* 24 (2): 293–295.
159. Powell, T., and Chuang, E. 2020. Covid and NYC: What we could do better. *American Journal of Bioethics* 20 (7): 62–66; see also: Abbott, J., D. Johnson, and M. Wynia. 2020. Ensuring adequate palliative and hospice care during Covid-19 surges. *JAMA* 324 (14): 1393–1394; Wynne, K.J., Petrova, M., and R. Coghlan. 2020. Dying individuals and suffering populations: Applying a population-level bioethics lens to palliative care in humanitarian contexts: Before, during and after the Covid-19 pandemic. *Journal of Medical Ethics* 46: 514–525.
160. Radbruch, L., F. M. Knaul, L. de Lima, et al. 2020. The key role of palliative care in response to the Covid-19 tsunami of suffering. *Lancet* 395 (10235): 1467–1469; see also: Berlinger, N., J. Abbott, A. Milliken, et al. 2020. Access to therapeutic and palliative drugs in the context of Covid-19. *The Hastings Center*, July 14.

161. Back, A., J.A. Tulsky, and R.M. Arnold. 2020. Communication skills in the age of Covid-19. *Annals of Internal Medicine* 172 (11): 759–760.
162. Gordon, Goodman, Achterberg, et al., Commentary: Covid in care homes.
163. Sullivan, and Curtis, A view from the frontline.
164. Blinderman, C. D., R. Adelman, D. Kumaraiah, et al. 2021. A comprehensive approach to palliative care during the coronavirus pandemic. *Journal of Palliative Medicine* 24 (7): 1017–1022. See also: Janwadkar, A.S., and T. M. Bibler. 2020. Ethical challenges in advance care planning during the Covid-19 pandemic. *American Journal of Bioethics* 20 (7): 202–204.
165. Richards, T., and H. Scowcroft. 2020. Patient and public involvement in covid-19 policy making. *British Medical Journal* 370: m2575.
166. Arya, A., S. Buchman, B. Gagnon, and J. Downar. 2020. Pandemic palliative care: Beyond ventilators and saving lives. *Canadian Medical Association Journal* 192: E400–E404; Heyland, D.K., P. Dodek, G. Rocker, et al. 2006. What matters most in end-of-life care; perceptions of seriously ill patients and their family members. *Canadian Medical Association Journal* 174: 627–633; Auriemma, C.L., M.O. Harhay, K.J. Haines, et al. 2021. What matters to patients and their families during and after critical illness: A qualitative study. *American Journal of Critical Care* 38 (1): 11–20.
167. King, C.S., D. Sahjwani, A.W. Brown, et al. 2020. Outcomes of mechanically ventilated patients with COVID-19 associated respiratory failure. *PLoS ONE* 15 (11): e0242651. See also: Karagiannidis, C., C. Mostert, C. Hentschker, et al. 2020. Case characteristics, resource use, and outcomes of 10.021 patients with Covid-19 admitted to 920 German hospitals: An observational study. *Lancet Respiratory Medicine* 8 (9): 853–862; Cardona, M., M. Anstey, E.T. Lewis, et al. 2020. Appropriateness of intensive care treatments near the end of life during the Covid-19 pandemic. *Breathe* 16 (2): 1–9.
168. Hollingsworth, H. 2021. More people choosing to die at home as hospitals limit visitations amid pandemic. *HuffPost*, February 7.
169. Powell, V.D., and M.J. Silveira. 2020. What should palliative care's response be to the Covid-19 pandemic? *Journal of Pain and Symptom Management* 60 (1): e1–e3.
170. Sheahan, and Brennan, What matters? 795.
171. Horowitz, J., and E. Bubola. 2020. Italy's coronavirus victims face death alone, with funerals postponed. *New York Times*, March 19; Johnson, W. 2020. The photograph that shocked a nation. *National Geographic*, July 24.
172. Silva, D.S., M.J. Smith, and R.E.G. Upshur. 2013. Disadvantaging the disadvantages: When health policies and practices negatively affect marginalized populations. *Canadian Journal of Public Health* 104 (5): 410–412.
173. Lewis, and Schuklenk, Bioethics met its Covid-19 Waterloo; Richards, and Scowcroft, Patient and public involvement in Covid-19 policy making.

Chapter 8
Prevention and Ethics

Abstract Covid-19 is not merely a national or regional threat but a global one. It requires coordinated action of the global community to mitigate the spread of the pandemic. Even more important is the question how to prevent the next pandemic. Humankind has been warned multiple times for emerging diseases and the risks of pandemics, although preparatory and preventive responses have been deficient. Preventive interventions are possible since it is known how and where infectious diseases emerge. Such interventions proceed on the basis of shared vulnerability and responsibility for global health. This chapter begins with discussing the need for primary prevention and preparedness in order to avoid that a pandemic will originate. Once an infectious diseases is spreading, secondary prevention attempts to mitigate its effects. Human behavior plays a crucial role in this stage and this will subsequently be discussed. The most promising tool to control a pandemic is vaccination which will be investigated in the final sections of this chapter. Ethical problems arise in the successive stages of development, production, distribution and application of vaccines.

Keywords Biosecurity · Conspiracy mentality · Human behavior · Inequity · Pandemic fatigue · Preparedness · Prevention · Vaccination · Vaccine certificates · Vaccine hesitancy

8.1 Introduction

The old saying 'prevention is better than cure' is often attributed to the sixteenth century Dutch scholar Desiderius Erasmus. In fact, concern with prevention is older since for much of its existence the curative capabilities of medicine were limited. The Hippocratic writings as well as classic Chinese and Indian medicine gave many recommendations for a healthy lifestyle. During earlier pandemics, physicians offered recipes and advices how to ward off contagion. Nowadays, prevention is one of the established goals of medicine, in addition to cure, care, and palliation. It has a broad scope, including prevention of diseases, injuries and disabilities but also

H. ten Have, *The Covid-19 Pandemic and Global Bioethics*, Advancing Global Bioethics 18, https://doi.org/10.1007/978-3-030-91491-2_8

preservation and promotion of health. It is frequently assumed that medicine is primarily aimed at cure: trying to heal diseases comes first. However, while cure and prevention are different activities, they share a larger common goal: the removal of disease. The difference is one of timing: prevention is directed at the stage before diseases happen, cure involves activities after diseases have manifested themselves. Early identification of a health problem through screening, for example, will allow rapid treatment, and thus enhances the probability of a cure [1].

A usual distinction in preventive interventions is based on the evolution of a health problem. Primary prevention aims at avoiding that a disease arises (for example, vaccination). Secondary prevention is focused on early detection of a disease in order to delay or arrest its progression (an example is population-based screening). Efforts to forestall that a disease becomes chronic or to counteract the progression of health problems (for example, increased pressure in the eyes) are considered as tertiary prevention which is often a component of curative healthcare. Promoting health has two dimensions: health protection undertaken by agencies to ensure that the health of citizens is not harmed (e.g., restricting the sale of tobacco, regulating the quality of food, and prohibiting the release of toxic substances in the air), while health promotion includes interventions to encourage healthy lifestyles and behaviors, and to create favorable environmental and social conditions.

Compared to other forms of healthcare, prevention is different because it usually involves healthy people, or at least persons who do not feel ill. Curative healthcare commonly starts with a complaint or request of a patient, while in preventive healthcare individuals are approached who have no request for help but the initiative for contact is taken by care professionals. A second characteristic is that prevention used to involve populations rather than individuals. Specific groups at risk may be contacted, particular sections of the population (such as babies and children) or the population at large. Furthermore, in curative care the patient will notice, mostly within a short timeframe, whether the treatment is effective or not because his condition will improve. Preventive efforts of individuals (e.g., a healthy diet) produce effects only in the long term. The results of preventive programs are frequently not observable by the individual participants; if prevention is successful, the potential disease will not occur. Since prevention aims at populations it is significant for groups of individuals who are at risks (reflected in lower mortality and morbidity rates), while many other people will not benefit since they would not have incurred the relevant disease anyway. That means that preventive activities are demanding efforts which are primarily beneficial for others and the community as a whole. This phenomenon is called the 'prevention paradox': a preventive measure that has advantages for the community has little to offer to each participating individual. Finally, prevention frequently includes interventions outside the proper domain of healthcare (for example, the use of safety belts in cars). Governmental policies attempt to influence social and economic determinants of health, the physical and psychological environments and the lifestyle of individuals. These measures can significantly impact the daily existence of human beings, also because they can be restricting individual freedoms. This makes prevention into a political topic: it implicates many aspects of social life so that the value of health must be balanced against other social values.

These characteristics of prevention are associated with a range of specific moral concerns, particularly questioning its moral justification. For example, how can it be defended that healthy people are involved in the healthcare system while they have not asked for this? Is suggesting that people without health complaints may be at risk and need screening not promoting medicalization? How is curtailing the freedom of individuals through restrictive measures and the intrusion of the government in private life justified in the name of health? Preventive interventions require the balancing of individual autonomy, the interests of other people, and the interests of society. In debates on the moral justification of such interventions two basic questions are at stake: what is the good promoted by prevention, and what means should be used? The advantages of prevention are reduction of mortality and morbidity; avoidable death and suffering do not occur. The quality of life, the well-being of individuals, and the control of life can be enhanced. There are also potential benefits for the healthcare system such as reducing expenses, saving resources, and avoiding overburdened services. Possible disadvantages are potential harm to individuals (creating anxiety and worries, risks of tests, mental problems due to lockdown), medicalization of society, and increasing healthcare costs (through identification of more cases of disease). Preventing one disease (for example, lethal infection, acute heart failure and cancer) may lead to subsequent diseases later in life (for example, Alzheimer's disease). The good accomplished by prevention (better health and longer life) is therefore relative but the underlying assumption is that most people will value living and health rather than disease and dying. The second question concerns the means that are used to accomplish the good of prevention. Individuals have only limited abilities to protect themselves against epidemic disease. Protection requires governmental interference. But then individual and collective interests can be at odds. To safeguard the public interest and the collective good, individual autonomy needs to be limited (with stay-at-home orders, quarantines or curfews). The question is how much restriction and coercion should be allowed. An example is the traditional debate about childhood vaccination. In many countries, vaccination is voluntary. Information is provided to persuade parents to have their children vaccinated. But when a growing group of parents do not follow the advice, and therefore endanger other people, the issue is how much influencing will be justified: nudging, manipulation, or coercion? Similar questions arise with vaccination against coronavirus disease. In order to reach herd immunity as quickly as possible, should vaccination be made mandatory?

As previous chapters point out, the above moral questions become salient in the Covid-19 pandemic. This chapter will focus on prevention in connection to the pandemic. Primary prevention in this context has two purposes: inhibiting that the disease will emerge, and countering its expansion into a pandemic. Secondary prevention aims at limiting or avoiding the damage once a pandemic has occurred. Current policy measures are primarily concentrated on secondary prevention, with efforts to limit the amplification of the infection by 'flattening the curve.' Much less attention is paid to impeding the emergence of infectious disease. As experiences discussed in Chap. 2 expose, after each pandemic or pandemic wave, societies prefer to return to normal as quickly as possible, waiting for the next pandemic or wave to occur.

What is striking in the Covid-19 pandemic is the lack of preventive efforts, not merely prior to but also during the pandemic. It was known for a long time that future pandemics would be inevitable, and countries had taken preventive measures (such as storing face masks). But over the last decade, many of these precautionary measures have been reduced. This is shown in the lack of testing equipment, materials, and manpower. Such unpreparedness demonstrates not only absence of preventive vision but is also the result of deliberate policies. In numerous countries, public health services were cut back so that basic approaches such as testing, contact tracing and isolating were simply impossible for a relatively long time. It furthermore illustrates the one-sided development of globalization; basic ingredients for testing and medication are nowadays produced in a limited number of countries such as China and India, not expecting that these countries will also be paralyzed in a global pandemic.

8.2 Preparedness and Prevention

Multiple times a "coming storm" has been predicted, particularly with a microbe combining high lethality and transmissibility [2]. Confronted with a current pandemic such as Covid-19, it should not be taken for granted that this will be the expected disaster. SARS-CoV-2 is not a very deadly virus compared to other potential viruses but it is disastrous because it overwhelms the healthcare system and prolongs its impact because public health is insufficient. The Covid pandemic could be a "rehearsal for a potentially more lethal pandemic" [3]. In light of these predictions, two responses are promoted: preparedness and prevention.

8.2.1 Preparedness

The idea to be ready for potential public health emergencies in the immediate future became influential after the anthrax mailings following the 9/11 terror attacks in the United States in 2001 [4]. Evidently, dangerous micro-organisms could be used to threaten populations and the best way for societies to protect themselves is to apply a range of monitoring and detection activities but also prepare themselves for the event and be ready with countermeasures so that there is an adequate capability to respond. The concept of preparedness also applies to unintentional threats with the result that emerging diseases are framed as a security concern, in keeping with the military vocabulary that has always been used to characterize infections as invasions. The context of biosecurity emphasizes that infectious diseases endanger societies, economic exchange and political stability, reinforcing the role of states. The frame of biosecurity introduces a specific perspective with characteristics that can be easily recognized in today's responses to Covid-19. First, it assumes a polarity: threats are foreign, coming from the outside, and they must be contained and controlled in order to keep citizens safe. One of the consequences is that antagonism rather than global relationship is promoted, minimizing the moral vocabulary of global solidarity. Second, securitization distorts priorities. Because it focuses on specific pathogens and the areas where they emerge, the conditions in which they are produced are not addressed. Concerns of inequality, injustice and poverty that often contribute to the spread of diseases cannot be discussed while the attention primarily goes to micro-organisms whereas global trade and biodiversity loss are not considered as security problems. Third, the frame of biosecurity regards assessment of risks as a technical issue. Evaluating risks is regarded as the prerogative of experts, not as the outcome of public debate about which risks are acceptable and to whom. It is argued that the common good should be protected but there is little space for dialogue, engaging communities, and transparency. Framing health and disease as security issues has prompted governments to introduce legislation and practices that restrict human rights, surveil the public sphere, and practically eliminate privacy. Finally, biosecurity requests a specific approach to biological threats. Since the future is filled with imminent catastrophes and the emergence of these threats is unpredictable, the best strategy is to take preemptive action before the actual threat occurs. That means stockpiling drugs, vaccines, and protective materials. During the 2009 swine flu outbreak for example, Western governments invested hundreds of millions of dollars to stockpile and distribute the antiviral drug Tamiflu, to little avail since the outbreak was much milder than predicted. The resources were not invested to improve the public health infrastructure in most countries, neither in providing clean drinking water, food or basic healthcare in developing countries to make them more resilient in case of diseases [5].

After the SARS epidemic in 2002–2003, the World Health Organization issues a plan to prevent similar outbreaks in the future. It recommends better preparation to respond effectively, and to make supplies of personal protective equipment and essential medicines [6]. Two years later, an evaluation of the responses to SARS concludes that humankind has been lucky this time since the SARS-CoV virus could be relatively easily contained since there was no asymptomatic transmission, so that the traditional public health measures of isolation, quarantine and contact tracing could control the dissemination of the virus. However, these measures were not perfect in some countries and it was unclear why the outbreak stopped [7]. Looking back at the Ebola Virus Disease outbreak (2013–2016), independent assessment panels conclude that the World Health Organization as the leading global health agency has insufficient capacity to respond to a public health emergency. Countries do not have the capacity to detect, report, and respond rapidly, while a robust global system to protect all people based on solidarity, shared vulnerability and responsibility is missing. Many recommendations are made to better prepare for future outbreaks [8].

Following the avian flu outbreaks in 2003 and 2004 (H_5N_1 influenza virus) affecting mainly birds and occasionally humans, and alerted by the SARS outbreak, the

World Health Organization, updating its *Influenza Pandemic Plan* of 1999, launches the *Global Influenza Preparedness Plan*, urging countries to make national bio-preparedness plans. Each country is expected to achieve international harmonization of preparedness measures [9]. Many countries make these plans in view of a deadly pandemic that experts warn will surely come. Over the years, drills and exercises with simulated pandemics are hold. For example, in 2012 the Robert Koch Institute in Germany tests a hypothetical scenario with a new virus, transferable from person to person, and with a mortality rate of 10%. Many of the effects of the current Covid pandemic on healthcare, society and the economy are predicted. The results are incorporated in a National Pandemic Plan [10]. In 2015, a report in the United States warns that the majority of states are not adequately capable to prevent, detect, diagnose and respond to infectious disease outbreaks. It urgently recommends to expand resources for public health infrastructure [11]. In 2016, the government of the United Kingdom organizes Exercise Cygnus to test the country's readiness for a pandemic with a scenario of a devastating influenza pandemic, affecting almost half of the population. The conclusion is that preparedness and response are insufficient to cope with a severe pandemic. The report however is never published [12]. A few years later, just before the Covid-19 pandemic, a similar simulation exercise takes place in the United States. This so-called Crimson Contagion shows that the federal government and the states are incapable to effectively respond to a severe influenza pandemic; funding and coordination are lacking, and production capacity for protective equipment, face masks, and ventilators is insufficient. The report is only published in September 2020 [13].

Although preparedness is advocated since a long time, plans have been developed, and exercises have taken place, the Covid-19 pandemic demonstrates that countries were not prepared. One reason might be that for a long time the focus was on biosecurity in a limited sense, fixated on the threat of bioterrorism. Another reason is that serious pandemics would not be reasonably expected. In the meantime, the world has seen epidemics of Ebola and Zika but they primarily affected a limited region and did not transform into global pandemics. The focus of preparedness efforts was also primarily directed at influenza, because the most recent deadly pandemic has been the Spanish flu. Anxieties concerning avian flu (H_5N_1 virus) since 2003 did not materialize since a pandemic has not occurred. Swine flu, caused by a similar virus as the Spanish flu (H_1N_1) developed into a pandemic in 2009 but it turned out to be milder than seasonal flu. The alarmist framing of the threat, and the gap between what was expected and what occurred might have relaxed preparedness efforts and produced "preparedness fatigue" [14]. As soon as a threat is not manifest or diminishes, priorities change with the result that resources for preparedness are not consistently allocated. Especially in Europe and North America there was "widespread complacency," with political leaders underestimating the danger [15]. Repeated warnings by national and international health organizations are ignored, and even the obvious evidence from simulation exercises showing lack of preparation has not instigated political action. This refers to a significant degree of contentment. The Global Health Security Index, first published in October 2019, concludes that no country is fully prepared for epidemics or pandemics. The countries that are best prepared are the United States, the United Kingdom, and the Netherlands [16]. Similar conclusions are advanced in a WHO study of national and regional preparedness capacities in 2018: more than 50% of the 182 analyzed countries have the highest levels of operational readiness capacities. Most countries in the European region reported high levels of capacity to prevent, detect, and respond to a public health event [17]. Both studies are based on self-reported data, showing that especially high-income and middle-income countries are satisfied with their capacities to cope with a health emergency. Anyway, the Covid-19 pandemic simply overwhelmed all countries, developed or underdeveloped, and particularly affected high-income countries that assumed they are best prepared. Overestimation of own capacities was joined with underestimation of the risks and effects of the virus.

8.2.2 Prevention

In 1994, the Centers for Disease Control and Prevention in the United States propose a prevention strategy for infectious diseases with four goals: surveillance and response; applied research; prevention and control; public health infrastructure. The strategy points out that research should not only be focused on virology and epidemiology but also on the role of behavioral factors that influence exposure to microbial threats. The importance of enhanced public communication of relevant information, as essential for the implementation strategies, is highlighted. Infrastructure requests attention for laboratory capabilities with surge capacity. But what is most important for prevention is monitoring and surveillance, not only nationally but globally: "Timely recognition of emerging infections requires early warning systems to detect such problems so they may be promptly investigated and controlled before they evolve into public health crises" [18]. The science of pandemic prevention includes three steps: early identification of epidemics, assessment of the probability that they develop into pandemics, and stopping lethal pathogens before they become pandemic [19]. Preventive approaches are feasible because it is known that microbes generally come from animals, and that some geographical areas pose greater risks that pathogens 'jump' to human because there is proximity between humans and animals.

The emphasis on preventive activities is related to the growing importance of the ecological perspective. Humankind is rapidly destroying biodiversity though it is essential for its health and survival. Global populations of mammals, birds, fish, amphibians and reptiles have dropped 68% since 1970. Declining species populations are a measure of deteriorating overall ecosystem health. Deforestation, overexploitation, reduction of wildlife habitat, air and water pollution lead to continual destruction of nature, accelerated by climate change [20]. Loss of biodiversity and disruption of ecosystems create the conditions for the emergence of diseases. Animals that are reservoirs and hosts of viruses are driven from their natural environments, and come into closer contact with humans. The natural barriers between animals in which viruses are circulating and humans are disappearing. However, it is not nature itself that is threatening but human interference with natural ecosystems that creates human health risks. Because of the decline of tropical forest some regions in West Africa, the Amazon basin, and South-East Asia are known as emerging disease hotspots. The cycle of viral spillover will only be broken when human activity will change, and our relationship with nature transformed [21]. The same considerations apply to the Covid-19 pandemic [22]. Although its origins are still unclear, the virus has most probably crossed from bats into humans, with perhaps an intermediary host. Future pandemics will in all likelihood also be zoonotic diseases associated with environmental change and human behavior. Prevention of disease spillover, and thus future pandemics, is closely connected to efforts to mitigate the global challenge of climate change. The transformation towards green economies will be economically expensive but compared to the current costs of the Covid-19 catastrophe relatively moderate.

The ecological perspective affirms that the first action to avoid pandemics is to prevent further deterioration of biodiversity. Global surveillance is important but in fact too late. It is estimated that 1.7 million viruses exist in mammals and birds (where most pandemics originate) but less than 0.1% have been described. The majority of possible pathogens is not even discovered [23]. The challenge is enormous because most of the risks are unknown. The Institute of Medicine report in 1992 presents a long list of emerging infectious diseases, with special attention to HIV, recognized in 1981. The WHO publishes in 2016 a blueprint for action to prevent epidemics, listing priority diseases with eleven pathogens for which few or no medical countermeasures are available (including two coronaviruses, SARS-CoV and MERS-CoV) [24]. A later updated list includes Covid-19 but also 'Disease X.' This refers to a pathogen that is currently unknown but that can cause a serious epidemic in the near future [25]. It evokes the nightmare scenario of a virus that combines the characteristics of the influenza virus (highly transmissible) and SARS-CoV (high lethality). In 2016, the initiative is taken to establish the Global Virome Project, a ten-year research project with the aim to develop a viral database and to detect zoonotic viruses with pandemic potential. It recognizes that prevention and responses thus far have been inadequate. Arguing that global health requires global action, the project assumes the shared vulnerability of the world, and the need for cooperation and solidarity, as well as equitable access to research data and benefits [26].

Pathogen surveillance is based on the image of war. Because emerging diseases cannot be eliminated, the enemies should be better understood so that they can be identified like terrorists before they do harm. We therefore need epidemic intelligence services, disease detectives, virus hunters, biodefense, and a "biological NATO" to defend the world against developing pathogens [27]. This military framework emphasizes a technical approach with data collection, anticipatory intervention and precise weaponry, assuming that war is inevitable, but it overlooks the possibilities for primary prevention. The significance of this framing is that it does not allow any space for considering shared moral responsibilities to prevent outbreaks. The notion of global health affirms that in many developing countries healthcare systems are fragile, and must be made more robust and resilient.

Moreover, they do not have adequate surveillance capacity, and need support because if one country is vulnerable, all countries will be. Prevention and preparedness is a shared responsibility, and requires global solidarity. The failure to strengthen health infrastructure and to ameliorate shared vulnerabilities makes the fact that hardly any lessons are learned from previous pandemics a moral failure. It signifies that no attention is given to global justice, equality, vulnerability, and solidarity. The moral implications of the notion of global health are not taken seriously so that technical and militarized approaches will not produce satisfactory preparedness and prevention for the next pandemic [28].

8.3 Human Behavior

Surveys show that most populations agree with the policy measures of their governments. At the same time, preventive behavior and adherence to the measures is a different issue. For example, in the Netherlands, the large majority of the population agrees that people with symptoms should stay at home; however, only 3 out of ten respondents indicate that they self-isolate with complaints or after returning from a high-risk area, and 2 out of ten do not follow the rules of quarantine if they are tested positive [29]. Most people agree that testing should be done when symptoms occur, but if they are symptomatic, only 28% has the intention to test, and only 11.5% indicate that they have actually done the test [30]. These findings suggest that people assume that policy measures should be followed, but that a lenient interpretation is allowed for themselves.

Once an infectious disease has started to spread, the focus is on secondary prevention. Widespread testing is used to identify infected persons, so that they can self-isolate or be quarantined in order to avoid that they infect others. Policy interventions aim at behavior modification such as physical distancing and masking to limit the possibilities to get infected and to further disseminate the virus. The effectiveness of secondary prevention depends on various factors, such as the availability of protective materials and testing, as well as the stringency with which policies are implemented. However, the crucial factor is human behavior. When the virus has no opportunities to infect humans, the pandemic will recede and eventually terminate. As argued in previous chapters, a pandemic disease is only partially a medical or epidemiological challenge; more often it is a social phenomenon that depends on how humans behave. Efforts to influence and modify behavior are confronted with the typical features of prevention. The majority of people do not feel ill; as individuals they do not notice any benefits while they follow the public health rules because these will primarily benefit the population as a whole. For preventive interventions to be effective, the focus should be on people, not on the disease. From an ethical perspective, prevention requires a specific approach. Lessons can be learned from criticisms of global health. Approaches to improve health often marginalize the social and human context of diseases, and proceed with a reductionist view: focusing on medical interventions, ICU care, drugs and vaccines without paying much attention to the underlying conditions that cause diseases to emerge, and disregarding preparedness as an epidemiological challenge [31].

During the first wave of the pandemic, adherence to restrictive policies is high. The lockdown of Wuhan demonstrates that it takes at least ten weeks of drastic measures to extinguish viral transmission [32]. Social behaviors and interactions are significantly changed; a limited number of people can enter shops; they are avoiding each other in the streets, and many people work from home. Especially older people isolate themselves. Although it takes some time before the effect of public health measures is visible, people observe that their behavior is flattening the Covid curve. The serious economic damage of restrictive measures is also not only the result of lockdowns but related to the behavior of people in response to the virus: consumption diminishes, even in commercial sectors that remain open. When governments are late and not stringent with intervention, consumer confidence retracts [33]. Economic damage can be mitigated if people behave more disciplined. The moral argument that the personal sacrifice of modifying behavior is the best way to protect oneself and others, and will be advantageous for society and economy in the longer run, seems persuasive for most people. When restrictions are relaxed after a few months, many feel relieved and assume that normal life can be resumed.

The second wave of Covid-19 after the summer of 2020 initiates a different dynamic. It becomes clear that the SARS-CoV-2 virus is not disappearing and takes advantage of less strict attitudes of the population. Although there is disillusion that the virus is still circulating, the feeling of crisis has abated. The effect of policy interventions is flattening; measures that have been effective in the first wave, have less impact in the second because people react differently. More information is available, people have different experiences, and often make a personal risk assessment. In some countries such as Italy, the second wave develops rather late because the population is more aware of the viral threat; face masks are generally used and the implementation of measures is more stringently controlled. In other countries, policies especially in the early phases of the second wave are fluctuating with intermittent suppression and relaxation, hoping that vaccines will soon be available. When the virus spreads rapidly, and hospitals are overwhelmed again, restrictive measures are anew inevitable, sometimes more drastic than in the first wave. A particular challenge is that this wave is not only more severe but also has a longer duration, with frightening new mutations of the virus, and serious obstacles in the distribution of vaccines that have been approved from the end of 2020. In the beginning of the next year, most countries are confronted again with a surge in infections and hospitalizations, often identified as the third and fourth waves of the pandemic. The rise of the Delta variant demonstrates that even in vaccinated populations the virus continues to spread and cause havoc, especially in unvaccinated populations but also with breakthrough infections among the vaccinated.

The characteristics of the subsequent waves of the pandemic raise the question of the persistence of public health efforts. Experts predict that prolonged or intermittent physical distancing and masking may be necessary for a long time, possibly into 2022 or longer [34]. It is even argued that Covid-19 will never disappear. It may become milder and endemic, like other human coronaviruses that cause a common cold. But the virus may also continue to mutate and become more severe, bypassing the immunity against some variants, resisting current vaccines, and leading to reinfections. According to this scenario, repeated vaccinations will be necessary as well as persistent use of face masks and physical distancing [35]. In this light, the challenge is how to prevent ‘corona fatigue.’ To sustain an adequate response to the pandemic, the collaboration of all citizens is necessary which requires solidarity, mutual trust, and confidence in the effectiveness of policies.

Pandemic fatigue is defined by the World Health Organization as “demotivation to follow recommended protective behaviors, emerging gradually over time and affected by a number of emotions, experiences and perceptions” [36]. It is a phenomenon that can be expected when there is a prolonged public health crisis: compliance with preventive behavior is declining over time, especially because it has a severe impact on everyday life. A substantial number of people, particularly in developing countries will not be able to comply because they live together with many others in crowded housing, or lack adequate housing, do not have formal jobs, and have to go out for making a living, when government efforts to provide economic relief, secure income and health insurance are absent. Parents, and especially women, are overburdened since they have to work from home and simultaneously to care for elderly relatives and for children when schools are closed. Many people are unable to work remotely [37].

The World Health Organization proposes five principles to counter demotivation: transparency, fairness, consistency, coordination, and predictability. Authorities should explain the reasons behind recommendations and restrictions. They should clarify that measures are not arbitrary or favoring specific groups in society. Policies must be consistent, based on scientific evidence, and leaders should follow their own recommendations. It is important to coordinate responses, so that mixed messages from experts, government representatives and healthcare workers are avoided. Finally, while the Covid pandemic is unpredictable, it is necessary to set objective epidemiological criteria and timelines for restrictions and their relaxation so that people know what to expect [38].

There are many instances in which these principles have not been followed. Numerous countries have seen behavior of authorities not in line with official policies. In the Netherlands, for example, the minister of justice and security, responsible for the implementation of policies, has not followed distancing guidelines at his wedding. The King went on holiday to Greece, just after the government strongly advices citizens not to travel to high-risk areas. Other examples are not difficult to find. Phil Hogan, European Commissioner for trade had to resign after he attended a dinner ignoring travel restrictions and a local lockdown. The prime ministers of Norway and Sweden were blamed for not following their own policies. The point is that such incidents undermine trust and create loss of confidence among the population. When authoritative personalities, often responsible for stringent measures, do themselves not follow the rules, how can they expect that the general public will comply? Persistence of public health efforts is furthermore frustrated because policies are not consistent within and among countries [39]. Recommendations for testing and masking have changed considerably, not merely on the basis of growing evidence, but because of different interpretations of risk and effectiveness in various countries. Some measures are modified or withdrawn since they lack an appropriate legal framework. In February 2021, after weeks of deliberation, the Dutch government for example, imposed a curfew. When this was contested in court, the judge immediately lifted it as unjustified on the basis of existing law. The government then has to make a special law to continue its implementation. This episode seriously corroded the credibility of public health measures. One of the issues not mentioned by the WHO is public involvement and engagement. People who are affected by policy measures should be involved and consulted so that community participation and solidarity is maximized. However, in a pandemic context the argument is used that public discussion will cause alarm and anxiety. Expert recommendations are usually confidential, and debate in Parliament takes place after policies have been implemented. This is no surprise, since already in the preparatory stage, the focus is on expert advice without much involvement of the population. Public debate, however, is ethically required because recommendations and policies are not merely based on scientific expertise but presuppose ethical values and specific moral preferences that are often not made explicit. One example is the assumption that scarcity (from masks to ICU beds) will be unavoidable, so that rationing of resources is foreseen. Analysis of national pandemic plans show that this view is simply accepted without examining possibilities to augment resources [40]. The logic of rationing is already accepted in the preparatory stage, long before the healthcare system is overwhelmed. The arguments for such acceptance are not clarified; brief references are made to the costs involved but how costs are related to loss of life and human suffering is not elaborated. Once the outbreak is there, value choices are also not visible. In most countries, the focus of policies is on care capacity, not on containment of the spread of the virus and prevention of transmission with serious implications for patients with non-Covid diseases. In practice, decisions to ration treatment and care are left to medical experts. There is usually little room for contributions of the general population. The principle of transparency requires that value judgments are discussed in a public consultation process, and that ethical dilemmas are clarified.

Collaboration, trust and solidarity as necessary components of compliance furthermore require reliable and effective communication, beyond simple and widespread promulgation of measures. Lessons from the Ebola Virus Disease indicate that fear, panic, denial and mistrust produce rejection of public health interventions. When communities are involved, they can change their approach (for example, in transforming burial practices) [41]. Transparent communication of risks, consistent messaging, and inspiring leadership will build trust and solidarity. Public trust is based on the perception that authorities are competent, fair, and open. However, it works both ways: if there is a high level of public trust, people will not feel personally responsible but assume that the government will manage the risks; if the level of trust is low, people will make their own risk assessment, and generally underestimate the risks. In both cases, compliance with government measures will be reduced [42]. An important determinant of compliance seems to be trust in science. People who have this trust, generally perceive Covid-19 as an actual risk, and will follow policy measures [43].

It is crucial what kind of information is provided, and how it is communicated. Reliable and transparent information is necessary to motivate citizens to protect themselves and others. The pandemic has produced enormous amounts of information and news media reporting that usually is a mix of truths and falsehoods. This so-called 'infodemic' can cause distress because information is threatening, contradictory and uncertain, and may lead to mistrust, panic purchases, and reduced compliance with preventive measures. Information distress is generally associated with better compliance, but it may also lead to avoidance of information (people become tired of Covid news) which is associated with worse compliance [44]. Communicating information needs to follow a fine line between making people anxious and worried so that they are motivated to implement measures, and creating so much anxiety and distress that people no longer attend to information, and feel that prevention is futile. The implication is that information should be provided to oppose information avoidance, with the awareness that pandemic fatigue has two dimensions: not only behavioral fatigue (feeling demotivated to follow restrictive measures) but also information fatigue (feeling tired of news about the pandemic).

Adherence to preventive policies, as evidenced in the importance of trust, as well as the WHO principles to counter pandemic fatigue, is not just a practical medical, technical or social issue but an ethical challenge. Two perspectives are advanced to explain why people comply with preventive restrictions. The rational choice perspective assumes that individuals comply when they perceive serious risks. This might be true for vulnerable groups but does not apply for groups that are not at risk. Some studies report that even the elderly who are disproportionally affected by Covid, do not comply better than other age groups. Those after 65 years of age comply less to masking mandates than younger people [45]. The normative perspective argues that a sense of duty and altruism motivates people to comply, not only because they want to follow governmental guidelines but also because they want to protect themselves and others, and are convinced that everyone has a duty to protect vulnerable citizens [46].

The last point illustrates that implementation of public health measures not only refers to the ethical notion of trust but also of solidarity. This is particularly clear in the context of global health [47]. Ironically, Covid particularly affects the wealthier parts of the world, certainly in its early stages. The policy measures used in these countries such as lockdowns and widescale testing are advocated for other countries while the main concerns are with hospital care and technological interventions. The context of less-resourced countries is different and not taken into account. The population is usually younger, healthcare less accessible, the economies fragile and informal, while older people do not reside in care homes. Lockdowns are simply impossible in poor and slum areas, and in households of extended families living together in small lodgings, with limited sanitary facilities, and no access to internet. Test kits are too expensive for people who barely survive, and who are directly threatened by hunger and other infectious diseases. In 2019, malaria made 229 million people ill, and caused 409,000 death, primarily in Africa. Covid-19 has impeded prevention and treatment campaigns in 2020 [48]. Reportedly, the dengue virus infects 390 million people each year, particularly in Asia; 96 million of them are diseased, and 22,000 died [49]. While the context of low-income countries is not taken into account, and they are supposed to implement the same public health measures as more affluent countries, they are not able to acquire sufficient protective equipment, and are not prioritized in the distribution of global resources such as medicines and vaccines. Without global solidarity, their populations cannot implement preventive measures, not because of pandemic fatigue but compelled by the circumstances of everyday life.

Trust and solidarity are ethical components of compliance with prevention strategies but an equally essential element is truth. The 'infodemic' associated with the pandemic disseminates not only reliable information but misinformation, rumors, and fake news inspired by conspiracy theories. This is not a new phenomenon; every pandemic in history has generated false information [50]. Contemporary digital media have amplified and accelerated the phenomenon so that viral misinformation spreads faster than SARS-CoV-2 [51]. Deceptive and intentionally false information negatively affects compliance with preventive measures, more than inaccurate and erroneous information that is often the result of uncertainty and conflicting evidence especially in the early stages of the pandemic [52]. In this context, so-called 'conspiracy theories' play a detrimental role. In fact, these are not theories but beliefs since they often lack coherence, and are invulnerable for evidence and falsification. Refutation or correction on the basis of factual evidence is impossible. As belief systems that do not allow any doubt, it is more appropriate to speak of conspiracy mentality or thinking. The fundamental assumption is that there is a hidden hand behind events, explaining how and why things are happening. The idea that social life is controlled by secret forces provides a rationale allowing believers to have some understanding of the apparent uncontrollability and unpredictability of the world. This rationale frequently is driven by ideological perspectives that consider the government as the ultimate enemy, that make a clear distinction between good and evil forces, and that accuse and blame specific groups and ethnicities. This kind of thinking is corroborated and magnified by being widely shared in the (online) media, and especially through video-clips and streamcast [53].

A conspiracy mentality specifically opposes behavior that is recommended by authorities such as governments and experts because it assumes sinister and secret motives to promote particular interests. This mentality undermines trust and solidarity but furthermore may lead to stigmatization, as evidenced in previous pandemics. The Covid-19 outbreak provides examples of racism and xenophobia, blaming people of Asian descent, but also discrimination of care providers.

The conspiracy mentality may instigate actions that obstruct pandemic policies. Conspiracy beliefs that recent 5G mobile phone technology is the cause of Covid-19, led to attacks on hundreds of telecommunication towers in several countries. Other beliefs promulgate the idea that Covid-19 is less dangerous than the seasonal flu, that corona statistics are manipulated, that public health policies are not aimed at reducing viral spread but an instrument of the powerful to curb the freedoms of individuals, that the virus is created by the Chinese government as a bioweapon, or by the pharmaceutical industry to promote the sales of drugs and vaccines. As a consequence, such beliefs encourage people to disregard public health measures, and to protest, sometimes violently, against government policies; they also reduce the intention to be vaccinated. Disinformation is widely shared in populations. In the Netherlands, one in ten citizens believe in conspiracies in the context of the pandemic [54]. In the United States in March 2020, 1 in 4 people believe that the Chinese government made the coronavirus, 1 in 4 that the Centers for Disease Control and Prevention exaggerate the danger of the virus, and 1 in 7 that the pharmaceutical industry created the virus. These beliefs have become more widely shared in July 2020 [55]. An analysis of 38 million news articles in traditional and online English-language media during the first four months of 2020 identified 1.1 million articles that disseminated misinformation in relation to the pandemic [56]. Powerful actors and celebrities are actively promoting misinformation and disinformation. One notorious example is the president of Tanzania declaring that Covid is a hoax; he ordered officials in May 2020 to stop reporting cases and notifying Covid-19 as a cause of death, and then announced that the country is Covid-free and has no need for vaccination [57].

Conspiracy thinking is very difficult to correct [58]. Presenting factual information and rational arguments will not change conspiracy beliefs. Opposite data often reinforces the misinformation beliefs, especially because the expert sources of such data are regarded as secret manipulating agents. The scientific practice of retracting misleading publications has a limited effect since these articles continue to have an influence once published. A notorious example is the publication in 1998 in the *Lancet* that causally linked autism and the measles, mumps, and rubella vaccine. Although the article was retracted and the author removed from the medical register in the UK, it led to a drop in vaccinations, and it continues to figure as an important document in conspiracy beliefs. Nonetheless, experiences with climate change, – a subject where denial and conspiracy beliefs are rampant during the last few decades – demonstrate that change in mentality is possible. In this context, misinformation has been more and more ignored, and direct behavioral interventions have been introduced, encouraging people to adopt ecologically friendly behaviors, for example to drive electric vehicles. Transforming alternative narratives into policy practices has turned climate deniers into a stubborn minority [59]. Although a conspiracy mentality is difficult to counter, from the perspective of ethics it is imperative to continue distinguishing between truth and falsehood. In the context of the Covid-19 pandemic this has at least two implications. The first is that pandemic policies should not be influenced by conspiracy beliefs. An example is vaccination policy. Initially, misinformation and disinformation about vaccines circulate widely, and many people are hesitant to use vaccines but when more and more people are vaccinated, hesitancy diminishes. The second strategy is to continuously debunk these beliefs. Public health experts should be more active with messages in the media to counter fake news and scams so that conspiracy beliefs are not only refuted but also more plausible explanations for what is happening will be disseminated. An example is the WHO Mythbusters webpage where common topics of misinformation are addressed and active strategies advanced to identify myths and debunk them [60]. Social media platforms such as Facebook have started in February 2021 to remove posts with false Covid-19 related information.

8.4 Vaccination

From the beginning of the pandemic, vaccines are regarded as the ultimate means to deliver humankind from the clutches of the virus. Enormous efforts are undertaken to develop vaccines as quickly as possible, with success in a relatively short time. Once available, it is challenging to produce and distribute vaccines to as many people as possible. Because quantities are limited, choices have to be made as to which sections of the population should be prioritized for vaccination. The vaccine story demonstrates again that a technical and biomedical approach is only a partial element of the Covid narrative. Of course, there first should be a vaccine, but its availability does not ensure that it will be applied. Social, economic, political, historical as well as cultural factors determine the utilization of vaccines in populations. The relevant ethical considerations in the subsequent phases of development, production, distribution, and application of vaccines are discussed in this section of the chapter.

8.4.1 Development

Usually, it takes more than ten years to develop a vaccine and make it available on the market. It lasted 43 years after the discovery of the Ebola virus before the first vaccine was approved in 2019 [61]. The mumps vaccine, licensed in 1967, was an exception; it took only four years to develop it. Covid-19 vaccines broke a new record. As soon as the genetic sequence of the virus is published by Chinese researchers in early January 2020, research institutes and pharmaceutical companies immediately start to develop hundreds of vaccine candidates. In July, the first results of phase 1 and 2 studies of two vaccines are published, indicating that they are safe and stimulate immunity against SARS-CoV-2 [62]. Ten months after the identification of the virus, the first phase-3 clinical trials are completed with positive results, showing vaccines to be effective. In December 2020, several vaccines receive emergency approval for general use. The speed with which vaccines have been developed and applied has impressed many. Already early in the pandemic, and more strongly after the first wave when it becomes apparent in many countries that behavioral restrictions alone cannot control the virus transmission, at the same time heavily impacting on social and economic life, authorities refer to a future vaccine as the best hope to return to normal life, and they promise to do everything possible to make them available, sometimes putting pressure on researchers, companies, reviewers, and licensing bodies to speed up the process. At the same time, the results of trials are widely reported; progresses and setbacks are instantly discussed in the public media nourishing hope and optimism that the pandemic ordeal can be ended soon.

The speedy development of vaccines makes other people suspicious. While it usually takes years to make vaccines, and for some infectious diseases successful vaccines never have been developed, why has progress been so rapid in the case of Covid-19? Fears that corners have been cut, and safety and ethical standards sacrificed for speed, have contributed to conspiracy beliefs and vaccine hesitancy. Scholars argue that there is no need for mistrust since there are good reasons to explain the fast development and availability of Covid-19 vaccines [63]. One reason is that vaccine development builds on fundamental biology research where basic technologies using genetic engineering are applied and tested. The tools, technologies, and knowledge to make a vaccine were ready [64]. The two mRNA vaccines that have been approved first, are new and never used before in humans, but they have been in development for two decades. Only a few hours after the genetic code of SARS-CoV-2 becomes public, biotech company Moderna fabricates the first version of its candidate vaccine. Other technologies using adenoviral vectors have also been explored for some time to make vaccines for Ebola and Zika. The research team at the University of Oxford for example, has been preparing for Disease X by creating a biotechnology framework in which the genetic material of a pathogen could be inserted to generate a vaccine. Previous research on coronaviruses, especially since SARS and MERS has identified the viral spike protein as crucial for the infectivity of the virus, and thus as the main target for vaccines. Scientific experience and technology could therefore be put to work, immediately after the genetic code of the virus was known. Scientists engaged in intense worldwide collaboration, sharing research data and clinical trial results. A major contributor to vaccine development has been funding. It is estimated that developing a new vaccine from preclinical testing through phase 2 usually costs between $31 and $68 million, while others indicate that bringing one vaccine from design to manufacture will require a capital investment from $500 million to $1 billion [65]. Global investment for vaccines has been rather small, even for the eleven priority diseases listed by the WHO in 2016. Lack of investment explains why for many infectious diseases no effective vaccines are available. Other explanations refer to the limited number of vaccine manufacturers, as well as the market ideology which does not expect significant revenues if a disease (e.g., Ebola or MERS) is not widely disseminated [66]. The landscape significantly changes with the global threat of Covid-19, especially now that wealthier countries are affected. Governments allocate enormous emergency funds for research, development and manufacturing of Covid-19 vaccines [67]. They also pre-order billions of doses before the vaccines are sufficiently tested. In a short period of time, the money committed to vaccine research induces many companies and laboratories to start investigating vaccine candidates, and to overcome the usual reticence of market thinking in this area. A third factor that promotes the fast availability of vaccines is the rapid recruitment of participants in clinical trials. Usually, testing of vaccine candidates takes a much longer time than the development of the candidates themselves. They need to be tested in the laboratory, in animals, and then in humans; first in a small number of healthy volunteers to determine whether it is safe (phase 1), then in a few hundred volunteers to examine the immune response, and dosage (phase 2), and finally in phase 3 with thousands of subjects in order to analyze effectiveness. The vast majority of candidates (on average 94%) never reach the end stage [68]. Normally, it takes time to enroll participants in trials. With Covid-19 the conditions are different. Many people from various parts of the world are willing to participate in trials, and because the pandemic is raging in waves, there are always areas where the incidence of infections is high. A fourth factor speeding up the development of vaccines is the adaptation of the evaluation mechanisms. Testing procedures that used to proceed in subsequent stages, are executed parallel or in overlapping phases. Also, the regulatory agencies, normally starting reviewing when all research data are completed, now apply a rolling review, using data as soon as it is acquired. Experts assure, however, that the same rigorous guidelines are used as in ordinary circumstances [69].

Though rationally explainable, the development of vaccines is also influenced by less rational factors. The political and public pressure to deliver quick results is enormous. Geopolitical competition is evident since it is a matter of national pride which country and which company will be the first to launch a vaccine. In China and Russia large groups of people are injected with vaccine candidates long before the clinical studies are completed. That the United Kingdom is the first Western country to approve the first Covid-19 vaccine on December 2, 2020, is perhaps not unrelated to its official withdrawal from the European Union less than two month later. Each player in the race follows his own approach, setting up his own clinical trials, and self-defining success criteria, without coordinated or comparative studies [70]. The pressure to deliver positive results may raise ethical concerns, particularly in regard to the balancing of harms and benefits [71]. In view of the increasing damage of the pandemic on health, society and economy, and the persistent inability of public health measures to control the spread of the virus, it seems that any vaccine will be better than nothing. That means that greater risks are acceptable than in ordinary circumstances. Usually, animal studies are required to determine safety and immune response before vaccine candidates are applied in humans but at present they are done in parallel with human testing. The shortening of the various phases of clinical testing and the limited number of participants in phase 3 trials are in normal circumstances not accepted as sufficient evidence that vaccines are safe [72]. When they are injected in millions of people unexpected safety issues can arise, and long-term follow-up will be needed to obtain reliable information. One may easily assume that the practical results (i.e., efficacy and lack of serious side effects) justify the abbreviated development process but it is then ignored that the emergency use is provisionary and that vaccines are still investigational. Since adverse effects that are not identified in clinical trials are expected when large populations are vaccinated, long-term surveillance and monitoring for side effects, safety and efficacy are required. Safety of vaccines is a crucial consideration since it is a preventive rather than curative intervention. Vaccination is applied in healthy people; side effects are therefore taken more seriously than in the case of new medication for patients with severe illness where the balance between benefits and risks is different. The pharmacovigilance infrastructure to monitor this, is not at hand in many places [73]. WHO experts have argued that randomized studies used in phase 3 should continue with blinded follow-up of placebo recipients in these trials in order to better determine the longer-term effects of vaccines [74]. This proposal is ethically problematic, once effective lifesaving vaccines are available. At least participants in the trial should be informed that an effective vaccine is available and that they could leave the study but they could also be persuaded to continue in order to increase scientific knowledge. This problem with placebo-controlled trials will become more important, not only for the continuation of studies but also for the initiation of new trials which should no longer compare a new vaccine with a placebo but with existing vaccines. In this manner, every research participant will be protected, and it will be possible to compare vaccines and determine which one is most effective [75].

The belief that more risks are acceptable because of the urgency of the situation is reflected in another way. The most successful vaccine platforms are new and never put on the market before Covid-19. Both platforms use genetic engineering; viral vector vaccines insert corona-like DNA in adenoviruses while mRNA vaccines directly inject nanoparticles with genetic material into the human body. Prior experiences with other coronaviruses such as SARS and MERS indicate the possibility that a vaccine can enhance disease [76]. The clinical trials show no evidence that this occurs but perhaps the studied population is too small to detect this effect [77]. Another issue concerns the efficacy of vaccines. This refers to the desired effect or benefit of a vaccine. Whether a vaccine is efficacious can be assessed in different ways: prevention of infection, prevention of disease, prevention of severe disease (and thus hospitalization, intensive care admission, and death), and prevention of transmission. An ideal vaccine will combine all preventive effects but in practice the effect is more limited. Studies show that for example, the Moderna vaccine prevents illness and serious disease while protection against infection and blockage of transmissibility are unclear [78]. Obviously, reducing the burden of disease is a major accomplishment and clearly beneficial for individuals (diminishing the probability of serious illness, hospitalization and death) and for society (reducing the pressure on hospitals and critical care units) but as long as vaccines do no limit the spread of the disease public health measures must be continued (at least until herd immunity is reached). The focus of vaccines on diminishing (severe) disease will be particular important for vulnerable populations most in need of vaccines, such as the elderly, people with comorbidities and compromised immune-systems but also people belonging to ethnic, poor and migrant communities. Participants from these populations are often not included in research trials. Other reasons to be cautious are that it is not clear how long the immune response provided by vaccines will last and whether the vaccines will provide protection against new mutations of the virus [79]. In general, efficacy under controlled research conditions is not the same as effectiveness in real-life conditions when the vaccine is more widely deployed [80].

The above considerations demonstrate that rushing the development of Covid vaccines is not without dangers [81]. The general ethical issues related to the development of vaccines concern the level of risk and harm that is acceptable given the urgency of the situation and the expected benefits of vaccine candidates, taking for granted that it is too early to know whether there will be long-term side effects, as well as how effective vaccines will be and for how much time. Furthermore, two specific ethical concerns have emerged. One is the relation of vaccines and the abortion debate. Many vaccines are developed and produced with the use of cell lines derived from aborted human fetuses. Some Catholic bishops have argued that vaccination therefore implies complicity with the moral evil of abortion. Vatican statements, however, explain that it is morally acceptable to use vaccines that are relying on such cell lines [82]. The cells are derived from abortion in the 1970s, and commonly used in medical research. The connection to abortion is very remote, and accepting a vaccine cannot be regarded as formal cooperation with abortion. At the same time, there is an urgent need to address the pandemic, and although vaccination is not a moral obligation, there is a duty to pursue the common good [83]. All approved vaccines use fetal cell lines for laboratory testing but the mRNA vaccines do not use them for development and production, in distinction to the viral vector vaccines (Johnson & Johnson, and Astra-Zeneca/Oxford). For this reason, the American bishops argue that choosing alternatives from Pfizer and Moderna is preferable since they are morally less compromised [84]. However, the possibility to opt for a particular vaccine is given the present scarcity of vaccines illusory.

A second specific concern is the ethics of human challenge studies in which vaccinated healthy volunteers are deliberately exposed to an infectious pathogen. These studies are often done to test new vaccines with the argument that they accelerate the time to develop vaccines since a lower number of participants is needed, and the time of observation reduced. The assumption is that controlled infection in a controlled setting will provide better insights in the efficacy of a vaccine than studies that rely on naturally occurring transmission within communities which is varying according to protective behavior and public health measures [85]. Challenge studies have been allowed for diseases that are treatable and self-limiting [86]. However, these two conditions do not apply in the case of Covid-19. Proponents argue that extraordinary circumstances ask for extraordinary solutions [87]. Enormous benefits may be gained from experimenting with a small number of volunteers who might be seriously harmed while the lives of many more people could be saved. Speeding up the development of vaccines with the risk of sacrificing a few individuals is acceptable because it might benefit so many others. These arguments highlight the exceptionalism discussed in the previous chapter: in times of emergency, the core ethical principles of research ethics no longer apply. The basic rationale however is contested. It is not clear that challenge studies will lead to faster data on efficacy, especially now that the pandemic is widespread and recruitment of phase 3 trials is rapid. Controlling the infection will itself be a challenge, and the manufacturing of a challenge strain of the virus in special biosafety conditions and the development of a challenge model will take time [88]. In order to reduce the risks, challenge studies usually enroll young, healthy adults. This weakens the applicability of the results to people with the highest risks of Covid-19 so that the social value of challenge studies is questionable [89]. In the case of Covid-19 a proper assessment of risks and benefits is not possible; knowledge of the disease is limited but infection may have severe effects even for younger subjects. The uncertainties also compromise the informed consent process. Finally, challenge studies may have negative impact on public confidence; they may enhance mistrust and conspiracy beliefs, notably when healthy people are harmed after being intentionally infected. The consequentialist framework promoting human challenge studies is thus not justified on the basis of its consequences. From the perspective of ethics, there are limits to what can be done to individual persons. In vaccine studies, individuals are asked to accept risks for the common good; they are the ones running the risks while the benefits are for the community. But their interests should be respected in the sense that serious harm should not be intentionally inflicted on them, whatever the possible benefits, the more so when the disease cannot be treated. Respect for human persons, their dignity and rights, has been the basis for research and healthcare ethics since the last Word War, particularly because of experiences of abuse and exploitation. It is not realistic to assume that such exploitative mechanisms will not exist in emergency circumstances when utilitarian thinking prevails [90].

8.4.2 Production

Having a vaccine is one thing, vaccination is another. The challenge is enormous since controlling the pandemic demands that the majority of people across the globe must be vaccinated. Depending on the number of injections and the need for booster shots, billions of vaccines will be needed to immunize global citizens. Anticipating approval, early manufacturing for most vaccines already started in parallel with clinical trials. Wealthy countries preordered vaccines long before the results of clinical trials are known (the USA and the European Union each ordered 700 million doses). Some countries pre-ordered more doses than necessary to vaccinate the entire population. Canada, for example, ordered nine doses per inhabitant while countries such as Brazil and India ordered enough vaccines to immunize just 23.9%, respectively 18.5% of the population [91]. Poorer countries were not able to preorder as long as the characteristics and costs of approved vaccine are not known [92]. The race to get hold of vaccines before they are sufficiently tested and approved, is risky because it implies betting on the most promising candidates. The European Union for example initially buys 400 million doses from AstraZeneca, and 300 million from Pfizer, but none from Moderna. Deception comes quickly because the Pfizer vaccine, the first Covid-19 vaccine approved, is confronted with serious production obstacles. In fact, at some point all manufacturers of approved vaccines are forced to admit that they cannot meet production arrangements. Just before its vaccine is approved by the European Medicine Agency (EMA), AstraZeneca informs the European Union that it can only deliver 60% of the ordered vaccines. The European Commission that has bought the vaccine on behalf of the 27 Member States is incensed; all countries have to revise their scheduled vaccination strategy and immunizing the entire population will be much later than expected since this vaccine is the cornerstone of the strategy in most countries. At that time, AstraZeneca has four production sites in Europe (two in the United Kingdom and two on the Continent). The suspicion is that the company has exported vaccines from the Continent to the UK that started an early and efficient vaccination program with only the AstraZeneca vaccine. Later, it becomes clear that vaccines produced on the Continent were indeed exported to the UK while no vaccines produced there were shipped to the Continent. The European Union responds with a measure to monitor and register the export of vaccines. The pharmaceutical company gives no explanation why the production of vaccines in the Continental facilities is delayed, whereupon the European Union requests public disclosure of the contract. The company blames the Union for its slowness in the negotiations (the United Kingdom contracted vaccine delivery in June and the EU in August 2020) and argues that the contract does not stipulate obligations to deliver but only 'an utmost effort' to deliver the vaccines in time. Experts, however, refer to other factors. One is that the approval of the vaccine by the EMA was late because it requested additional information due to irregularities in the clinical trial data (the British authorities licensed the vaccine on 30 December 2020, while the European Commission granted conditional marketing authorization on 29 January 2021). The contract with the pharmaceutical company was concluded later than in the UK and USA because the EU negotiated a lower price and better conditions for liability. The details, however, are unknown, since in the published contract crucial information is blackened [93]. In distinction to other companies such as Pfizer, the experience of AstraZeneca with vaccine production is limited. Moreover, it is complicated to upscale manufacturing vaccines from the laboratory to mass fabrication in bioreactors. At the same time, several companies are engaged in the same effort, and basic ingredients, equipment, glass vials, and syringes have become scarce. Early March 2021, Johnson & Johnson told the EU that it is facing supply issues. It has promised to deliver 55 million doses of its vaccine in the second quarter of the year, but it will possibly not be able to provide. This vaccine is eagerly awaited since only one dose is needed. Because of delays in the provision of vaccines, several countries engage in other initiatives. Deviating from the joint purchase and distribution procedure in the EU, Hungary and Slovakia procure Russian and Chinese vaccines. These products are considered with suspicion in Western countries because they have been administered before rigorous evaluation is concluded, although later studies show that the Sputnik vaccine is safe and 91.6% effective [94]. The vaccine is already used in countries such as Egypt, Mexico and Argentina, and since March under review by the EMA while contracts for production in various European countries have been concluded [95].

That countries have been too optimistic about mass production of vaccines should have been clear from the beginning. Traditionally, vaccine production is not very attractive for pharmaceutical companies because of risks of litigation and low profits. Recently, the interest in vaccine research and production has revived due to the emphasis in global health policies on prevention of diseases, rather than cure. The use of vaccines has been spurred by the establishment in 1999 of the Global Alliance for Vaccines and Immunization (GAVI) aiming to increase access to vaccines in developing countries. Concerns with biosecurity also stimulated research into new vaccine approaches [96]. The global expertise to develop and produce vaccines is nonetheless limited. For prevention of infectious diseases, the world is dependent on only a few major companies, suppliers of raw materials, and production facilities. The capability to produce (coronavirus) vaccine is chiefly concentrated in the US, India and China. The largest manufacturer in the world is the Serum Institute of India (contracted with AstraZeneca to produce 1 billion doses of its vaccine). Only two major pharmaceutical companies (Sanofi Pasteur and

GlaxoSmithKline) have been manufacturing a broad range of vaccines in the past [97]. For some time, studies have warned that vaccine production capacity is insufficient: "Unfortunately, the global industry is not yet at a level where it can respond to meet the full pandemic vaccine need in a timely and equitable manner …" [98]. For ten out of 16 emerging infectious diseases with pandemic potential, the twenty largest pharmaceutical companies in the world have no research and development activities so that preparation for new pandemics is inadequate [99]. The Covid-19 crisis reveals that many pharmaceutical companies are more interested in securing financial gains than in producing goods and services such as vaccines. Financial reserves of companies are not invested in research or productive capabilities but distributed in increasing payouts to shareholders. To finance these payouts, the price of drugs has increased exponentially and the business model has changed from producing and selling of products to buying up competitors and biotechnology companies to diminish competition and acquire intellectual property rights [100]. Against this backdrop, Covid-19 has been a blessing for the pharmaceutical industry. The share prices of many companies have skyrocketed and profits boomed [101]. Governments have allocated billions of dollars without demanding strict conditions but given the urgency of the pandemic and the suffering of populations, pharmaceutical companies are in a much stronger negotiating position than governments. It is in the company's interests that an atmosphere is created in which governments are pressured into contracts with secret clauses, gigantic prepayments and few binding obligations. The example of Israel illustrates that vaccination is not simply a matter of vaccine production. For several months, the country is leading the world-wide vaccination race. Early March 2021 it has vaccinated 44% of its population (with 9 million doses given). Apparently, there is no shortage of vaccines for this country. The country made a deal with Pfizer, paying a much higher price than the USA and EU, and promising to provide the company with the anonymized age, gender and demographic data of vaccinated people, enabling it to carry out a real-life study of the effectiveness of its vaccine [102]. The example illustrates that vaccines are produced for the highest and most compliant bidder. Companies decide who will receive the produced doses first, and global health is clearly not a priority consideration in these decisions.

Two ethical principles are advanced in the context of vaccine production: sharing of knowledge and sharing of benefits. Both require transparency that is missing because protocols and details of statistical analyses of the studies are not public. In September 2020, clinical trials of the AstraZeneca vaccine are halted due to safety concerns but no further information is released. This has worried scientists because taxpayers have supported the development of vaccines and transparency is needed to bolster public confidence. Scientists asking for trial protocols to study them are ignored. In the Covid-19 pandemic, academic research institutions are increasingly sharing knowledge. An outstanding example is the Recovery trial determining the effects of steroids on Covid-19 mortality: the protocol and statistics are fully published. Another example is the use to preprint servers, making the data and findings of research publicly accessible and open to review prior to acceptance for publication in scientific journals. Another argument is that manufacturing of vaccines can be boosted by removing barriers to production with dose sharing, technology transfer and voluntary licensing. More radical proposals suggest to waive intellectual property rights and create a global patent pool of vaccine knowledge which can be used by domestic manufacturers with reasonable compensation to the companies. In March 2020, Costa Rica is the first country to propose the creation of a global pooling mechanism. Together with 36 member states it launched the WHO Covid-19 Technology Access Pool as a global platform to share knowledge, intellectual property and data in order to accelerate the development and manufacturing of products that are necessary to fight Covid-19 and to make them equally accessible for all countries [103]. The proposal has not received much support from countries and industries, and no contributions to the pool have been made [104]. A similar proposal to suspend intellectual property rights for Covid-19 vaccines until everybody across the word is protected is later submitted by India and South Africa to the World Trade Organization and backed by the African Union and the WHO. The initiative, supported by more than 100 countries, is blocked by wealthy countries, notably the UK, Germany, and Switzerland with the argument that it is a disincentive to innovation [105].

It is remarkable that the argument of exceptionality, so often advanced for policy measures during the pandemic is not used in this discussion. The intellectual property rights regime is firmly anchored in global governance. Nonetheless, it is continuously subjected to ethical and legal criticism because the inherent conflicts between economic interests and public health. Patent rights particularly infringe on the right to health as basic human right. In 2001, flexibilities have been introduced in order to give priority to public health issues, if necessary. However, very few countries have used these flexibilities. Many countries have resorted to bilateral and regional trade agreements with powerful nations demanding stricter enforcement of patents, closing avenues to protect public health and making manufacturing and import of generic drugs more difficult [106]. Nevertheless, mechanisms in global trade regulations exist to stimulate domestic manufacturing capacity in case of emergencies. An example is the Medicines Patent Pool (MPP) initiated by UNITAID in 2010 for the production of low-cost generic versions of antiretroviral drugs for HIV/AIDS. This initiative has been successful since it improved access to medication; patents on all treatments for HIV recommended by the WHO are now licensed to the MPP. This pool has been expanded in 2015 to include treatments for hepatitis C and tuberculosis, and in 2018 to include essential medicines listed by the WHO [107].

Arguments for technology transfer and sharing patents with the WHO, and thus the global community, are pragmatic and moral. Inadequate production will lead to slow vaccination of populations, in particular in developing countries, and this will provide more opportunities for the virus to mutate with the risk that new viral variants will become less susceptible to existing vaccines. In the current circumstances only a tiny minority of the world population will be vaccinated. As long as the majority is not protected against the virus, everybody is in fact not fully protected. Pooling patents is not the only mechanism to achieve larger production and equitable access to vaccines. What is also needed is access to manufacturing knowledge which is often not included in patents. This requires transfer of know-how across companies and widespread technology transfer, and thus open access to information that is usually protected by trade secret law [108]. That governments can do more than funding the development and production of vaccines is demonstrated in the historical example of mass production of penicillin that saved thousands of soldiers in World War II. The U.S. government invested heavily in large-scale production, and intervened to construct production plants while major pharmaceutical companies were reluctant to cooperate. Technology and expertise were made available to all companies [109].

Moral arguments emphasize that the basic concepts and technologies to make vaccines have been developed in universities supported with government-funded research, and also that massive public funding has diminished the risks of development and production for companies while liability for adverse effects is covered by governments [110]. It is not fair when most of the profits go to some pharmaceutical companies while citizens have to bear most of the financial risks. This raises the question who actually owns the vaccines: are they the private property of companies, so that producing them is their monopoly, or should they be regarded as public property? If the products of international research and public funding are global public goods, market mechanisms do no longer apply [111]. In the Covid-19 pandemic, there are not merely concerns about the urgent need for vaccines but also about equitable and affordable access to vaccines. The much acclaimed right to health in fact demands universal access to Covid-19 vaccines [112]. A pooling mechanism addresses these concerns and demonstrates global solidarity and shared responsibility [113]. The spirit of solidarity is manifested in the rapid sharing of information and research results among scientists, collaborating for example to share the genetic sequence of the virus and to explore the structure of viral proteins. It is not acceptable when such a spirit is not expressed when the results of collective efforts must be distributed. While scientists have learned a different way of working in a pandemic, governments should redefine their approaches too [114]. It is also a matter of justice when some countries not only pre-order vaccines and thus secure preferential access, but even more when they prevent other countries from manufacturing, and thus expanding the global supply [115]. The shortage of manufacturing facilities has furthermore collateral damage; other vaccines, for example against influenza, measles and rubella can no longer be produced in sufficient quantities [116].

8.4.3 Distribution

Once a vaccine is developed and produced, the next step is distributing it. The main ethical challenge at this stage is global justice, in this context framed as 'vaccine equity.' In December 2020, it became clear that wealthy countries (representing 14% of the world population) have secured 53% of the most promising vaccines [117]. In February 2021, almost 130 countries had not received a single dose of vaccines while 75% of the total 210 million doses had been distributed in ten countries [118]. This is opposed by the statement of Antonio Guterres, secretary-general of the United Nations that a Covid-19 vaccine must be regarded as global common good, a "people's vaccine" that should be accessible to everybody [119]. Despite appeals to global solidarity, most countries are concerned with securing vaccines for their own population first as particularly illustrated in the scramble of wealthy countries to pre-order the largest share of vaccines. From the perspective of ethics, this is not only unfair but also potentially harmful. As long as the present inequity exists it will take years before everybody on the globe is protected, creating an 'immunity divide' that allows the virus to continue to circulate across the globe. This will increase the changes that the virus will mutate, and that existing vaccines will no longer be effective against new viral variants. Equal distribution of vaccines is therefore not only a matter of justice but also of self-interest. Otherwise, the desired return to normality and the recovery of global trade and travel will be delayed.

Inequity is associated with two phenomena: vaccine nationalism and vaccine diplomacy. Vaccine nationalism means that countries prioritize their own interests and make sure that vaccines are first of all available to their citizens [120]. It is not a new phenomenon. In the 2009 H_1N_1 pandemic the first vaccine was manufactured in Australia but not immediately exported because the country wanted to vaccinate its own citizens first [121]. The global supply of vaccines was bought up by wealthy countries while the worst affected resource-poor countries had no supplies [122]. The worldwide race to obtain Covid-19 vaccines starts early in 2020 when the US President attempts to buy the German company CureVac, later issuing an Executive Order that vaccines produced in the US may only be exported if Americans themselves are sufficiently vaccinated [123]. Such export bans, not only for vaccines but also raw materials, are issued by many other countries. Lack of solidarity is manifest in Europe when a vaccine alliance of Germany, France, Italy and the Netherlands in June 2020 jointly makes a deal with AstraZeneca to buy its vaccine, at that time non-existent, bypassing the European Commission that has committed to negotiate with pharmaceutical companies on behalf of all member states. In the aftermath of Brexit, a vigorous row develops between the European Union and the United Kingdom when it transpires that 34 million doses of vaccine have been exported from the EU to other countries, including 9 million doses to the UK while the last country has blocked the export of vaccines produced in its territory. The EU then imposes control mechanisms and in some cases blocks the export of vaccines, arguing that there should be 'reciprocity' between countries producing vaccines [124]. Efforts to procure and reserve vaccines for national citizens are opposed to global solidarity, especially when practiced by the most powerful and wealthy nations since they lead to, what is called 'vaccine apartheid.': people in the Global South will be vaccinated years later than those in Western countries [125]. They not only have to wait longer, but also pay higher prices. South Africa and Uganda for example have paid three times more for the AstraZeneca vaccine than the EU [126]. Many developing countries have no production facilities, and distribution infrastructure is insufficient. mRNA vaccines requiring extremely low temperatures are impossible to use in these conditions. While many clinical vaccine trials have been conducted in for example Argentina, Brazil and South Africa, citizens in these countries have not received significant quantities of vaccines. Another example of vaccine apartheid is the policy of Israel, quickly vaccinating its Jewish population but not Palestinian Arabs. A wry manifestation of it is the decision of four African countries to refuse any vaccines at all, some arguing that they have no Covid, have it under control, or have their own indigenous treatment [127]. The ethical argument against this backdrop is that a global approach is unrealistic since it does not recognize that every nation-state has a primary responsibility to first protect its own citizens. Nonetheless, countries have also a responsibility to provide global assistance, and the main challenge is how to balance national and international obligations. If projections of vaccine production are correct, there should be enough vaccines to inoculate the most vulnerable individuals in all countries [128]. It is ironic that many liberal regimes aim at vaccinating their own population first whereas authoritarian regimes export their vaccines long before their populations are inoculated.

Related to vaccine nationalism is vaccine diplomacy. Countries race to be the first to develop and procure vaccines but also to be the first to start vaccinations. The media continuously report on the progress of vaccination campaigns around the world and praise the frontrunners in the race (Israel, the United Kingdom, Chile) while the slow rollout, particularly in Europe, is severely criticized. Countries such as China and Russia deploy their vaccines before clinical trials are completed while scientific information about results is not publicly available. They also use their vaccines as instruments of influence, shipping them to developing countries. The nationalism of Western countries has created room for vaccine diplomacy. These countries let the distribution of vaccines mostly to pharmaceutical companies while the Chinese and Russian governments decided to market the vaccines themselves, setting up manufacturing plants in other countries such as Indonesia, Malaysia and the United Arab Emirates [129]. While its vaccines are not trusted by Western governments, China quickly exports them to many developing countries where they are welcomed with much publicity, even though its own population is very slowly vaccinated [130]. China poses itself as a leader in global health, and repeats the UN language that vaccines are a public good. In the West, the Russian Sputnik vaccine is initially rejected as unreliable but later studies conclude that it is highly effective and safe. Only a small minority of the Russian population is inoculated, but the vaccine is aggressively promoted in foreign countries. Later, ethical questions emerge concerning the voluntariness of research subjects in the clinical trials (notably conscripted soldiers). In the European Union, member states have agreed to a joint vaccination strategy in order to have a strong negotiating position with the pharmaceutical industry, but because of vaccine shortages several countries (for instance, Hungary and Slovakia) undercut this solidarity and purchase Russian and Chinese vaccines. In Slovakia, the government had to resign after a secret deal to acquire the Sputnik vaccine has been revealed. The policies of Russia and China create a lot of envy, and are condemned as 'vaccine propaganda' [131]. They are clearly aimed at expanding geopolitical influence (with unclear and often secret conditions for delivery and distribution of vaccines), at the same time they help to mitigate the pandemic especially in Africa and Latin America.

In order to counter vaccine nationalism, and based on experiences in the H_1N_1 pandemic of 2009, the WHO together with the Global Alliance for Vaccines and Immunizations (Gavi) and the Coalition for Epidemic Preparedness Innovations (CEPI) took the initiative in April 2020 to establish Covax (the Covid-19 vaccines global access facility). The basic aim is to have global equitable access to Covid-19 vaccines. The facility invests in research and development, and will procure vaccines so that 20% of all people can be vaccinated, protecting the most vulnerable populations around the globe. Covax will create a common vaccine pool with different commitments of countries. Developed countries will provide funding and buy vaccines through the facility whereas developing countries will obtain vaccines for lower prices or receive donations. The notions of equity and solidarity underlying this initiative are reiterated in the Vaccine Equity Declaration of January 2021, as the only way to avoid, in the words of the WHO Director-General, "a catastrophic moral failure" [132]. The first doses are delivered to Ghana in mid-February but the success of Covax will depend on the provision of sufficient funding. It is also substantially hampered by the national policies of wealthy countries that not only bought vaccines through the facility but preemptively engaged in bilateral contacts with the pharmaceutical industry and rapidly secured more doses than Covax. Some wealthy countries regarded the facility as a backup stockpile, and bought vaccines to supplement their national stockpile, leaving even less vaccines for resource-poor countries. Against this backdrop, it is not expected that Covax will accomplish its aim of providing two billion doses of vaccines worldwide in order to inoculate the first 20% of all populations by the end of 2021 [133]. At the end of October 2021, it has only distributed 406 million vaccines among 144 countries, while globally 6.8 billion doses have been administered.

The need for equal distribution of vaccines is justified with different arguments. Moral arguments emphasize global justice and solidarity. Rather than vaccinating all people in some countries as advocated in vaccine nationalism, global solidarity among countries will require the vaccination of at least some people in all countries, preferably people at higher risk. Moreover, it requires an upfront commitment to share vaccines as soon as they are available, and not as left-overs after first use by some countries. Pledges to share excess supplies have been made at the G7 meeting (in February 2021) by leaders of countries that have bought many more doses than are needed for their population [134]. The mentality still is: our own people first, and subsequently charity can be practiced. However, it is not clear if and when this will happen since there are basic uncertainties: it is not known how long the immunity from vaccines will be and whether new variants of the virus will emerge so that regular booster injections will be needed to maintain immunity. Many pledges have not resulted in actual donation. For example, the Netherlands promised to donate 27 million doses to developing countries but has only donated 2.15 million vaccines (as of the end of October 2021).

Another type of argument articulates public health rationality. Models predict that if the first 2 billion doses of vaccines are received by 50 high-income countries, global mortality will decrease with 33% (compared to the situation without vaccines). If all countries receive the first doses proportional to their populations, 61%

of all deaths will be averted [135]. The message of this argument is that equal distribution eventually is beneficial to everybody. In fact, there is no conflict between national and common interests. When vaccines are shared among the world population, global interconnections, and thus trade, travel, and tourism, will resume faster than when every country first secures its own population. There is also a collective economic interest since expenses for global vaccination are insignificant compared to the trillions that have already been spend to stimulate economies, and that will be even more necessary after Covid has receded.

Implementing equal distribution demands strategies based on specific criteria. Covax initially suggests proportionality as a criterion. Countries should not vaccinate more than 20% of the population until all countries have vaccinated 20% of their population. The facility should allocate vaccines until this proportion has been reached in all countries. However, not all countries have equal needs and are confronted with similar problems. Some countries are evidently more severely affected by Covid-19 than others, so that vaccines could have a much greater impact on reducing harm there [136]. A different framework proposes to allocate vaccines on the basis of need and vulnerability. Vaccination should proceed in stages with priority for healthcare workers, elderly people, and the most vulnerable individuals (e.g. high risk individuals with comorbidities and people in socio-economic groups that are disproportionately affected). The aims of this strategy are threefold: reduction of death and disease burden, protect the functioning of essential services in society, and protect people who bear significant risks and burdens of Covid-19 to safeguard the welfare of others [137]. Another framework proposed ís the 'fair priority model' that distributes vaccines first to countries where the greatest number of life years can be saved. The aims are to limit the health consequences and the economic damage of the pandemic [138]. Finally, a multi-value ethical framework uses three principles for global allocation: ability to provide care, ability to implement the distribution of vaccines, and reciprocity (e.g., when countries have participated in clinical trials) [139]. The problem with all frameworks is that they in fact prioritize rich countries. Since these countries have more healthcare workers than poor countries, the WHO proposal allocates a larger share of available vaccines to high income countries. The fair priority framework assumes that wealthier countries first control the pandemic (having enough vaccines to keep the rate of transmission, the Rt, below 1) before they release vaccines for other countries. Another ethical problem with this framework is that saving the most life years unfairly discriminates elderly and disabled people. The multi-value framework risks to reinforce the disadvantaged status of poor countries [140].

8.4.4 Application

Administration of vaccines after they have been developed, produced and distributed is complicated. Governments are forced to set up vaccination programs in a very short period. It the first time that the world faces a challenge to vaccinate the entire global population. Practical difficulties relate to the fact that the available vaccines are different, and difficult to compare. mRNA vaccines for example require storage in ultracold temperatures (only later adapted to more standard conservation conditions of −25 to −15 °C). All countries have to build cold-chain infrastructures but this is impossible in many developing countries. They are also too expensive for these countries so that the application of specific vaccines will depend on the local setting. These vaccines are highly effective, so that a choice has to be made for using less effective vaccines in numerous countries [141]. However, all available vaccines have a much higher efficacy than the 50% efficacy threshold set by the WHO. Covid vaccination is a "stress test" for healthcare systems, and almost every country has to resolve deployment challenges [142]. They must mobilize human resources, set up systems for efficient administration of vaccines, identify eligible individuals, establish immunization registries, recall patients for second doses, arrange transportation to rural and remote areas, provide personal protection equipment for vaccinators, and create safe waste management. Only few low- and middle income countries can make use of existing capacity for annual vaccination programs [143]. In high income countries, public health departments that have been underfunded for decades are already overwhelmed with increased demands for testing and contact tracing.

While the public expectation is that as soon as a vaccine is approved, it will be available, in most countries the roll-out is generally slow. Again, there is a competitive element. While the United Kingdom starts vaccination in early December, most other countries struggle. Around Christmas 2020, the first vaccinations are given in a kind of public relations show, to provide confidence to the population but actual mass programs start much later. The Netherlands is the last country in the European Union to start vaccinations. The government is relying on existing structures: general practitioners who used to administer annual influenza vaccines, nursing home doctors usually inoculating residents, and local and regional municipal health services. Trust in the existing health system, and lack of central direction and control prove disappointing. The first vaccines available cannot be used in general practice (because of ultracold storage requirements). In nursing homes delays are expected since all residents need to be informed and give permission, while the municipal health services have not yet created an ICT system for registration, and lack sufficiently trained personnel (who are supposed to start mass vaccination of millions of people on August 2021). To many people, these difficulties reflect the earlier blunders with procuring protective materials and setting up testing capacity. In Germany, logistical problems are less stringent: huge vaccination centers are established across the country, with mobile teams and assistance of the military. In general, programs in Continental Europe are slowly implemented, not only due to late approval but also due to problems with supply of vaccines. In other countries, progress is more rapid after regulatory approval of the Pfizer vaccine, in the beginning of December 2020 [144]. In early January 2021, 12 vaccination doses per 100 inhabitants are administered in Israel, increasing to 114 in March 2021. In the United Arab Emirates, the doses increased from 8 to 78 in the same period, in Chile from 0.05 to 49, in the UK from 2 to 48 and in the US from 1.2 to 40, while in France and Germany the increase is from near zero to 14 [145]. The slow roll-out leads to frustration and 'vaccine stress' in many populations dissatisfied with prolonged lockdowns and curfews. Pessimists estimated that it will take a long time before the majority of populations will be protected [146]. However, an early start of vaccinations does not guarantee a more successful outcome in the longer term. In Israel, the peak of the number of vaccinated people is in January 2021 but by the end of October only 65% of the population is fully vaccinated. The same applies for the United Kingdom and the United States which have a peak in March-May, respectively April but since then the number of vaccinations has declined (with 66% resp. 56% of the population fully vaccinated by the end of October). France and Germany, on the other hand have most vaccinations in June-July and 67%, resp. 65% fully vaccinated at the end of October. At that time Spain and Italy have 80% resp. 71% of the population fully vaccinated [147].

Mass immunization programs face ethical challenges. Initially, when vaccinations start up, available doses are scarce so that decisions must be made about how to distribute this resource within and across populations. Ethical issues of prioritization and policy implementation are on the table in this stage. When programs are established, sufficient quantities of vaccines are manufactured. Countries that have preordered and purchased vaccines are now faced with a different problem: the majority of the populations is vaccinated but a minority is not, so that the goal of herd immunity is not in sight. In this later stage of vaccination campaigns more stringent measures are proposed and introduced to increase the number of vaccinated people. Societies where the majority of citizens are vaccinated are also opening up and many of them require a corona-pass to participate in social activities. Ethical issues at this stage concern vaccine hesitancy, voluntary or mandatory administration, and the use of vaccine certificates. At the global level, the major ethical challenge is vaccine inequity. While developed countries start to provide booster shots of vaccines, in many developing countries most people have not even received a first dose of a vaccine.

8.4.4.1 Prioritization

Prior to approval of vaccines, and assuming that vaccines will be scarce, at least initially, numerous proposals and recommendations are made to decide who will be the first to receive a vaccine [148]. Such decisions are ethical, i.e. they are based on ethical principles and values rather than scientific information, even more so because, at least initially, the basic characteristics of vaccines are unknown, such as efficacy in specific populations and patients, safety and adverse effects, the duration of immunity, and effect on the transmission of the virus. When more scientific evidence emerges, priority schedules need to be dynamic and adapted, using the ethical framework outlined.

All prioritization proposals start with identifying relevant ethical principles. The World Health Organization refers to human well-being, equal respect, equity in access, reciprocity, and legitimacy [149]. The National Academies of Sciences, Engineering, and Medicine (NASEM) in the U.S. use a more simple framework of three ethical principles: maximum benefit, equal concern, and mitigation of health inequities. The Norwegian Institute of Public Health identifies five values important for prioritization: equal respect, welfare, equity, trust, and legitimacy while the Health Council of the Netherlands suggests two principles: utility (maximizing health benefits) and fairness (equal respect) [150]. The goals of vaccination may also differ: reduce or eliminate viral transmission, minimize mortality, reduce morbidity, diminishing the burden of the healthcare system, protect social infrastructure, and limit social and economic disruption. On the basis of ethical values and goals, different priority groups are listed, assuming that the distribution of the vaccine must proceed in subsequent phases, depending on the availability of vaccines.

Nearly all proposals focus on giving priority to people with the highest risk of serious disease and death. Generally, this includes older persons although the age limit is different between countries, for example, all people older than 60 years in the Netherlands, over 65 year in Belgium, people older than 70 years in Sweden, while in several countries it is proposed to start with the oldest people first and then progress to vaccinating younger categories, as in the UK. First priority is also given to people with underlying conditions who have higher risks (when sometimes a specific list of such conditions is provided). Both preferences (age and medical risk) are justified with the emphasis on the moral value of human life and well-being. These priorities will achieve the goal of reducing death and serious illness. Indirectly, they will contribute to the other goal of protecting the healthcare system since the burden of care will diminish. A practical consideration is that the ongoing clinical studies of vaccines at that time estimate the impact on morbidity and mortality, not transmission of the virus. More controversial is the inclusion in this first phase of special groups that are at increased medical risks. Some groups are disproportionately affected by Covid-19 because they are disadvantaged by race, ethnicity, occupation and socio-economic circumstances. This point is specifically articulated by NASEM out of contemporary concerns with racial inequality and discrimination. The framework proposes that allocation criteria take into account the vulnerability of these groups. Several instruments are available (for instance the Social Vulnerability Index) to determine who are the 'worst off' or most disadvantaged. Rather than explicitly prioritizing disadvantaged groups (e.g., allocation on the basis of race or ethnicity), the proposal is that for each population group in the various prioritization phases, high vulnerable areas are identified and vaccines are first given to people in these areas [151]. The Norwegian document, recognizing the disproportionate risks of disadvantaged groups, does not propose them as a separate priority category. However, it recommends geographical prioritization as the second priority category, focusing on areas with high infection rates. This focus is not explicitly related to reducing health disparities and it is not clear whether the worst-off populations will be concerned [152]. Another special group are residents of long-term care facilities such as nursing, care and disability homes. They are given top priority in many proposals, often together with vaccination of staff working with them [153]. The same questions arise here as with triage of treatment. Other groups also live in crowded settings such as homeless shelters, prisons, and migrant facilities. Although they generally are not given high priority, NASEM includes them in the category of "older adults living in congregate or overcrowded settings" in the high priority Phase 1b, including nursing homes in the same list as prisons and homeless shelters [154].

Finally, healthcare workers are generally included in the first phase of prioritization. Three arguments are used to justify their inclusion: the need to maintain the functioning of healthcare systems, their exposure to high risks, and their risk of transmitting the virus to others. This raises questions similar to those in triage of treatments (see Chap. 7). Who exactly will qualify for vaccination: healthcare workers providing care to Covid patients, other healthcare workers (in transportation and administrative services or in nursing homes), people in occupations of high social value such as teachers, school staff and childcare workers, workers that have frequent physical contacts with clients, and people in critical occupations with increased risks of transmission? In its allocation framework, NASEM gives the highest priority (Phase 1a) to high risk health workers (not only those in direct patient care) and first responders (emergency services, police, and firefighters).

The Norwegian proposal assign a lower priority (category 3) to occupational groups, notably healthcare personnel and people in critical societal functions but they should move to a higher category when the risk of viral transmission and thus the infection burden is high [155]. The Health Council of the Netherlands advises to vaccinate care workers if they are in direct contact with people in medical risk groups who cannot be vaccinated for medical reasons, providing indirect protection [156]. It points out that healthcare workers as a group may have an increased risk of infection but not necessarily an increased risk of severe morbidity or mortality compared to vulnerable patients. These workers also have more opportunities than others at this moment in time to reduce potential risks with the use of personal protective equipment and preventive measures in healthcare facilities. At least, provisional data in the Netherlands indicate that the number of positive tests in healthcare workers is not higher than in people working in education and informal care, and lower than in law enforcement and contact professions [157].

Ethical Strategies

Prioritization is an ethical exercise. It requires asserting and weighing values that determine the kinds of good that should be accomplished. Compared to treatments, vaccines have a greater societal value. They are preventive rather then curative, with a wider set of goals. Vaccination has a direct effect in that it protects the individual from death and serious illness when infected because the immune system is activated. But vaccines do not only benefit individuals, they also protect others and they contribute to the common good when the majority of a population is vaccinated. The indirect effect of individual inoculation is that the social burden of disease is reduced, the pressure on the healthcare system is diminished, and gradual reopening of society will be possible. To reach these social goals it is imperative that as many individuals as possible agree to be vaccinated. Individual and social interests are therefore closely aligned. However, in the current state of affairs, it is not known whether vaccines are equally effective and safe for all groups of people, and whether they prevent transmission of the virus. Empirical knowledge will be obtained when millions of people are vaccinated. Efficacy and safety are now well established, although the emergence of new mutations of the virus may create potential difficulties. Controversies exists concerning viral transmissibility. Potentially, three vaccination strategies are possible: reducing morbidity and mortality, reducing transmission of the virus, and preventing social disruption [158]. The direct effects of one strategy may also have indirect effects for the other strategies. When vaccination programs start, it is evident that the approved vaccines are efficacious in reducing severity of illness and the risk of death. This made the choice of strategies easier. Many countries accept the first strategy, aiming at saving as many lives as possible, and diminishing the burden of serious illness, thereby preferring individual protection of the most vulnerable people. In some other countries proposals combine aims, for example in the United States the aim of reducing death and illness is combined with the aim of preserving the continued functioning of society, reflected in the priority of key workers, and rejection of age as a general criterion. In some developing countries, the choice between vaccinating those with the greatest risk of dying and those with the greatest risk of becoming infected and transmitting the virus is different. For example, in Indonesia and Kenya younger people are vaccinated before the elderly. Following health care workers and support staff, members of the public workforce (such as public servants, police, and teachers) will have priority. Rather than protecting individuals, this strategy first aims at group protection (reducing viral transmission in the population) and recovery of social and economic life. It is justified because in these countries the overall mortality in hospitals is lower than in high-income countries (since the average age of the population is lower, and comorbidities are less frequent [159].

Vulnerability

That prioritization cannot simply be based on medical criteria but requires ethical assessment of individual and social considerations is reflected in the common use of the notion of vulnerability. Almost all proposals agree that vulnerable individuals and groups should have priority but they differ in what vulnerability means. It can be related to age, health, occupation, or social conditions. If being vulnerable means increased risk of severe morbidity and mortality, priority is given to the elderly and people with underlying health conditions. However, age is not an uncontroversial criterion. Vulnerability does not necessarily increase with age, and not all elderly people are equally vulnerable. For example, older men are more vulnerable to Covid-19 than older women. Healthy older persons who can shelter at home are less vulnerable than those in crowded housing [160]. Another example are minority populations; they are affected by the virus at an earlier age than the majority population. While numerous proposals and guidelines use age as a simple cut-off point for vaccine allocation, it is rejected by some as an unreliable predictor of risk. The argument is that age makes people vulnerable if it is combined with multiple comorbidities [161]. NASEM recommends therefore to give priority to "people of all ages with comorbid and underlying conditions that put them at significantly higher risk" [162]. This argument is countered with empirical studies showing that given priority to individuals older than 65 saves the most lives. Age is the strongest predictor of vulnerability, more than underlying conditions [163]. Scientists also argue that the distinction between saving the most lives and saving the most years of life that is made in discussions about prioritization of treatment are not relevant in prioritization of Covid-19 vaccination: immunizing the oldest individuals saves not only the maximum number of lives but also most future lives because the risk of death is exponentially increasing with age [164]. Similar worries are expressed about underlying conditions. It is evident that health is endangered by such conditions and that they are associated with higher vulnerability to Covid-19. But lists of these conditions are too static. Rare diseases and many disabilities are often not included while they may lead to increased risks for developing serious illness from Covid-19 [165].

If vulnerability is related to occupational risk, healthcare workers are prioritized, and possibly people with other professions and occupations. Being vulnerable here means more exposure to possible infection as well as heavy workload with physical and emotional stress. Less than in the previous category, vulnerability is related to the risk of illness and death but more to the burden of executing the work and the risk that vital social services can no longer be provided.

Finally, vulnerability can depend on socio-economic circumstances, for example living in crowded conditions so that residents of care homes but also prisoners should be the first to receive vaccines. But there are also disadvantaged populations that as a result of structural inequalities, racism, and discrimination are unfairly treated and confronted with poverty, unemployment, health problems, and lack of access to healthcare. It is clear that these populations, especially racial and ethnic minorities, are particularly vulnerable to Covid-19 and show disproportionate morbidity and mortality. Prioritization should take into account that the coronavirus is not impacting equally on everybody but harms some populations significantly more than other, so that health inequity should be mitigated. Like the pandemic itself, structural vulnerability is manifested in waves, first in nursing homes and long-term care facilities, then minority groups, and finally correctional facilities [166].

Policy Deviations

While priority schedules are recommended by scientific and ethical councils before vaccines become available for use, many deviations can be observed when the vaccines are actually administered. Discrepancies are noticeable at the policy and practical levels. Proposals of advisory bodies are not necessarily followed by governmental authorities and translated in clear policies. The CDC guidelines for example differ from the NASEM recommendations. They locate healthcare personnel and long-term care facility residents in priority phase 1a, moving frontline essential workers to phase 1b. More importantly, it includes in this phase all people aged 75 years and older, whether or not they have underlying conditions [167]. This substantially broadens the eligible population, from 19 to 20 million individuals with multiple comorbid conditions as estimated by NASEM to 49 million people of 75 years and older [168]. States also adapt their own policies and deviate from federal guidelines. In Texas and Florida first priority is given to all adults 65 years and older as well as persons with underlying conditions who will thus receive vaccines sooner than in the federal recommendations while front-line essential workers have to wait somewhat longer. In the Netherlands, policies are fluctuating and selection criteria changing. The Health Council recommends to start with vaccinating people with the highest medical risk. This includes nursing home residents but practical problems emerge since the first available vaccine (Pfizer vaccine) needs to be stored in ultracold conditions, is delivered in large packages, and must be used within a few days after de-freezing which makes it difficult to deploy in smaller settings. The Ministry then decides to first vaccinate people working in care and nursing homes to provide indirect protection to residents since they can go to vaccination centers with refrigerator equipment. As soon as this decision is taken, health professionals start protesting and put pressure on politicians with the argument that it is important to prevent collapse of the care system. Subsequently, the Ministry concedes and changes the priority order, allowing vaccination of healthcare workers prior to the most vulnerable citizens. After that decision, general practitioners protested, and they are then also included in the priority group.

With the rise of the third wave of the pandemic in Spring 2021, and the fear that new mutations of the virus will multiply and become more dangerous given the slow pace of vaccination programs, some countries (e.g. UK and Denmark) decide to delay the second dose of vaccines. Available vaccines at that time require two shots to obtain optimal immunity (the Pfizer vaccine needs a booster after 3 weeks, the Moderna vaccine after 4 weeks). Delaying the second shot (up to 3 months) will give the opportunity to vaccinate more people, providing at least some protection and mitigating the transmission of the virus. Instead of keeping a large stock for the second dose, available vaccines should be immediately used, in the hope of course that the supply problems will be solved soon [169]. However, the approval of the vaccines is based on evidence from clinical trials with two doses after a specific interval, and the effects of delay are unknown. Deviations from this evidence is questionable. The decision to change the application regime of vaccines is ultimately an ethical one: preferring maximum outreach, i.e. protecting more people with the vaccine, even if the level of protection is lower, or maximum protection, i.e. providing the highest level of protection to a smaller population. Given the high efficacy of the two dose regime, clinical trials comparing the efficacy of a single dose with that of two doses is no longer ethical [170]. Only observational studies can indicate the efficacy after one dose, and they indeed later confirm that the risk of disease, hospitalization and death is diminished with 60–70% in the age group of 70+ within 4 weeks after the first dose [171]. Studies in the UK show that delaying the second dose of the Pfizer vaccine with 12 weeks even produces a much stronger antibody response [172]. However, it is still uncertain how long the immune response will last after receiving a vaccine. It is also pointed out that immune response is not the same as proven protection against Covid-19 [173]. The emergence of the Delta variant of the coronavirus has complicated matters further since one vaccine dose is less effective against this variant than two doses, so that the UK has to reverse its policy to much quicker administration of the second dose [174].

Practice Deviations

During the rollout of vaccines many irregularities occurred compared to the idealized prioritization schedules. In several countries, politicians are the first to be vaccinated. In the United States government leaders receive the vaccine before other citizens, although prioritization frameworks do not grant them special status [175]. In Peru, hundreds of prominent citizens secretly secure the vaccine month before it becomes available for the rest of the population. Vaccine producer Moderna announces that it will give its workforce priority to receive its vaccine. These examples are not exceptional, and reflected at all levels of society. In the city of Halle (Germany) the mayor gives priority to himself and his collaborators. In Florida, two younger women disguise themselves as seniors and receive the first dose of the vaccine. In the Netherlands, several nursing homes distribute the vaccines to all employees whether or not they are involved in direct care for residents. In August 2021, Germany was shocked to learn that a nurse has injected thousands of people with saline rather than vaccine [176].

The scarcity of vaccines and the enormous demand for fast inoculation generate phenomena such as vaccine tourism (travel agencies offering vaccine holidays in Dubai), vaccine applications (people presenting themselves to their general practitioner as extremely vulnerable and in urgent need of vaccination), and vaccine line crashing or interloping (people who do not meet the criteria travelling to areas where vaccines are available, for example poor and indigenous regions where more vaccines are provided to mitigate inequity) [177]. These examples of jumping the queue illustrate at the national level the earlier discussed experiences at the global level of countries securing priority access to vaccines. In populations eager to receive vaccines they elicit outrage and angry responses because usually healthy and young people are inoculated at the expense of vulnerable seniors.

Dissatisfaction with the slow rollout produces proposals for faster immunization of larger portions of the population. One is to focus on herd immunity rather than immunizing priority groups. This will reduce everyone's risk of infection. Prioritizing community benefit rather than direct individual benefit of course has the risk that many vulnerable people will die or get seriously ill. This can be prevented when vaccination is massive and fast which is not the case in most countries [178]. It is argued that the current approach is complicated since it requires administrative efforts, identification of those that meet the criteria, and scheduling of appointments. The experiences in Israel show that simpler priority criteria lead to faster vaccination, with only broad allocation groups (such as all people 60 and older) rather than an order of priority within and between groups [179]. However, such simplification of the logistics is ethically defensible when there is no scarcity of vaccines. It is only possible in a limited number of countries, as illustrated in the decision of the US administration to lift all vaccine eligibility requirements by May 1, and make 90% of adults eligible by April 19, 2021 [180].

8.4.4.2 Expanding the Success of Programs

When most Western countries in the second part of 2021 manage to vaccinate the majority of their populations, the tempo of vaccinations slows. Instead of scarcity of vaccines, the unvaccinated become one of the major concerns. In many countries more stringent policies are introduced such as mandatory vaccination and use of vaccine certificates.

Vaccine Hesitancy

The success of vaccination campaigns depends on the willingness of people to be vaccinated [181]. In 2020, before vaccines are available, a significant part of populations does not wish to be vaccinated or does not want to commit yet: in the United States 68%, in France 76%, and in the Netherlands 72%. During 2020, willingness to vaccinate has declined globally, based on concerns about the speed of development, the fact that mRNA vaccines have not been used before, and conspiracy beliefs [182]. Now that vaccines have become available and experiences are shared, vaccine acceptance is growing in many countries (in January 2021 rising to 51% in the US, 42% in France, and 83% in the Netherlands. However, in Asia acceptance rates are decreasing [183]. The willingness to be vaccinated is related to public trust in the government and medical authorities, as well as confidence in the reliability of information on vaccines [184]. An important contributor to skepticism, hesitancy or refusal of vaccines is the earlier mentioned 'infodemic' and the spread of misinformation inspired by conspiracy beliefs. The number of individuals determined not to be vaccinated is relatively small; most people have an attitude of wait and see. When vaccines apparently are very effective and safe, and more information is known about longer-term effects, enthusiasm for vaccination is growing. The lack of trust that makes people reluctant can therefore be overcome with clear and transparent information [185]. At the same time, a more aggressive campaign will be needed to counter anti-vaccine disinformation and conspiracy beliefs [186].

But there are other factors that must be taken into account. In some countries a history of deception and misinformation makes at least some segments of the population distrustful of authorities. In France, health authorities distributed blood products in 1991, knowing that they were contaminated with HIV. In 2009, the government ordered massive doses of vaccine against H_1N_1 influenza which were unnecessary, raising suspicions about the involvement of the pharmaceutical industry. In 2019, a court case was opened against the company Servier since it has promoted Mediator, a medicine for diabetes and weight loss that caused the death of several thousands patients. Because of its adverse effects, the drug was withdrawn in other European countries but in France (the main market) only in 2009. The company is accused of prioritizing profits over the safety of patients [187]. Similar experiences exist in the United States where a history of racism has made Black, Hispanic and Indigenous communities distrustful of medical, scientific and political authorities [188]. These are also the populations most seriously affected by the pandemic.

Several factors during the rollout of vaccinations have not contributed to increasing trust. One is shifting policies as the result of supply problems but also of political lobbying. The effect is that many people are uncertain when they will receive a vaccine. Another is the uncertainty about effectiveness and safety. Initially, it is not clear which type of vaccine will be appropriate for which categories of recipients. A significant factor are the continuing troubles with the AstraZeneca vaccine and the deficient communication of the producing company. In September 2020, the clinical trial with this vaccine is paused because of possible safety risks. While it is later restarted in Brazil, South Africa and the UK, information is kept secret and the trial protocol not shared with other researchers [189]. After approval, there are continuous problems with supply in Europe since lower quantities are delivered than promised, leading to a clash between the company and the European Commission and necessitating European countries to adapt their vaccination strategies. In the early stages it is not clear whether the vaccine is effective in people older than 65 (since this age group is not sufficiently included in the first clinical trials) with the result that several countries only authorize the vaccine for people below a certain age limit (55 years in Belgium and Italy, 60 years in Poland, and 65 in Germany and France). In February 2021, South Africa puts the rollout of this vaccine on hold because it does not protect against the coronavirus mutation in this country [190]. In March 2021, several countries suspend the administration of the vaccine for safety reasons since it is associated with rare blood clots, potentially occurring in one of 100,000 vaccinated people, especially in women younger than 55 [191]. Countries resuming vaccination, now only indicate the vaccine for people older than 60 years. At the same time the company publishes the results of a large trial showing that the vaccine is 79% effective in preventing symptomatic illness, and 100% against critical disease and hospitalization, arguing that there is no increased risk of thrombosis. The US Data and Safety Monitoring Board within hours replies that outdated and incomplete information has been used [192]. It is clear that these troubles diminish trust in the company and its vaccine. In several countries, health care professionals and patients are requesting not to be vaccinated with the AstraZeneca vaccine but with other vaccines. People do not show up for scheduled vaccination, so that in some countries stocks of unused vaccines are growing [193].

In August 2021, the FDA in the US approved the Pfizer vaccine, so far available under Emergency Use Authorization, for the prevention of Covid-19 in individuals older than 16 years. That means that the vaccine meets the high standards for safety, effectiveness, and manufacturing quality, required for all medication. The hope is that this approval will increase confidence, so that especially people with a wait-and-see attitude will be convinced to take the vaccine. Another option to increase the number of vaccinated persons is to expand the age categories for vaccination.

Many countries now offer Covid vaccines to adolescents from 12 to 17 years of age. This use remains under Emergency Use Authorization.

Voluntary or Mandatory Administration

The aim of community protection will not be reached if a substantial number of citizens is not vaccinated [194]. Against this backdrop the question is raised whether Covid-19 vaccination should be mandatory. The European Court of Human Rights accepted in April 2021 that compulsory vaccination of children can be necessary in a democratic society in order to protect individual and public health. The argument is that governments are responsible for preventive measures to protect society against lethal infections. This responsibility sometimes requests restricting individual freedom in order to protect individuals from serious and deadly harm with a relatively simple intervention such as vaccination. It also requires intervention when others may be harmed by individual choices not to be vaccinated. In the case of Covid-19 vaccines, this argument becomes stronger when more knowledge is obtained regarding protection in the longer term, as well as serious side effects. Most vaccines, however, have received Emergency Use Authorization rather than full approval precisely because less safety and efficacy data are provided [195].

While early on some countries (for example, Brazil) have mandated inoculation, in most countries vaccination started to proceed on a voluntary bases. However, the slowing of the vaccination rate as well as the emergence of the Delta variant of the virus have promoted the idea of mandated vaccination in several countries. Hospitalizations and deaths are increasing, particularly among non-vaccinated groups. Italy made vaccination mandatory for all care personnel. France did the same in July. In September 2021, the US President Biden made vaccination required for all federal workers and for private companies with more than 100 employees (or otherwise tested weekly). Large companies such as Facebook and Google require vaccination of employees before they can work in their offices.

In many other countries, rejecting direct mandatory immunization, policies are implemented to put indirect pressure on people to get inoculated. Incentives (for example payment) can make vaccination more attractive, while disincentives discourage people from refusing to be vaccinated [196]. The corona-pass (QR code) is nowadays used in many countries for travel and entering public spaces. If a person decides to be vaccinated, he or she will be allowed certain privileges such as visiting a restaurant or festival, or the freedom not to wear a mask or practice physical distancing. Vaccination will enable the government to relax restrictive measures, and such relaxations will first be granted to specific groups that are vaccinated, for example in nursing homes. The underlying idea is that if people are free to make their own decisions, they should also accept the consequences. This view is particularly expressed by liberal political parties, arguing that precedents already exists, for example in the Netherlands where a bill proposes that childcare institutions have the right to refuse non-vaccinated children. From a utilitarian perspective, the aim of maximizing the good of all citizens and preventing harm to others is better reached through payment incentives than mandatory measures [197]. From a communitarian point of view on the other hand, it may be argued that different treatment of vaccinated and non-vaccinated citizens reinforces social inequity.

These proposals to incentivize vaccination are criticized because they do not address the reasons for avoiding vaccination. Monetary payments will not help since aversion to vaccines is not motivated because they are inconvenient. Non-monetary payments such as exemption from physical distancing and masking can be counterproductive since social norms and solidarity advocated by public health policies may be undermined when not everybody has to follow them. The main problem with reluctance and hesitancy to vaccinate is lack of trust which requires effective and honest communication and persuasion [198].

Special problems arise regarding vaccination of healthcare workers [199]. If a mandate for the general population is rejected, can it be imposed on specific sub-populations, notable professionals who work with patients? In Italy, Greece and France, the governments decreed that if healthcare workers refuse to be vaccinated, they are transferred to another function, or have to stay home without salary. A significant argument here is that vaccination prevents harm to others (patients and colleagues). Also, hospitals and care institutions have a moral duty to provide a safe environment for patients and workers but at the same time they are not allowed to discriminate. Mandating vaccination will create problems of its own. When a substantial number of care workers is not vaccinated, a mandate and its consequences will reduce the number of professionals and thus impact the capacity of healthcare systems [200]. It will also make at least some people more distrustful towards vaccines and government policies, and thus corroborate conspiracy thinking. The issue of trust is important since many healthcare systems have failed to sufficiently protect healthcare workers and patients against disease transmission. Mandating vaccination is only an option of last resort when less coercive and intrusive measures have been fully implemented and a safe working environment is provided [201].

Opposition to mandatory vaccination is based on four arguments. First is the importance of individual autonomy. Many vaccination programs are voluntary, and successful because they appeal to personal responsibility. Studies indicate that less people are accepting to be vaccinated if it is mandated by governments or employers [202]. Second, respect for autonomy implies respect for convictions even when they are not shared. Some people do not assign the highest value to health, for example because of religious beliefs, and governments have to recognize such different value systems. The third argument is that mandating vaccination is infringing the usual relationship between physician and patient. If general practitioners for example are asked to vaccinate their patients, they can only do so on a voluntary basis. Vaccination violates bodily integrity and cannot be done without explicit consent of the patient. Fourth, it is argued that a vaccine mandate is disproportional, and it will be difficult to execute and to enforce. Especially when it is combined with penalties (such as fines or job dismissal) it will reinforce the idea that citizens are manipulated. It will also undermine the moral justification for vaccination, emphasizing that it is an altruistic choice.

Whether coercion will be effective in reducing Covid cases and illness, needs to be seen. Indonesia makes inoculations mandatory for all adults in February, with fines up to 5 million rupiah (300 euros) for refusal. Vaccinations start in January and increased until mid-March, declining since then due to shortage of vaccines, and rising again in June. By the end of October 2021, 41% of the population is fully or partially vaccinated. In India, where vaccination is not mandated, at the same point in time 52% is fully or partially vaccinated [203].

The effects of vaccines on transmissibility of the virus are not clear at the moment. The contagious Delta variant makes the distinction between the vaccinated and unvaccinated permeable. Vaccinated people are protected against serious illness but it seems that they can still be infected and transmit the virus to others. The latest studies apparently indicate that vaccination is not sufficient to prevent transmission in settings where people live together and have prolonged exposure [204]. Inoculated or not, it remains important for everybody to continue to implement the public health measures. The promise that vaccines will bring relief from these measures and a return to 'normal' life, turns out to be illusory, and frustration is projected on unvaccinated people who are blamed for the continuation of restrictions. It is true that most hospitalized patients nowadays are not vaccinated but the causes are more complicated than simple vaccine refusal. In September 2021, nine out of ten corona patients in Dutch hospitals are not vaccinated but half of them have a non-Western migration background. They usually do not trust the government and public authorities, and are often disadvantaged economically and socially, with difficulties in communication and health literacy [205]. In such circumstances, vaccine mandates will not be helpful. From the perspective of ethics, it is preferable to strongly engage people about the motives and consequences of decisions not to be vaccinated.

Vaccine Certificates

The question what to do with the unvaccinated and the desire to safely reopen economies and societies has reactivated the earlier proposal of immunity passports (see Chap. 5) in relation to vaccination [206]. The basic idea is the same: when a person is vaccinated, he or she will receive a document or smartphone badge confirming inoculation, allowing activities that otherwise will be restricted, such as travel, family reunions, going to church and concerts, and restarting economic undertaking, so that 'normal life' can return to a certain degree. Vaccination passports are imaginable when a large proportion of the population is vaccinated. Israel is the first country to issue such passports, many other countries have implemented the idea, and several business sectors (e.g., airlines) have started to require proof of vaccination for access to their services. In March 2021, after intense lobbying of tourist destinations such as Spain, Portugal and Greece, the European Commission has proposed a Digital Green Certificate to allow citizens to travel and go on holiday in the summer, although it is anxious not to call it a vaccine passport [207].

The idea of certificates is supported by many populations since it promises liberation from the public health restrictions that have been imposed since the beginning of 2020 [208]. Previously proposed immunity passports are based on evidence of past infection, and they are not recommended because of doubts concerning the accuracy and reliability of antibody tests and the duration of immunity. Vaccination passport are different; immunity is no longer the result of infection, and test are no longer needed. These passports provide an incentive to get vaccination, while immunity passport may have a perverse incentive to seek out infection [209]. The ethical questions, however, are similar. Making a difference between people on medical grounds adds another dimension to already prevailing inequities between human beings. The chances to get qualified for a certificate are unfairly distributed [210]. High income countries are well provided with vaccines, while certain groups within countries (especially the young in developed countries) will have to wait much longer before they receive a certificate. Granting special rights to privileged groups will further exacerbate existing inequities. Vaccination certificates will for example allow tourists to travel to developing countries where the majority of the population is not vaccinated. As long as vaccines are not widely distributed and administered, certificates only further increase existing global inequities [211]. Besides inequity, the potential for discrimination may be another ethical problem [212]. Because the benefits of certificates are not equally distributed, there will be a class of people who will not have access to services, and will continue to endure public health measures, at least as long as sufficient supplies of vaccines are not available. This concerns people who cannot be vaccinated for medical reasons, or decide not to be vaccinated. Certificates may thus identify two different types of citizens. The objectives of certificates may differ from exempting individuals from restrictive measures (for example, the need to quarantine after travel) to allowing access to facilities such as restaurants and cinemas. The last objective is more problematic since it controls which individuals can participate in social life. In public health, that option should be used that is the least impairing individual liberties [213]. To avoid discrimination, the uses of certificates should be clearly defined. Another ethical concern is that certificates introduce a disguised and indirect vaccine mandate. The assumption is that vaccination is a matter of choice (in distinction to immunity passports since people ordinarily do not choose to be immune). But this freedom of choice is compromised when a vaccinated status has important consequences for social life. It will depend of course on what kind of consequences and how strict they will be implemented. Health certificates are already used, and the often mentioned example is the certificate for yellow fever vaccination requested for entry into several countries but they only restrict some travel possibilities [214]. A related ethical concern is that the introduction of vaccination certificates may set a precedent for future pandemics. Especially in digitalized form, they can be used to monitor the health status and movements of individuals. The example is China where citizens are required to download an App indicating health status and travel history, that is scanned by authorities (with personal data sent to police). Such use reinforces the biosecurity state which is controlling the virus through monitoring and scrutinizing what human beings do [215].

Finally, there are practical challenges. It is currently not certain whether vaccines prevent transmission, and how long protection will last so that a false sense of security is present [216]. This is amplified by the fact that the privileges of the certificate for many people are so important that a black market has occurred. Like many paper yellow cards in the past were falsified, fake QR codes can easily be obtained nowadays [217]. Policy-makers rely on the digital technology but the effectiveness of the corona-pass depends on whether and how it will be used by human beings. Will it be consistently checked when people want to enter bars or restaurants? Another problem is that global standards and cooperation are absent, so that a very diverse corona-pass landscape exists with multiple complications for cross-border activities. Available vaccines are very different in various countries, so it is important to determine what vaccines will qualify for a certificate. There is furthermore a serious risk that virus mutations will occur, so that certified travelers will be exposed to virus strains for which their vaccine is less responsive. This underlines the need for a global approach. As long as the majority of the world population is not vaccinated, it will be necessary to uphold public health measures such as masking and physical distancing [218]. Vaccine certificates will provide only temporary relief and will be unnecessary when everybody is inoculated quickly. It is therefore important to have a sunset clause; when certificates are introduced it should be clear when they will be terminated. Denmark was one of the first countries to join the EU Digital COVID Certificate program in June, after it has launched its own pass in March. Early September 2021 it abolished the use of the certificate within the country, with 76% of the population fully vaccinated.

8.4.4.3 Global Vaccine Inequity

At the end of October 2021, 38% of the world population is fully vaccinated, while 49% has received at least one dose. A total of 6,92 billion doses has been administered; in Europe 844 million, in North America 666 but in Africa 184 million. While for example in Portugal 87% of the population has been fully vaccinated, in Nigeria only 1.3% [219]. These figures demonstrate vaccine inequity. As discussed earlier in this Chapter, this ethical problem has existed from the start, even before vaccines have been produced, it has only further grown during the pandemic. Despite the existence of Covax, donations of vaccines have been low. The international community has not moved to change the global regime of intellectual property right to enhance the production of vaccines [220]. Furthermore, the use of vaccine certificates in many countries, primarily facilitates the movement of people in wealthier countries. In the course of 2021, the rapid dissemination of the Delta variant, and studies documenting the steady decline of antibody levels among the vaccinated, indicate a growing risk of breakthrough infections. These findings add a new dimension to the inequity problem: the possible need for booster programs to increase protection against the virus [221]. The FDA authorizes the use of booster shots for several vaccines in the autumn. Israel and the United Kingdom are the first countries to provide a third vaccine dose; many others are following.

When the majority of people in low- and middle-income countries is not vaccinated at all, it is morally questionable when citizens in high-income countries who are already protected, receive a third dose [222]. The global vaccine supply is still limited, and producing vaccines takes time. Although the effectiveness of vaccines diminishes over time, but they are still highly effective in preventing and reducing serious and fatal illness while the burden of the disease nowadays is mostly on unvaccinated people, within countries but even more so across the globe. The expected benefits of booster vaccinations are small compared to the benefits of vaccinating people who did not receive a vaccine so far. Booster campaigns will not change the fact that in high-income countries the majority of hospitalizations and deaths are among unvaccinated individuals while they seriously impact on the prevention of Covid morbidity and mortality in other countries that have not shared in the distribution of vaccines. Booster shots are first of all promoted by pharmaceutical companies interested in selling more of their products to wealthier countries. From the perspective of global bioethics, greater good can be accomplished by making vaccines available to middle- and low-income countries, rather than using them as boosters. An additional argument is that potential harm for all countries will be diminished when the vaccines are more equitably distributed. As long as the pandemic is not controlled globally, risks are increasing that new and dangerous virus variants will emerge that may eventually evade the immunity provided by vaccination everywhere.

8.5 Conclusion

Humankind cannot argue that it has not been warned for the Covid pandemic. Experiences with four recent disease outbreaks (SARS 2002–2003; avian flu 2003–2004; swine flu 2009–2010, and Ebola 2013–2016) have made clear that preparedness and prevention of emerging infectious diseases are insufficient, not only at the global but also national levels. Lack of a culture of anticipation is demonstrated by the fact that specific recommendations have been made each time to improve capacities to detect, report and respond in an early stage to threatening infections but they have not been taken seriously. Many of the horrible dilemmas in the current pandemic could have been avoided if these recommendations were taken into account. This chapter argues that it is an ethical imperative to intensify efforts of primary prevention even if it is too late now for the Covid-19 pandemic. When the pandemic is over, return to 'normal' life should not mean that usual practices are resumed but it should show that the lessons have been learned for disease prevention and public health. One lesson is that health is not merely a national but a global concern. Because the world is interconnected, threats to human health arising anywhere are threatening for everyone. Another lesson is that health is not only the result of medical care but dependent on social and environmental conditions so that human health cannot be separated from animal and ecological health. The third lesson is that health is not equally distributed; some individuals, groups and populations are more vulnerable to disease than others. The Covid pandemic illustrates that these lessons are easy to ignore. Each country focuses on its own citizens, implements its specific policies, and tries to procure as much vaccines as possible, overestimating domestic capacities to cope with a global threat. The denial of global interdependencies forgets that viruses do not know borders or nationalities, and that they can only be overcome with cooperation.

A further lesson relates to science and technology. Scientists across the globe are intensively cooperating to analyze and innovate diagnostic tests, and to make these more accurate and easier to use. Numerous clinical trials are ongoing to assess potential medication for Covid-19. Researchers study the social, psychological and economic effects of different types of public health policies, and epidemiologists produce modelling studies to predict the effect of various policies. Finally, new vaccines have been successfully developed in a very short period of time. The scientific contributions in this pandemic have been greater and more reliable than ever before in previous pandemics. Nevertheless, how good and fast they are, they are not sufficient to address the pandemic. This is not because scientific knowledge is inherently uncertain, and even more so when a unknown virus affects the planet, but because of the inevitable role of human beings. They are victims as well as vectors of the infectious disease. This role is discussed in this chapter at two levels of response to Covid-19. First, public health measures to prevent and mitigate the spread of the virus will only be effective with compliance of the majority of the population. Second, vaccines will produce protection of the entire population if most people are willing to be inoculated. These conclusions will apply at national levels, but even more at the global level. As soon as policies are different across the world, with lockdowns in some countries, and re-opened societies and economies in others, the virus will continue to disseminate, and, more worrying, will have a chance to mutate. At the global level, confirmed cases of Covid-19 continuously grow during 2020, peaking in the beginning of January 2021. In 2021, the pandemic is accelerating globally, with more infected and deceased people than in the entire previous year. So far two waves occurred in 2021 (with peaks in April and August) bringing the total number of confirmed cases at 244 million, with almost 5 million deaths (as of October 27) [223]. Vaccination is progressing in some countries, but in most it is slow. It is estimated that in advanced economies, the majority of the adult population will be vaccinated by mid-2022, in middle-income countries in late 2022 or early 2023, but in the poorest countries only in 2024, if this goal will be accomplished at all [224].

Although vaccines are generally regarded as the ultimate solution for the Covid-19 crisis, the immediate future is riddled with uncertainties concerning how vaccines will work in real-life conditions, how safe they are, whether they are effective against mutations, how long vaccine induced immunity will last, and whether there is a need for booster shots. The expectation is that with the early massive rollout in several countries (especially Israel, Chile, UK and USA) studies will clarify these uncertainties. Now that millions of people are vaccinated, evidence confirms that vaccines (Pfizer, Moderna, and AstraZeneca) substantially prevent disease, reduce morbidity and mortality. It takes some time before these effects are noticeable: first reduction of hospitalization and then reduction of deaths. Vaccines also seem to provide some protection from infection (based on the finding that they are associated with 33% lower chance to test positive), and thus may indirectly benefit unvaccinated citizens [225]. However, other vaccines used in many other countries are less effective, but no vaccine guarantees that nobody will be infected with the Covid-19 virus. Many studies are done to determine whether vaccines can prevent the transmission of the virus. This is especially important from an epidemiological and public health perspective since blocking transmissions will ultimately control the pandemic if sufficient numbers of people are vaccinated. However, it is difficult to determine as long as public health measures such as lockdowns and physical distancing are maintained [226]. When such measures are abolished, as happened in several countries later in 2021, the Delta variant seems able to break through the protection offered by vaccines and to make vaccinated persons able to transmit the virus. This is necessitating these countries to reintroduce at least some public health measures for the general population.

The dream to end the pandemic is connected to the idea of herd immunity. Community protection through herd immunity is the final goal of vaccination; the virus will no longer have any chances to circulate and be transmitted. As discussed in previous chapters, the concept of herd immunity is misused in the earlier stages of the pandemic, when some countries (e.g., UK and Sweden) considered to give the virus free range in order to acquire high levels of immunity within the population without the need for stringent public health measures. Experts quickly point out that such strategy will lead to enormous numbers of casualties while obtaining sufficient levels of immunity will take years. Proposing this strategy illustrates a way of ethical thinking that gives priority to opening the economy over maximal saving of lives, while vulnerable people will be 'protected' by completely separating them from all social contacts [227]. Never before in the history of public health has this concept been used to respond to an outbreak, and it is generally rejected as an unethical strategy. Herd immunity is only a proper goal of a vaccination strategy [228]. The pandemic will end and relaxation of public health measures is safe when a majority of the world population is immune (due to either natural immunity after disease or due to vaccination), although estimations of the required percentages vary. Usually it is assumed that approximately 60–70% of the population should be immune but the percentage required for herd immunity depends on the effectiveness of the vaccines, the basic reproduction number of the virus, and the implementation of physical distancing measures. If these last measures are continued and the vaccine is 90% effective, the threshold for herd immunity is at least 55%. If public health measures are lifted, the threshold with the same vaccine rises to 67%. When the basic reproduction number increases because of more infectious new virus variants, a higher percentage of populations should be immune [229]. The Delta variant has made the goal of herd immunity more elusive, requiring almost 90% of the population to be immune [230].

Such levels of immunity are necessary to protect society as a whole, and in particular people who cannot be vaccinated due to medical reasons. To reach herd immunity, in all nations people should be encouraged and motivated to get vaccination. It should be pointed out that individual decisions not to be vaccinated have social consequences. If herd immunity is not reached and a substantial number of people is not vaccinated, reservoirs of the virus will remain to circulate and outbreaks will recur. This is the case at national levels but also across the globe. But with new and more contagious variants of the virus, it becomes more and more unlikely that the goal of herd immunity can ever be reached. Within countries that already have a high vaccination rate, frustration may lead to ever more strenuous efforts to convince or mandate the last percentages of unvaccinated people, but 90% immunized people seems a faraway goal, as long as vaccines are not 100% effective and do not stop transmission. Even more, across the globe, the goal of herd immunity is remote; the virus continues to circulate and new escape variants may emerge.

For these reasons, the reliance on vaccines as the sole means of ending the pandemic is therefore problematic. The main advantage is that the risk of being seriously ill and dying is substantially reduced, so that hospitalizations will diminish. Even if the risk of being infected is lower, the dominant variants still infects people who have been vaccinated. This brings us back to the importance of traditional public health measures. When is it safe to relax these health measures? In countries such as Israel and the United States restrictive measure are relaxed when a substantial part, but not the majority of the population is vaccinated. They are immediately confronted with a new wave of Covid cases, and have to reinstate many public healthy measures. This illustrates the tension between individual (clinical) and population (epidemiological) perspectives. When individuals are vaccinated and thus protected against serious disease and death, they feel liberated from restrictive pressures, and their behavior may possibly change. But as long as the majority of populations everywhere is not vaccinated, the risk of infections has not disappeared. We do not know how long immunity lasts. Infectious disease specialists and epidemiologists are worried that the effect of existing vaccines will be over within one year. We do not know how the virus will evolve. New mutations may emerge that are more transmissible and deadly, as well as less susceptible to vaccines as long as the majority of the world population is not vaccinated [231]. Another concern is that SARS-CoV-2 will never disappear because it becomes endemic [232]. The pandemic may be over but the coronavirus will stay. Public and political expectations that Covid-19 is only a disastrous episode that may soon be ended, so that life as usual can be resumed, probably are deceptive since we have to learn to live with the virus and to find a different way of life. Perhaps that the serious impact of the virus will be mitigated (with less hospitalizations and death), but that only applies to those who are vaccinated. Since the virus (probably in new mutated forms) will continue to circulate across the world, continued implementation of public health measures is the only means to prevent infections. This is a paradoxical result because vaccination has always been presented as the best way to get rid of these policy measures.

The implication of these worries is that ending the pandemic is not the result of relying on technological tools such as vaccines and QR codes but ultimately a matter of human behavior and ethics. Concerns with the common good and collective interests, rather than an individualistic and nationalistic mindset, will open the only way for humanity to control the pandemic, and to find a balance in living with the virus. While science provides preventive knowledge and tools, these will only be effective in an ethical context which motivates people to use them. The pandemic finally teaches us the significance of global responsibility and solidarity. The often used phrase that we are all in this together is not just a hollow slogan but perfectly illustrates that the only way out is through global cooperation.

References

1. Broadbent, A. 2019. *Philosophy of medicine*. New York: Oxford University Press, 51 ff.
2. Wolfe, N. 2011. *The viral storm. The dawn of a new pandemic age*. New York: St. Martin's Griffin *The viral storm*, 16.
3. Khan, A.S. 2020. *The next pandemic. On the frontlines against humankind's gravest dangers*, ix. New York: PublicAffairs.
4. The history of bio-preparedness goes back to the beginning of the Cold War in the late 1940s and 1950s. Fee, E., and T. M. Brown. 2001. Preemptive biopreparedness: Can we learn anything from history? *American Journal of Public Health* 91 (5): 721–726.
5. Ten Have, H. 2019. *Wounded planet. How declining biodiversity endangers health and how bioethics can help*. Baltimore: Johns Hopkins University Press, 93 ff.
6. World Health Organization. 2004. *SARS risk assessment and preparedness framework*. Geneva: WHO.
7. World Health Organization. Regional Office for the Western Pacific. 2006. *SARS: How a global epidemic was stopped*. Manila: WHO Regional Office for the Western Pacific.
8. World Health Organization. 2015. *Report of the Ebola Interim Assessment Panel*; Moon, S., D. Sridhar, M.A. Pate, et al. 2015. Will Ebola change the game? Ten essential reforms before the next pandemic. The report of the Harvard-LSHTM Independent Panel on the Global Response to Ebola. *Lancet* 386: 2204–2221.
9. "Every country is strongly urged to develop or update a national influenza preparedness plan..." World Health Organization. 2005. *WHO global influenza preparedness plan: The role of WHO and recommendations for national measures before and during pandemics*. Geneva: World Health Organization, 1.
10. Ribes, M. 2020. Covid-19 retrospective, a disaster that should have been averted. *Bioethics Observatory*.
11. Trust for America's Health and Robert Wood Johnson Foundation. 2015. *Outbreaks: Protecting Americans from infectious diseases*. Washington.
12. Pegg, D. 2020. What was Exercise Cygnus and what did it find? *The Guardian*; See also: Horton, R. 2020. *The Covid-19 catastrophe*. Cambridge: Polity, 25 ff.
13. Department of Health and Human Services (HHS). 2020. *Crimson Contagion 2019 Functional Exercise After-Action Report*. Washington.
14. Brown, J. 2018. *Influenza. The hundred-year hunt to cure the deadliest disease in history*, 172. New York: Simon & Schuster.
15. Horton, *The Covid-19 catastrophe*, 35.
16. Global Health Security Index. 2019. *Building collective action and accountability*.
17. Kandel, N., S. Chungong, A. Omaar, and J. Xing. 2020. Health security capacities in the context of Covid-19 outbreak: An analysis of International Health Regulations annual report data from 182 countries. *Lancet* 395: 1047–1053.
18. Centers for Disease Control and Prevention. 1994. *Addressing emerging infectious disease threats: A prevention strategy for the United States* (Executive Summary). MMWR 43 (No. RR-5), 8.

19. Wolfe, *The viral storm*, 207.
20. World Wide Fund for Nature. 2020. *Living Planet Report 2020 – Bending the curve of biodiversity loss*. Gland: WWF.
21. Daszak, P. 2020. We are entering an era of pandemics – It will end only when we protect the rainforest. *The Guardian*.
22. Vidal, J. 2020. 'Tip of the iceberg': Is our destruction of nature responsible for Covid-19? *The Guardian*.
23. Daszak, We are entering an era of pandemics.
24. World Health Organization. 2016. *An R&D Blueprint for action to prevent epidemics. Funding & coordination models for preparedness and response*.
25. ———. 2018. *Prioritizing diseases for research and development in emergency contexts*.
26. Bellagio Initiative on the Global Virome Project. 2016.
27. Langreth, R. 2021. Five steps to prevent the next pandemic. *NDTV*.
28. Smith, M.J., and R.E.G. Upshur. 2015. Ebola and learning lessons from moral failures: Who cares about ethics? *Public Health Ethics* 8 (3): 305–318; Smith, M.J., and R.E.G. Upshur. 2020. Learning lessons from Covid-19 requires recognizing moral failures. *Journal of Bioethical Inquiry* 17: 563–566.
29. RIVM Corona Gedragsunit. 2020. *Naleven van quarantaine en isolatie advies*.
30. ———. 2020. *Analyse thuisblijven, testen en quarantaine*.
31. Holst, J. 2020. Global health – Emergence, hegemonic trends and biomedical reductionism. *Globalization and Health* 16: 42.
32. Horton, *The Covid-19 catastrophe*, 15.
33. Neuteboom, N., P. Golec, and S. Phlippen. 2020. *De Nederlandse economie tijdens Covid-19*.
34. Kissler, S.M., C. Tedijante, E. Goldstein, Y.H. Grad, and M. Lipsitch. 2020. Projecting the transmission dynamics of SARS-CoV-2 through the postpandemic period. *Science* 368: 860–868.
35. Greshko, M. 2021. Covid-19 will likely be with us forever. Here's how we'll live with it. *National Geographic*; Philips, N. 2021. The coronavirus will become endemic. *Nature* 390: 382–384.
36. World Health Organization. 2020. *Pandemic fatigue – Reinvigorating the public to prevent COVID-19. Policy framework for supporting pandemic prevention and management*. Copenhagen: WHO Regional Office for Europe, 7.
37. Badre, J. 2021. How can we deal with 'pandemic fatigue' *Scientific American*.
38. World Health Organization, *Pandemic fatigue*, 22 ff.
39. See Chapter 4.
40. Kotalik, J. 2005. Preparing for an influenza pandemic: Ethical issues. *Bioethics* 19 (4): 422–431.
41. Smith, and Upshur, Ebola and learning lessons from moral failures, 309.
42. Wong, C.M.L., and O. Jensen. 2020. The paradox of trust: Perceived risk and public compliance during the Covid-19 pandemic in Singapore. *Journal of Risk Research* 23 (7–8): 1021–1030.
43. Plohl, N., and B. Musil. 2021. Modeling compliance with Covid-19 prevention guidelines: The critical role of trust in science. *Psychology, Health & Medicine* 26 (1): 1–12; Bargain, O., and U. Aminjonov. 2020. Trust and compliance to public health policies in times of Covid-19. *Journal of Public Economics* 192; 104316.
44. Siebenhaar, K.U., A.K. Köther, and G.W. Alpers. 2020. Dealing with the Covid-19 infodemic: Distress by information, information avoidance, and compliance with preventive measures. *Frontiers in Psychology* 11: 567905.
45. Daoust, J.-F. 2020. Elderly people and responses to Covid-19 in 27 countries. *PLoS ONE* 15 (7): e0235590.
46. Murphy, K., H. Williamson, E. Sargeant, and M. McCarthy. 2020. Why people comply with Covid-19 social distancing restrictions: Self-interest or duty? *Australian & New Zealand Journal of Criminology* 53 (4): 477–496.

47. Cash, R., and V. Patel. 2020. Has Covid-19 subverted global health? *Lancet* 395 (10238): 1687–1688.
48. World Health Organization. 2020. *World malaria report 2020. 20 Years of global progress & challenges*. Geneva: WHO.
49. ———. 2020. *Dengue and severe dengue*. Geneva: WHO.
50. See Chapter 2.
51. Zarocostas, J. 2020. How to fight an infodemic. *Lancet* 395: 676; Larson, H. J. 2018. The biggest pandemic risk? Viral misinformation. *Nature* 562: 309.
52. Hameleers, M., T.G.L.A. van der Meer, and A. Brosius. 2020. Feeling 'disinformed' lowered compliance with Covid-19 guidelines: Evidence from the US, UK, Netherlands and Germany. *The Harvard Kennedy School Misinformation Review* 1; Marinthe, G., G. Brown, S. Delouvée, and F. Jolley. 2020. Looking out for myself: Exploring the relationship between conspiracy mentality, perceived personal risk, and Covid-19 prevention measures. *British Journal of Health Psychology* 25: 957–980.
53. Meek, J. 2020. Red pill, blue pill. *London Review of Books*: 19-23; Kramer, J. 2021. Why people latch on to conspiracy theories, according to science. *National Geographic.*
54. Visser, M. 2020. Eén of de tien Nederlanders gelooft dat er rond corona vieze spelletjes worden gespeeld. *Trouw.*
55. Romer, D., and K.H. Jamieson. 2020. Conspiracy theories as barriers to controlling the spread of Covid-19 in the U.S. *Social Science & Medicine* 263: 113356.
56. Evanega, S., M. Lynas, J. Adams, and K. Smolenyak. 2020. *Coronavirus misinformation: Quantifying sources and themes in the Covid-19 'infodemic.'*
57. Tampa, V. 2021. It's time for Africa to rein in Tanzania's anti-vaxxer president. *The Guardian*. On Feb 20, 2021, the Director-General of the WHO called on Tanzania to report Covid-19 cases, share data, and implement public health measures. WHO. 2021. *WHO Director-General's statement on Tanzania and Covid-19.* Early March 201, it became clear that the President was in a hospital in Kenya in a critical condition and he died later that month, possibly from Covid-19. His successor, Samia Suluku, reversed the Covid denial policies. Akinwotu, E. 2021. Tanzania's missing president is in Kenya with Covid, says opposition leader. *The Guardian.*
58. Lewandowsky, S., U.K.H. Ecker, C.M. Seifert, et al. 2012. Misinformation and its correction: Continued influence and successful debiasing. *Psychological Science in the Public Interest* 13 (3): 106–131; Chan, M.S., C.R. Jones, K.H. Jamieson, and D. Albarracin. 2017. Debunking: A Meta-Analysis of the Psychological Efficacy of Messages Countering Misinformation. *Psychological Science* 28 (11): 1531–1546; Romer, and Jamieson, Conspiracy theories as barriers to controlling the spread of Covid-19 in the U.S.
59. Lewandowsky, Ecker, Seifert et al, Misinformation and its correction.
60. World Health Organization. 2020. *Coronavirus disease (Covid-19) advice for the public: Mythbusters.*
61. Mullard, A. 2020. Covid-19 vaccine development pipeline gears up. *Lancet* 395 (10239): 1751–1752.
62. Folegatu, P.M., K.J. Ewer, P.K. Aley, et al. 2020. Safety and immunogenicity of the ChAdOx1 nCoV-19 vaccine against SARS-CoV-2: A preliminary report of a phase ½, single-blind, randomized controlled trial. *Lancet* 396 (10249): 467–478.
63. Davis, N. 2020. How has a Covid vaccine been developed so quickly? *The Guardian*; Kramer, J. 2020. They spend 12 years solving a puzzle. It yielded the first Covid-19 vaccines. *National Geographic*; Haque, A., and A. B. Pant. 2020. Efforts at Covid-19 vaccine development: Challenges and successes. *Vaccines* 8 (4), 739; Mullard, Covid-19 vaccine development pipeline gears up; Gerberding, J. L., and B. F. Haynes. 2021. Vaccine innovations – Past and future. *New England Journal of Medicine* 384 (5): 393-396.
64. Rhodes, J. 2021. *How to make a vaccine. An essential guide for Covid-19 & beyond.* Chicago/London: The University of Chicago Press; Allen, A. 2020. For billion-dollar Covid vaccines, basic government-funded science laid the groundwork. *Scientific American.*

65. Gouglas, D., T.T. Le, K. Henderson, et al. 2018. Estimating the costs of vaccine development against epidemic infectious diseases: A cost minimization study. *Lancet Global Health* 6: e1386–e1396; Plotkin, S.A., A.A.F. Mahmoud, and J. Farrar. 2015. Establishing a global vaccine-development fund. *New England Journal of Medicine* 373 (4): 297–300.
66. Plotkin, Mahmoud, and Farrar, Establishing a global vaccine-development fund.
67. The US Government committed $10 billion to vaccine research, the European Union $8 billion. See: Ball, P. 2021. What the lighting-fast quest for Covid vaccines means for other diseases. *Nature* 589: 16–18.
68. Gouglas, Le, Henderson, et al., Estimating the costs of vaccine development against epidemic infectious diseases, e1386.
69. See, for example, Folegatu, Ewer, Aley, et al., Safety and immunogenicity of the ChAdOx1 nCoV-19 vaccine against SARS-CoV-2; Haynes. B. F. 2021. A new vaccine to battle Covid-19. *New England Journal of Medicine* 384 (5): 470–471; Park, A. 2020. Covid-19 vaccines are coming: Here's what to expect. *Time*.
70. Torreele, E. 2020. The rush to create a Covid-19 vaccine may do more harm than good. *British Medical Journal* 370: m3209.
71. Iserson, K.V. 2021. SARS-CoV-2 (Covid-19) vaccine development and production: An ethical way forward. *Cambridge Quarterly of Healthcare Ethics* 30: 59–68.
72. The first vaccines receiving emergency authorization are based on clinical trials with 44,000 volunteers (Pfizer/BioNTech vaccine) and 30,000 volunteers (Moderna vaccine); half of the volunteers received a placebo.
73. Castells, M.C., and E.J. Phillips. 2021. Maintaining safety with SARS-CoV-2 vaccines. *New England Journal of Medicine* 384 (7): 643–649.
74. WHO Ad hoc expert group on the next steps for Covid-19 vaccine evaluation. 2021. Placebo-controlled trials of Covid-19 vaccines – Why we still need them. *New England Journal of Medicine* 384 (2): e2(1-3).
75. Weintraub, K. 2020. Continuing Covid-19 vaccine trials may put some volunteers at unnecessary risk. Is that ethical? *USA Today*.
76. Graham, B.S. 2020. Rapid Covid-19 vaccine development. *Science* 368: 945–946.
77. Rubin, E.J., and D.L. Longo. 2020. SARS-CoV-2 vaccination – An ounce (actually, much less) of prevention. *New England Journal of Medicine* 383 (27): 2677–2678.
78. Baden, L.R., H.M. El Sahly, B. Essink, et al. 2021. Efficacy and safety of the mRNA-1273 SARS-CoV-2 vaccine. *New England Journal of Medicine* 384 (5): 403–416; Haynes, A new vaccine to battle Covid-19, 470–471.
79. Haque, and Pant, Efforts at Covid-19 vaccine development; Kossakovski, F. 2020. What we've learned about how our immune system fights Covid-19. *National Geographic*.
80. Hodgson, S.H., K. Mansotta, G. Mallett, et al. 2021. What defines an efficacious Covid-19 vaccine? A review of the changes assessing the clinical efficacy of vaccines against SARS-CoV-2. *Lancet Infectious Diseases* 21 (2): E26–E35.
81. Haseltine, W. A. 2020. The risks of rushing a Covid-19 vaccine. *Scientific American*; Zhang, S. 2020. A vaccine reality check. *The Atlantic*.
82. Pontifical Academy for Life. 2005. Moral reflections on vaccines prepared from cells derived from aborted human foetuses. *The Linacre Quarterly* 86 (2–3): 182–187; Pontifical Academy for Life. 2017. *Note on Italian vaccine issue*.
83. Congregation for the Doctrine of the Faith. 2020. *Note on the morality of using some anti-Covid-19 vaccines*.
84. Glenza, J., and M. Pengelly. 2021. Catholics in New Orleans and St Louis told to avoid Johnson & Johnson vaccine. *The Guardian*; Peiser, J., and M. Boorstein. 2021. U.S. Bishops splinter on the morality of taking coronavirus vaccines. *Washington Post*. That this is not only a concern of the bishops is evident in Michigan where legislators request that recipients of the Johnson & Johnson vaccine must be told that it is developed with the use of a stem cell line derived from an aborted human fetus. Luscombe. R. 2021. Michigan must tell Johnson & Johnson vaccine recipients that it was developed using stem cells. *The Guardian*.

85. Hodgson, Mansotta, Mallett et al., What defines an efficacious Covid-19 vaccine?, E32-E33; Parker, L. 2020. To find a vaccine for Covid-19, will we have to deliberately infect people? *National Geographic*; Deming, M. E., N. L. Michael, M. Robb, et al. 2020. Accelerating development of SARS-CoV-2 vaccines – The role for controlled human infection models. *New England Journal of Medicine* 383: e63.
86. Grady, C. 2004. Ethics of vaccine research. *Nature Immunology* 5 (5): 465–468.
87. Eyal, N., P.G. Lipsitch, and P.G. Smith. 2020. Human challenge studies to accelerate coronavirus licensure. *Journal of Infectious Diseases* 221: 1752–1756; Eyal, N. 2020. Why challenge trials of SARS-CoV-2 vaccines could be ethical despite risk of severe adverse events. *Ethics & Human Research* 42: 24–34; Plotkin, S.A., and A. Caplan. 2020. Extraordinary diseases require extraordinary solutions. *Vaccine* 38: 3987–3988.
88. It will take 1 to 2 years according to Deming, Michael, Robb et al. Accelerating development of SARS-CoV-2 vaccines, e63(3).
89. Kahn, J.P., L.M. Henry, A.C. Mastroianni, et al. 2020. For now, it's unethical to use human challenge studies for SARS-CoV-2 vaccine development. *Proceedings of the National Academy of Sciences of the United States of America* 117 (46): 28538–28542.
90. Solbakk, J.H., H.-B. Bentzen, S. Holm, et al. 2021. Back to WHAT? The role of research ethics in pandemic times. *Medicine, Health Care and Philosophy* 24: 3–20; Holm, S. 2020. Controlled human infection with SARS-CoV-2 to study Covid-19 vaccines and treatments: Bioethics in Utopia. *Journal of Medical Ethics* 46: 569–573; Elliott, C. 2020. An ethical path to a Covid vaccine. *New York Review of Books*.
91. Mullard, A. 2020. How Covid vaccines are being divvied up around the world. *Nature*.
92. Mason, B. 2020. *Covid-19 vaccine review – Pre-orders, approvals, and prioritizations*.
93. Chin-A-Fo, H., and C. van de Wiel. 2021. Heeft de EU zich door farmaceuten laten bespelen? *NRC*.
94. Logunov, D.Y., I.V. Dolzhikova, D.V. Shcheblyakov, et al. 2021. Safety and efficacy of an rAd26 and rAd5 vector-based heterologous prime-boost COVID-19 vaccine: An interim analysis of a randomised controlled phase 3 trial in Russia. *Lancet* 397: 671–681.
95. Oltermann, P., and A. Giuffrida. 2021. Russia's Sputnik V Covid vaccine gaining acceptance in Europe. *The Guardian*.
96. Almond, J.W. 2007. Vaccine renaissance. *Nature Reviews. Microbiology* 5: 478–481.
97. Buchholz, K. 2021. Where coronavirus vaccines will be produced. *Statista*.
98. Smith, J., M. Lipsitch, and J.W. Almond. 2011. Vaccine production, distribution, access, and uptake. *Lancet* 378 (9789): 436.
99. Access to Medicine Foundation. 2021. *Access to Medicine Index*.
100. Fernandez, R., and T.J. Klinge. 2020. *The financialization of Big Pharma*. Amsterdam: SOMO.
101. Kollewe, J. 2021. From Pfizer to Moderna: Who's making billions from Covid-19 vaccines? *The Guardian*.
102. Goenka, A. 2021. Israel's vaccine rollout has been fast, so why is it controversial and what can other countries earn? *The Conversation*; Homs, O. 2020. How has Israel launched the world's fastest Covid vaccination drive? *The Guardian*.
103. Boseley, S. 2021. WHO chief: Waive Covid vaccine patents to put the world on 'war footing.' *The Guardian*.
104. Safi, M. 2021. WHO platform for pharmaceutical firms unused since pandemic began. *The Guardian*.
105. Boseley, S. 2020. US and UK 'lead push against global patent pool for Covid-19 drugs.' *The Guardian*. President Biden has reversed the American position, and the United States is now supporting the proposal. Inman, P. 2021. Drop Covid vaccine patent rules to save lives in poorest countries, UK and Germany told. *The Guardian*, June 12.
106. Ten Have, *Wounded planet*, 126 ff.
107. Abbas, M.Z. 2020. Treatment of the novel Covid-19: Why Costa Rica's proposal for the creation of a global pooling mechanism deserves serious consideration. *Journal of Law and the Biosciences* 7 (1): 1–10.

108. Price, W.N., A.K. Rai, and T. Minssen. 2020. Knowledge transfer for large-scale vaccine manufacturing. *Science* 369 (6506): 912–914; Crager, S. E. 2014. Improving global access to new vaccines: Intellectual property, technology transfer, and regulatory pathways. *American Journal of Public Health* 104 (11): e85–91.
109. Neushul, P. 1993. Science, government, and the mass production of penicillin. *Journal of the History of Medicine and Allied Sciences* 48: 371–395.
110. Allen, For billion-dollar Covid vaccines, basic government-funded science laid the groundwork; see also: Caplan, A., and D. Reiss. 2020. Fair compensation for rare vaccine harms. *The Hastings Center.*
111. Maboloc, C.R. 2021. Global ethics and the right to universal access to Covid-19 vaccines. *Eubios Journal of Asian and International Bioethics* 31 (3): 169–171.
112. 't Hoen, E. 2020. Protect against market exclusivity in the fight against Covid-19. *Nature Medicine* 26 (6): 813.
113. Abbas, Treatment of the novel Covid-19, 4.
114. Editorial. 2020. Everyone wins when patents are pooled. *Nature* 581: 240.
115. Boseley, US and UK 'lead push against global patent pool for Covid-19 drugs.'
116. Khamsi, R. 2020. If a coronavirus vaccine arrives, can the world make enough? *Nature* 580: 578–580.
117. McVeigh, K. 2020. Rich countries leaving the rest of the world behind on Covid vaccines, warns Gates Foundation. *The Guardian.*
118. Kashyap, A., and M. Wurth. 2021. Rich countries must stop 'vaccine apartheid'. View. *Euronews*; Associated Press. 2021. 'Wildly unfair': UN says 130 countries have not received a single Covid vaccine dose. *The Guardian.*
119. United Nations. 2020. *Everyone, everywhere must have access to eventual Covid-19 immunization*, Secretary-General says in video message to global vaccine summit. Press release, June 4.
120. Karp, P. 2020. Former WHO board member warns world against coronavirus 'vaccine nationalism'. *The Guardian.*
121. Khamsi, If a coronavirus vaccine arrives, can the world make enough?
122. Fidler, D.P. 2010. Negotiating Equitable Access to Influenza Vaccines: Global Health Diplomacy and the Controversies Surrounding Avian Influenza H5N1 and Pandemic Influenza H1N1. *PLoS Med* 7 (5): e1000247.
123. White House. 2020. *Executive Order on ensuring access to United States Government Covid-19 vaccines.* December 8.
124. Boffey, D. 2021. Vaccine row: EU has exported 34 m doses – Including 9 m to the UK. *The Guardian*; Herszenhorn, D. M., and J. H. Vela. 2021. European Commission to propose tougher vaccine export rules. *Politico*; Boffey, D., and J. Elgot. 2021. EU to widen criteria for possible Covid vaccine export bans. *The Guardian.*
125. Davies, M., and R. Furneaux. 2021. World is on course for a coronavirus vaccine 'apartheid', experts warn. *Independent.*
126. Sullivan, H. 2021. South Africa paying more than double EU price for Oxford vaccine. *The Guardian.*
127. The countries concerned are Burundi, Eritrea, Madagascar, and Tanzania.
128. Lie, R.K., and F.G. Miller. 2020. Global allocation of coronavirus vaccines. *The Hastings Center.*
129. Safi, M., and M. Pantovic. 2021. Vaccine diplomacy: West falling behind in race for influence. *The Guardian.*
130. At the end of February 2021, more people outside of China have been vaccinated with Chinese vaccines than citizens of China itself. Vlaskamp, M. 2021. Chinees vaccin verovert de wereld. *De Volkskrant*; Parkinson, J., C. Deng, and L. Lin. 2021. China deploys Covid-19 vaccine to build influence, with U.S. on sidelines. *The Wall Street Journal.*
131. Retmann, A. 2021. EU blasts UK and Russia in 'vaccine propaganda' war. *EUObserver.*

132. World Health Organization. 2021. *Call to action: Vaccine equity declaration*. Geneva: WHO; see also: Beaumont, T. 2020. 'Landmark moment': 156 countries agree to Covid vaccine allocation deal. *The Guardian*.
133. Wouters, O.J., K.C. Shadlen, M. Salcher-Konrad, et al. 2021. Challenges to ensuring global access to Covid-19 vaccines: Production, affordability, allocation, and deployment. *Lancet* 397: 1024–1034.
134. Dearden, N. 2021. System change, not charity, will end the vaccine apartheid. *Al Jazeera*.
135. Goalkeepers Report. 2020, September. *Covid-19. A global perspective*. Bill & Melinda Gates Foundation, 16.
136. Herzog, L.M., O.F. Norheim, E.J. Emanuel, and M.S. McCoy. 2021. Covax must go beyond proportional allocation of covid vaccines to ensure fair and equitable access. *British Medical Journal* 372: m4853.
137. World Health Organization. 2020. *WHO SAGE values framework for the allocation and prioritization of COVID-19 vaccination*. Geneva: WHO.
138. Emanuel, E.J., G. Persad, A. Kern, et al. 2020. An ethical framework for global vaccine allocation. *Science* 369 (6509): 1309–1312.
139. Liu, Y., S. Salwi, and B. Drolet. 2020. Multivalue ethical framework for fair global allocation of a Covid-19 vaccine. *Journal of Medical Ethics* 46: 499–501.
140. Hassoun, N. 2020. How to distribute a Covid-19 vaccine ethically. *Scientific American*.
141. Ledford, H. 2021. Why Covid vaccines are so difficult to compare. *Nature* 591: 16–17.
142. Lee, T.H., and A.H. Chen. 2021. Last-mile logistics of Covid vaccination – The role of health care organizations. *New England Journal of Medicine* 384 (8): 687.
143. Anderson, T. 2021. Covid-19 vaccines: Resolving deployment challenges. *Bulletin of the World Health Organization* 99: 174–175.
144. The Pfizer vaccine is approved in the UK on December 2, in Canada on December 9, in the US on December 11, and in the EU on December 21, 2020.
145. Our World In Data. 2021. *Statistics and Research. Coronavirus (Covid-19) vaccinations*.
146. It is estimated that for example in the Netherlands with the current pace of vaccinations it will take until November 2022 before more than 60% of the population is vaccinated. See: Covid-19 vaccine tracker. This estimation was much too pessimistic since 68% of the Dutch population was fully vaccinated in mid-October 2021.
147. Our World In Data. 2021. *Statistics and Research. Coronavirus (Covid-19) vaccinations*.
148. Subbaraman, N. 2020. Who gets a Covid vaccine first? Access plans are taking shape. *Nature* 585: 492–493; Russell, F. M., and B. Greenwood. Who should be prioritized for Covid-19 vaccination? *Human Vaccines & Immunotherapeutics*; Persad, G., M. E. Peek, and E. J. Emanuel. 2020. Fairly prioritizing groups for access to Covid-19 vaccines. *JAMA* 324 (16): 1601–1602.
149. World Health Organization. 2020. *WHO Sage values framework for the allocation and prioritization of Covid-19 vaccination*. Geneva: WHO.
150. National Academies of Sciences, Engineering, and Medicine 2020. *Framework for Equitable Allocation of COVID-19 Vaccine*. Washington, DC: The National Academies Press; Feiring, E., R. Førde, S. Holm, O. F. Norheim, B. Solberg, C.T. Solberg, and G. Wester. 2020. *Advice on priority groups for coronavirus vaccination in Norway*. Report 2020. Oslo: Norwegian Institute of Public Health; Health Council of the Netherlands. 2020. *Covid-19 vaccination strategies*. The Hague: Health Council of the Netherlands.
151. National Academies of Sciences, Engineering, and Medicine, *Framework for Equitable Allocation of COVID-19 Vaccine*, 96-97, 132-134. See also: Schmidt, H., P. Pathak, T. Sönmez, and M. U. Űnver. 2020. Covid-19: How to prioritize worse-off populations in allocating safe and effective vaccines. *British Medical Journal* 371: m3795.
152. Feiring, et al., *Advice on priority groups for coronavirus vaccination in Norway*, 16.
153. For example, in England, they are the first priority group, followed by all people older than 80 years, and frontline health and social care workers. Public Health England. 2021. *Covid-19 vaccination first phase priority groups*.

154. National Academies of Sciences, Engineering, and Medicine, *Framework for Equitable Allocation of COVID-19 Vaccine*, 119–120; Siva, N. 2020. Experts call to include prisons in Covid-19 vaccine plans. *Lancet* 396: 1870.
155. Feiring, et al., *Advice on priority groups for coronavirus vaccination in Norway*, 16-17.
156. Health Council of the Netherlands, *Covid-19 vaccination strategies*, 41.
157. Health Council of the Netherlands, *Covid-19 vaccination strategies*, 42.
158. Health Council of the Netherlands, *Covid-19 vaccination strategies*, 40 ff.
159. Al Jazeera Staff. 2021. Young people first: Indonesia's Covid vaccine strategy questioned. *Aljazeera;* Surendra, H., I. R. F. Elyazar, B. A. Djaafra, et al. 2021. Clinical characteristics and mortality associated with Covid-19 in Jakarta, Indonesia: A hospital-based retrospective cohort study. *The Lancet Regional Health – Western Pacific.*
160. Persad, Peek, and Emanuel, Fairly prioritizing groups for access to Covid-19 vaccines, 1601.
161. Berlinger, N., M. Wynia, T. Powell, et al. 2021. Ethical challenges in the middle tier of Covid-19 vaccine allocation: Guidance for organizational decision-making. *The Hastings Center.*
162. National Academies of Sciences, Engineering, and Medicine, *Framework for Equitable Allocation of COVID-19 Vaccine*, 125–126.
163. Bubar, K.M., K. Reinholt, S.M. Kissler, et al. 2021. Model-informed Covid-19 vaccine prioritization strategies by age and serostatus. *Science* 371 (6532): 916–921.
164. Goldstein, J.R., T. Cassidy, and K.W. Wachter. 2021. Vaccinating the oldest against Covid-19 saves both the most lives and most years of life. *Proceedings of the National Academy of Science of the United States of America* 118 (11): e2026322118.
165. Bowen, L. 2021. Whose underlying conditions count for priority in getting the vaccine? *Scientific American.*
166. Solis, J., C. Franco-Paredes, A.F. Henao-Martinez, et al. 2020. Structural vulnerability in the U.S. revealed in three waves of Covid-19. *American Journal of Tropical Medicine and Hygiene* 103 (1): 25–27.
167. Centers for Disease Control and Prevention. 2020. *CDC's Covid-19 vaccine rollout recommendations.* Updated March 25, 2021.
168. National Academies of Sciences, Engineering, and Medicine, *Framework for Equitable Allocation of COVID-19 Vaccine*: 118–119.
169. Cohn, J. 2021. Delay second doses? A guide to the latest Covid-19 vaccine debate. *HuffPost*; Kramer, J. 2021. Experts torn over changing vaccine doses to speed up lagging rollout. *National Geographic.*
170. Shevzov-Zebrun, N., A. Caplan, and B. Parent. 2021. Should Covid vaccination schedules deviate from the status quo – As a last resort? *The Hastings Center.*
171. Bernal, J. L., N. Andrews, C. Gower, et al. 2021. Early effectiveness of Covid-19 vaccination with BNT162b2 mRNA vaccine and ChAdOx1 adenovirus vector vaccine on symptomatic disease, hospitalization and mortality in older adults in England. *MedRxiv preprint*, March 2.
172. Sample, I. 2021. Delay in giving second jab of Pfizer vaccine improves immunity. *The Guardian*, May 14.
173. Koff, W.C., T. Schenkelberg, T. Williams, et al. 2021. Development and deployment of Covid-19 vaccines for those most vulnerable. *Science Translational Medicine* 13: eabd1525. See also: Kramer, Experts torn over changing vaccine doses to speed up lagging rollout.
174. Geddes, L. 2021. *Five things we know about the Delta variant (and two things we don't.* GAVI, June 15.
175. Hughes, M.T., J. Kahn, and A. Kachalia. 2021. Who goes first? Government leaders and prioritization of SARS-CoV-2 vaccines. *New England Journal of Medicine* 384 (5): e15. (1–2).
176. Cook, M. 2021. Germans shocked by fake vaccinations. *BioEdge,* August 15.
177. See for example: Stillwell, C. 2021. 'Vaccine interlopers' exploit chaotic policies to skip line in rural Tennessee. *The Guardian.*
178. DeVita, M.A., and L.S. Parker. 2021. Ethics supports seeking population immunity, not immunizing priority groups. *Hastings Center.*

179. Cylus, J., D. Panteli, and E. van Ginneken. 2021. Who should be vaccinated first? Comparing vaccine prioritization strategies in Israel and European countries using the Covid-19 Health System Response Monitor. *Israel Journal of Health Policy Research* 10: 16.
180. Glenza, J. 2021. 'Oversupplied' US faces pressure to send Covid vaccine doses to less wealthy countries. *The Guardian*; Luscombe, R. 2021. Biden says up to 90% of adults will be eligible for Covid vaccine by 19 April. *The Guardian*.
181. Vaccine hesitancy is defined by the WHO as: "a delay in acceptance or refusal of vaccination despite availability of vaccination services.' MacDonald, N. E., and SAGE Working Group on Vaccine Hesitancy. 2015. Vaccine hesitancy: Definition, scope and determinants. *Vaccine* 33: 4161-4164.
182. Wouters, O.J., K.C. Shadlen, M. Salcher-Konrad, et al. 2021. Challenges in ensuring global access to Covid-19 vaccines: Production, affordability, allocation, and deployment. *Lancet* 397: 1030.
183. Henley, J. 2021. Covid vaccine acceptance rising across Europe but falling in parts of Asia. *The Guardian*; Henley, J. 2021. A quarter of people in France, Germany and the US may refuse Covid vaccine. *The Guardian*.
184. Lazarus, J.V., S.C. Ratzan, A. Palayew, et al. 2021. A global survey of potential acceptance of a Covid-19 vaccine. *Nature Medicine* 27 (2): 225–228.
185. Wallis, C. 7 ways to reduce reluctance to take Covid vaccines. *Scientific American*.
186. Hotez, P.J. 2021. *Preventing the next pandemic. Vaccine diplomacy in a time of anti-science*. Baltimore: Johns Hopkins University Press; Hotez, P. 2021. Covid vaccines: Time to confront antivax aggression. *Nature* 592: 661; Coutinho, R. 2021. *Vaxx. Hoe vaccinaties onze wereld beter hebben gemaakt*. Amsterdam: Ambo/Anthos.
187. Willsher, K. 2021. Vaccine scepticism in France reflects 'dissatisfaction with political class.' *The Guardian*.
188. Whatley, Z., and T. Shodiya. Why so many Americans are skeptical of a coronavirus vaccine. *Scientific American*.
189. Editorial. 2020. Covid vaccine confidence requires radical transparency. *Nature* 586: 8.
190. Ledford, H. 2021. Why Covid vaccines are so difficult to compare. *Nature* 591: 16–17.
191. Associated Press. 2021. Canada suspends use of AstraZeneca Covid vaccine for those under 55. *The Guardian*; Marsa, L. 2021. Can AstraZeneca dispel doubts about its shots? *National Geographic*.
192. Boseley, S. 2021. Covid: AstraZeneca vaccine 79% effective with no increased blood clot risk – US trial. *The Guardian*; Boseley, S. 2021. US agency questions AstraZeneca's Covid vaccine trial data. *The Guardian*.
193. Oltermann, P. 2021. Scepticism over Oxford vaccine threatens Europe's immunization push. *The Guardian*.
194. Pence, G.E. 2021. *Pandemic bioethics*. Peterborough: Broadview Press, 121 ff.
195. Gostin, L.O., D.A. Salmon, and H.J. Larson. 2021. Mandating Covid-19 vaccines. *JAMA* 325 (6): 532–533.
196. Lazarus, Ratzan, Palayew. et al., A global survey of potential acceptance of a Covid-19 vaccine, 226.
197. Savulescu, J. 2021. Good reasons to vaccinate: Mandatory or payment for risk? *Journal of Medical Ethics* 47: 78–85.
198. Savulescu, Good reasons to vaccinate: Mandatory or payment for risk?, 83.
199. Pennings, S., and X. Symons. 2021. Persuasion, not coercion or incentivization, is the best means of promoting Covid-19 vaccination. *Journal of Medical Ethics*.
200. Hoffman, D.N. 2021. Vaccine mandates for health care workers raise several ethical dilemmas. *The Hastings Center*, August 10.
201. Flood, C.M., B. Thomas, and K. Wilson. 2021. Mandatory vaccination for health care workers: An analysis of law and policy. *Canadian Medical Association Journal* 193 (6): E217–E220; Bowen, R.A.R. 2021. Ethical and organizational considerations for mandatory Covid-19 vaccination of health care workers: A clinical laboratorian's perspective. *Clinica*

Chimica Acta 510: 421–522; Mello, M. M., R.D. Silverman, and S.B. Omer. 2020. Ensuring uptake of vaccines against SARS-CoV-2. *New England Journal of Medicine* 383 (14): 1296–1298.
202. Gur-Arie, R., E. Jamrozik, and P. Kingori. 2021. No jab, no job? Ethical issues in mandatory Covid-19 vaccination of healthcare personnel. *BMJ Global Health* 6: e004877.
203. Our World In Data. 2021. *Statistics and Research. Coronavirus (Covid-19) vaccinations.*
204. Singanayagam, A., S. Hakki, J. Dunning, et al. 2021. Community transmission and viral load kinetics of the SARS-CoV-2 delta (B.1.617.2) variant in vaccinated and unvaccinated individuals in the UK: A prospective, longitudinal, cohort study. *Lancet Infectious Diseases*, October 28.
205. Van Kempen, J., and B. Soetenhorst. 2021. Hoe ga je verder als land als niet iedereen de inenting wil? *Het Parool*, September 3.
206. Pence, *Pandemic bioethics*, 155 ff.
207. Henley, J. 2021. Covid: EU unveils ‘digital green certificate’ to allow citizens to travel. *The Guardian.*
208. Nehme, M., S. Stringhini, and I. Guessous. 2020. Perceptions of immunity and vaccination certificates among the general population: A nested study within a serosurvey of anti-SARS-CVoV-2 antibodies (SEROCoV-POP). *Swiss Medical Weekly* 150: w20398.
209. Phelan, A.I. 2020. Covid-19 immunity passports and vaccination certificates: Scientific, equitable, and legal challenges. *Lancet* 395: 1595–1598; Voo, T.C., A.A. Reis, B. Thomé, et al. 2021. Immunity certification for Covid-19: Ethical considerations. *Bulletin of the World Health Organization* 199: 155–161.
210. Hassoun, N. 2021. How to make ‘immunity passports’ more ethical. *Scientific American.*
211. Beduschi, A. 2020. *Digital health passports for Covid-19: Data privacy and human rights law.* University of Exeter.
212. Kofler, N., and F. Baylis. 2021. Covid-19 vaccination certificates: Prospects and problems. *The Hastings Center.*
213. Osama, T., M.S. Razai, and A. Majeed. 2021. Covid-19 vaccine passports: Access, equity, and ethics. *British Medical Journal* 373: n861.
214. Phelan, Covid-19 immunity passports and vaccination certificates, 1597.
215. The Royal Society. 2021. *Twelve criteria for the development and use of COVID-19 vaccine passports.*
216. Osama, Razai, and Majeed, Covid-19 vaccine passports; Kofler, and Baylis, Covid-19 vaccination certificates.
217. Humphreys, G. 2021. Opening up with Covid-19 passes. *Bulletin of the World Health Organization* 99: 546–547.
218. Schlagenhauf, P., D. Patel, A.J. Rodriguez-Morales, et al. 2021. Variants, vaccines and vaccination passports: Challenges and chances for travel medicine. *Travel Medicine and Infectious Disease* 40: 101996.
219. Our World In Data. 2021. *Statistics and Research. Coronavirus (Covid-19) vaccinations.*
220. Liu, J., and R. Chung. 2021. Capitalist philanthropy and vaccine imperialism. *The Hastings Center*, September 10.
221. Dolgin, Covid vaccine immunity is waning.
222. Schaefer, G.O., R.J. Leland, and E.J. Emanuel. 2021. Making vaccines available to other countries before offering domestic booster vaccinations. *JAMA* 326 (10): 903–904.
223. WHO. 2021. *Coronovirus (Covid-19) Dashboard.*
224. Economist Intelligence Unit, 2021. *Q1 global forecast 2021 Coronavirus vaccines: expect delays.*
225. Mallapaty, S. 2021. Are Covid vaccination programmes working? Scientists seek first clues. *Nature* 589: 504–505; Dagan, N., N. Barda, E. Kepten, et al. 2021. BNT162b2 mRNA Covid-19 Vaccine in a Nationwide Mass Vaccination Setting. *New England Journal of Medicine* 384: 1412–1423; Bernal, Andrews, Gower, et al., Early effectiveness of Covid-19

vaccination with BNT162b2 mRNA vaccine and ChAdOx1 adenovirus vector vaccine on symptomatic disease, hospitalization and mortality in older adults in England.

226. Mallapaty, D. 2021. Can Covid vaccines stop transmission? Scientists race to find answers. *Nature*.
227. Aschwanden, C. 2020. The false promise of heard immunity. *Nature* 587: 26–28; Symons, X. 2020. Is seeking 'herd immunity' ethical? *BioEdge*, May 9.
228. "Never in the history of public health has herd immunity been recommended as a strategy for responding to an outbreak, let alone in a pandemic against a novel respiratory virus." It is a "flawed goal without a vaccine." Haque, and Pant, Efforts at Covid-19 vaccine development: challenges and successes, 4.
229. Hogan, A.B., P. Winskill, Q. Watson, et al. 2020. Report 33. In *Modelling the allocation and impact of a covid-19 vaccine*. London: Imperial College.
230. Grover, N. 2021. Delta variant renders herd immunity from Covid 'mythical.' *The Guardian*, August 10.
231. ———. 2021. Covid: New vaccines needed globally within a year, say scientists. *The Guardian*; Lavine, J.S., O.N. Björnstad, and R. Antia. 2021. Immunological characteristics govern the transition of Covid-19 to endemicity. *Science* 371: 741–745; Phillips, The coronavirus will become endemic.
232. Phillips, The coronavirus will become endemic.

Chapter 9
Bioethics After Covid

Abstract This chapter will analyze the implications of Covid-19 experiences for bioethical discourse. Mainstream bioethics operates with the ethical principles of respect for autonomy, beneficence and non-maleficence, and justice. Within this framework, ethical dilemmas are identified between personal freedom and public health, for example in regard to quarantine, face masks, and testing. The framework also encourages balancing benefits and harms of policy interventions and potential medical treatments and vaccines, as well as weighing different values such as health, freedom of movement, and employment. This chapter argues that from the perspective of global bioethics, a wider framework of ethical considerations should be used, especially vulnerability, connectedness and community, solidarity and cooperation. As a global phenomenon, the pandemic cannot be only interpreted from an individual point of view; it problematizes social and communal relations and requires a social and global ethical perspective.

Keywords Binarity · Common interest · Controllability · Ecological perspective · Exceptionality · Global health · Global governance · Relationality · Solidarity · Vulnerability

9.1 Introduction

After the first case of coronavirus disease is reported, it takes 3 months for the virus to spread rapidly over the world, with cases confirmed in 172 of the 194 member states of the WHO. Almost 2 years later (November 2021), all countries are affected with worldwide 244 million infected people and almost 5 million deaths. Instead of relenting, the pandemic has expanded over the globe in three waves. While cases and deaths are currently declining in most Continents, they are increasing again in Europe. In all countries people are tired of the pandemic, and governments are less vigilant and cautious. Citizens in most countries are confronted with public health measures since the beginning of 2020 but globally the same experiences continue: overwhelmed hospitals and intensive care units, postponed regular care, physical

H. ten Have, *The Covid-19 Pandemic and Global Bioethics*, Advancing Global Bioethics 18, https://doi.org/10.1007/978-3-030-91491-2_9

distancing and masking. New concerns are virus variants and the slowing pace of vaccination. The long struggle between humans and SARS-CoV-2 is exhaustive and many wonder what the future will bring.

This chapter begins with examining the context of global health and the ecological perspective on health and disease, articulating that the health of people across the world is interconnected. Next, it is argued that the scope of ethical debate is restricted since moral concerns are pre-structured by the emphasis on controllability, exceptionality, and binarity. The argument then moves to the elaboration of a more encompassing bioethical perspective, starting with the relationality of human existence, bridging the commonly assumed opposition between individual and common interests, and national and global approaches. The final part of the chapter will examine some of the potential consequences of the pandemic experience for the evaluation of policies, the imperative of prevention, the transformation of global governance, and the direction of globalization.

9.2 The Context of Global Health

As pointed out in Chap. 1, infectious diseases have been for some time high on the list of threats to global health. In the 1990s, the notion of emerging infectious diseases signaled that pandemics can be expected but not predicted, and that preparedness and preventive efforts are crucial for the immediate future. This notion itself materialized in the context of modified thinking about the concept of health. Based on the ideas of miasmatism and hygiene, sanitary reforms have been undertaken in many countries. In the nineteenth century, European countries started to cooperate in the domain of healthcare, confronted with recurrent epidemics of cholera and the need to control the spread of infections. For the first time, international sanitary conferences were convened where delegates deliberated about the best ways to counter infectious diseases [1]. Such efforts were certainly not disinterested since many European countries had extensive colonial empires and were afraid that so-called 'tropical diseases' could be imported; they were furthermore concerned about possible disruption in trade and commerce. In the early twentieth century, several international institutions are established for data collection, epidemiological studies, and policy implementation, culminating in the creation of the World Health Organization in 1946. While in the past pandemics have been approached from a limited, and often Eurocentric perspective, the international dimensions of health and disease (and thus the *pan* in the term pandemics) are now taken seriously, evidenced in changing terminology. 'International health' involves cooperation between a limited number of countries and only a few disciplines with the assumption that the expertise and provision of aid is the prerogative of some developed countries whereas the primary concern is with diseases of the developing world. The term 'world health' indicates a broader specter: health does not merely play a role in the relations between separate countries but is a concern for people everywhere, thus transcending national boundaries. The term also recognizes that health is significantly determined by social and environmental conditions beyond borders. To address health demands therefore global cooperation with the assistance of a broad range of disciplines, not only healthcare, epidemiology and virology but also statistic modelling, social and behavioral sciences, management and policy science, economics, and political sciences. This broader perspective, as well as the awareness that many health challenges are related to processes of globalization that expose considerable health inequalities among and within countries is nowadays summarized as 'global health' [2]. This term is increasingly used since 2000 [3].

The notion of global health basically affirms that health is a common concern for humankind. Its focus is on the health of global citizens, i.e., people everywhere on the globe. Health issues are not limited within national borders since our health cannot be separated from the health of everyone else. This is evident when a pandemic disease strikes; it affects people regardless of where they are located. This focus implies that collective action and cooperation are required, involving many actors and disciplines. Another implication is the recognition that human beings share not only similar vulnerabilities but also the capacity to address challenges, for example with the potential of knowledge and technology, even if this is not equally developed across the world. Finally, it implies that health is not the same everywhere. Health threats and hazards can affect some populations more severely than others while determinants of health (such as poverty) have more negative impacts in some places rather than other. Nonetheless, since health is interconnected in the global world, localized threats can easily become global. Contemporary economic globalization leads to inequitable distribution of benefits and risks, and monitoring global health is a way to minimize inequities through global health policies. Therefore, health disparities are a concern for all health policy makers, where ever they occur. The significance of these characteristics of global health is that global governance is unavoidable [4]. However, the perpetual challenge, as again illustrated in the policy responses to the Covid-19 pandemic, is that the implementation of a common framework for prevention and control of disease is often hampered by efforts to protect national security and trade [5].

Although global health as a concept has different interpretations and definitions, it shares many characteristics with public health [6]. The notion of global health introduces a new perspective in two ways: it draws attention to the interactions between health and the environment, and it highlights the importance of a broader global ethics framework for health interventions. The first point is emphasized in the interconnectedness of health. Not only is the health of all human beings linked so that a disease somewhere has implications, and often risks for people anywhere, but human health is also linked to social conditions, so that for example poverty and structural health inequalities should be eradicated in the world since they are associated with increased morbidity and mortality and can threaten the gains in health in richer countries. More recent is the awareness that health is dependent on environmental circumstances. Degradation of the environment and loss of biodiversity jeopardize human health but also the provision of safe water, clean air, and adequate nutrition as basic conditions for healthy living. The idea that the environment is a crucial determinant of health, not only for individual persons but for the entire population of the planet, is nowadays expressed in the wider concept of 'planetary health,' underlining that human health cannot be separated from the health of the earth [7]. On this basis it is argued that policies and governance should not be concerned with an abstract image of the globe but with the predicament of the new era of the Anthropocene in which humans themselves are rapidly deteriorating the conditions under which they exist, endangering health and civilization, not in the distant future but in the realities of today. The removal of disease and the improvement of health as the larger purpose of healthcare are impossible without preserving biodiversity, and reducing environmental and socio-economic hazards. In sum, health should be considered from an ecological point of view, besides a medical and social one. The second element is that the perspective of global health advocates a broader approach to ethics. As argued in Chap. 1, the interconnection between medical, social and environmental concerns has generated global bioethics as an ethical framework that takes these concerns into account, and works with a larger set of moral viewpoints such as vulnerability, solidarity, cooperation, social responsibility, equality, diversity, and future generations.

9.3 The Ecological Perspective: Global Vulnerability

Global interconnectedness implies that citizens in one country will be exposed to diseases when they emerge in other countries. The ecological perspective therefore stresses the need for solidarity. This is not just an ethical requirement, but a medical necessity. Closing borders, restricting travel, and concentrating on national interests have only a limited effect on the dissemination of Covid-19. Public health as a common good is essential for the well-being and survival of humanity. It will require cooperation and collective action, especially in the early stages of a pandemic disease when the caseload is still manageable and secondary prevention is possible. In an ecological perspective, vulnerability to infectious diseases is not confined to specific individuals, populations or nations.

Concepts such as 'planetary health' articulate the connection between health care and earth care. They make clear that the focus of bioethics should go beyond individual health, and that effective policies should be based on collective action. Given the interconnection between health and biodiversity, the notion of sheltering at our common home is most appropriate in the current circumstances. The earth is our home, as stated in the preamble of the Rio Declaration (in 1992) and in the encyclical letter *Laudato si'* of Pope Francis (in 2015). Like all homes, this common home has dark sides, clearly noticeable today. But humans have no other dwelling place. Our actions can destroy this place but they can also turn it into a home where everyone feels safe and secure. Facing the Covid-19 pandemic is an opportunity to make ourselves at home in the world and to preserve our common home.

The consequence of interconnectedness is global vulnerability. Processes of globalization have resulted in a world that not only creates more and new threats, but they have also undermined the traditional protection mechanisms (social security and welfare systems, family support systems), eroding the abilities of individuals and communities to cope with threats. Entire categories of people are disenfranchised, powerless and voiceless. The fact that the world has become increasingly interconnected and interdependent has created a sense of mutual vulnerability. In the words of the Director-General of the World Health Organization: "Vulnerability is universal" [8]. Being vulnerable is often the result of a range of social, economic and political conditions, and therefore beyond the power and control of individuals. Because it is related to globalization, vulnerability can no longer be framed, as it usually is in mainstream bioethics, as an individual affair. The background of this common framing is the ethical principle of respect for autonomy. Fragility or weakness reduce or impair the capacity of persons to make autonomous decisions (for example, the capability to give informed consent) so that they no longer have the ability to protect their own interests. Such vulnerable individuals need special protection. From the perspective of global bioethics, however, vulnerability may be caused by structural social, economic and political determinants that disadvantage people; they are made vulnerable in specific contexts and situations, so that vulnerability is first of all related to the ethical principles of justice, solidarity and equality, rather than individual autonomy. Vulnerability is then the fall-out of processes of globalization that have made everyday existence precarious, exposing it to more hazards and threats, and decreasing the capacities to cope. In this perspective, relating vulnerability to autonomy diverts attention away from the circumstances that make subjects vulnerable.

Vulnerability as a global phenomenon has significant implications for the bioethical debate. The first is the recognition that humans are social beings. The common idea that individual persons are autonomous and in control is challenged. Since the human condition is inherently fragile, all human beings are sharing the same predicament. Because their bodily existence is vulnerable, and embedded in environments such as the virosphere, humans have developed institutions and social arrangements to protect themselves. Vulnerability is neither an individual accomplishment nor a threat; it means that we are open to the world; that we can engage in relationships with other persons; that we can interact with the world. It is not a deficit but a positive phenomenon; it is the basis for exchange and reciprocity between human beings. We cannot come into being, flourish and survive if our existence is not connected to the existence of others. The notion of vulnerability therefore refers to solidarity and mutuality, the needs of groups and communities, not just those of individuals. The second implication is that vulnerability mobilises a different response: if vulnerability is a symptom of the growing precariousness of human existence and is exacerbated in certain conditions, the social and environmental context can no longer be ignored in bioethical analysis. On the contrary, bioethics should focus on the distribution and allocation of vulnerability at global level. Instead of focusing on individual deficits, analysis should criticise the external determinants that expose individuals to possible damage and harm. It also means that individual responses are insufficient; what is needed is a collective response, in other words social and political action [9]. However, such a collective response should have a global outreach. While most countries implement their own policies, often not coordinated or harmonized with those of other countries, vulnerability at the global level will continue to exist or even increase, and this will prolong vulnerabilities at national levels. Current vaccination strategies illustrate this point as often highlighted by the World Health Organization stating that nobody is safe until everybody is safe.

The perception of global vulnerability has motivated the idea that infectious diseases are emerging and re-emerging. This idea became more widely disseminated around the same time as the notion of global health. Since the 1992 report of the Institute of Medicine in the United States, emerging diseases became a fashionable topic of research and policy [10]. The report articulates three assumptions that are reiterated numerous times since its appearance. One is that new microbial diseases will emerge, although it cannot be predicted when and where. Another assumption is that environmental changes account for most emerging diseases [11]. These changes are anthropogenic, i.e., caused by human behaviour [12]. Infectious diseases are not just the effect of a specific pathogen but they have environmental and social causes; diseases emerge because the natural environments of pathogens have been transformed by human interference. A third assumption is that a paradigm shift is necessary from response to prevention. Experiences with previous pandemics have learned that vigilance is lessening once the disease threat is over, but a continuous effort will be needed. The discourse of emerging diseases therefore has negative and positive implications. First of all, it is a warning against human exploitation of surrounding nature. It also cautions against overconfidence in the accomplishments of post-war medicine which has made effective medication as well as vaccines available for major infectious diseases. Secondly, the increasing awareness that humans, animals, and biodiversity are connected, makes positive action possible. We know that most infectious diseases emerge from animals, so that a broader concept of One Health should be applied, requiring the collaboration of human, animal and ecological health professionals, which is especially important for prevention of zoonotic diseases. Most emerging diseases are transmitted to humans through contact with animals [13]. Pathogens emerge from their natural environment in the wild (particularly primates, bats, and rodents). Domestic animals bridge the gap between humans and wild animals. Another insight is that infectious diseases are associated with global civilization: urbanization, deforestation, fast and easy transportation, consumption of wild game, and industrial livestock practices, while rising temperatures will increase their incidence across the world [14]. Furthermore, we know that diseases emerge in certain hotpots, so that global surveillance can detect them in an early stage. Last but not least, prevention at the macro-level is possible for most epidemic and pandemic diseases. It is a mistake to assume that pandemics are not preventable, and that they occur randomly [15]. This will be a challenge since micro-organisms are everywhere, and unavoidably connected with life on the planet. New pathogens will continuously emerge, and existing pathogens will mutate and adapt to their environment. Whether they will evolve in epidemics or pandemics depends on how humans respond. Disease outbreaks are not just natural events but symptoms of a weak and less vigilant public health system [16].

9.4 Structuring Ethical Concerns

Previous chapters discuss a host of ethical issues related to the Covid pandemic and the efforts to control it. The imposition of strict measures such as lockdowns and curfews request a balancing of benefits and harms. These measures have serious effects and some argue that the cure (these measures) is worse than the disease; others point out that waiting longer with stringent measures leads to more harm. Providing care to patients without sufficient protective equipment and testing creates a dilemma between helping less patients in the safest way and employing personnel exposed to higher risks. The focus on hospital and intensive care minimizes the attention to the deleterious consequences of Covid-19 on businesses, education, and social and cultural activities. Stringent isolation of older patients in nursing homes produces reduced quality of life for many residents. Care for the dying is difficult to provide, palliative care is reduced and burial rituals are cancelled. Public health measures are most effective if the majority of the population implement them but appeals to solidarity and responsibility have only limited effect, resulting in a continuous balancing between voluntary cooperation and enforcement. Solidarity at the global level is lacking since the most powerful countries do not share resources such as protective materials, medication and vaccines.

How ethical concerns are formulated and addressed is the result of a specific manner of formatting. For instance, caring for infectious patients is interpreted as professional duty leaving aside considerations of personal risk or risk to family members and relatives but also the responsibilities of healthcare facilities to provide a safe environment. Another example are policy measures such as physical distancing and masking that often move from appeals to voluntary responsibility to mandatory requirements with the argument that the collective interest overrides the interests of individuals, emphasizing compliance with the measures rather than adherence to them on the basis of persuasion and motivation. A third example is the argument that in emergency circumstances priority should be given to treatment of Covid patients since that will save most lives while treatment of patients with other conditions is scaled down or cancelled. The formatting of ethical concerns is performed with three fundamental notions: controllability, exceptionality, and binarity.

9.4.1 Controllability

One of the striking features of the pandemic is the predominance of the war metaphor. Since the virus is an omnipresent threat to everyone, a massive common effort is needed to fight it. There are only two options: victory or defeat. The entire society must be mobilized. All hopes are established on a technical solution to the Covid crisis, overcoming the vagaries of human behavior by simply injecting a vaccine. In the meantime, the emphasis should be on hospital care and the best possible treatment. In this context, there are only heroes, victims and villains, and dissent cannot be tolerated. After this world war is over, efforts should be undertaken to prevent future outbreaks. The arms race between viruses and humans demands the building of a critical defense system at the global level. Like powerful countries have established extensive military systems to prevent nuclear war and have concluded international treaties to limit proliferation of nuclear weapons and to prohibit chemical and biological weapons, taking the war against viruses seriously implies a similar global system with surveillance and public health capabilities as well as international regulations than can be verified to ensure global security, concluded in a pandemic treaty [17].

The driving force of these efforts is the belief in controllability. Nowadays, viruses can be quickly identified, their genomes sequenced, diagnostic tests produced and vaccines developed. The spread of the virus can be controlled with rigorous public health measures, first of all physical distancing. When people feel ill, testing and contact tracing prevent the further dissemination of the virus. Hospitalized patients receive the best possible treatment and care to limit the impacts of the disease. Finally, with massive vaccination the expansion of Covid-19 can be halted and rolled back. Controllability, according to German philosopher Hartmut Rosa is a characteristic of modernity. Modern social existence as developed since the eighteenth century is different from previous human experiences because it has an "incessant desire to make the world engineerable, predictable, available, accessible, disposable (i.e. *verfügbar*) in all its aspects" [18]. The point is that emphasis on controllability transforms our relationship to the world. If the world is fully controlled, we feel abstracted from it. We enjoy theater, music, and games because they are not completely engineered and controlled. The same applies to love and health; we can follow many recipes and recommendations to minimize risks, but the outcome is never guaranteed. In short: "human life and human experience are defined by uncontrollability" [19]. The dimension of life that escapes control makes it most valuable. But the drive to control separates humans from the world in which they are situated, and regards the world as a resource to be exploited, a collection of objects to master, a treasury of facts and data to discover and to make useful, and an assemblage of obstacles to overcome in order to advance human flourishing. Everything is seen as a challenge. Against this backdrop, we encounter the world, in the words of Rosa, as a "point of aggression" [20]. This is exactly the perspective of the military metaphor in the pandemic. The virus is an outside enemy that needs to be controlled, and ultimately destroyed. The four dimensions of controllability are reflected in the approach of the viral threat. First it is made visible, using science to identify the virus and mathematics to quantify the impact; second it is made accessible through the development of a diagnostic test so that it can be followed how the virus spreads; third it is made manageable with the help of public health measures but most of all through vaccines; finally the threatened world is made controllable by making it useful and more efficient through digital surveillance, remote work and education, and economic restructuring.

The difficulty according to Rosa is that the desire for control is intimately connected to uncontrollability. The more the world is controlled, the more it eludes us, presenting new challenges. For example, processes of globalization and neoliberal policies promoted the idea that the world is a global market which is self-regulating and solving problems such as poverty and underdevelopment. At the same time, these processes and policies have produced environmental degradation and increasing inequality which are now threatening global security and nearly impossible to control. If we control some diseases, and for example eradicate smallpox and cholera, other diseases become more prominent and new ones will emerge, jeopardizing the idea that the phenomenon of disease can be controlled. The paradoxical connection between control and uncontrollability is observable in the pandemic. There is a strong conviction that science and technology are the optimal means for control that will bring relief. Recent studies on the history of infectious diseases articulate that the best way to eliminate the ever-present danger of these diseases is with the tools that medical science provides (medicines and vaccines). All other approaches (simply labelled as 'non-pharmaceutical interventions') are of secondary use [21]. Highly personal accounts of virus hunters and disease detectives relate how they travel the globe to identify microbes and to contain outbreaks. They warn for impending frightening disasters but their own work shows that something can be done to predict and prevent them, and that we can win the war [22]. At the same time, these works illustrate that it will be a perpetual battle since viruses are everywhere and most of them are unknown. As discussed in previous chapters, the sciences of virology and epidemiology are useful but as such no guarantee that viruses can be controlled since human behavior is not fully predictable and manageable. Even when effective tools such as vaccines are available, problems with production, distribution and deployment impede getting hold of the pandemic. Health security as the ultimately aim of control is therefore always precarious. The two options of the war metaphor (defeat or victory) do not allow for a third, i.e. that the virus will stay with us and that we have to find ways to live with it.

The quest for control and the discourse of war is difficult to criticize since it seems the most rational and efficient way to make the pandemic controllable. It reflects the rationalization and intellectualization of modern cultures that have made the world knowable, predictable, calculable, and manipulable. Modernity is ruled by instrumental rationality using means-end calculations and planning, referring to abstract and universally applied rules, laws and regulations. Science and rational governance dominate social life. This has liberated modern humans from irrational forces, from mystery and magic. However, this process has led to 'disenchantment' [23]. Rationalization has created a bureaucratic system that is formalized and impersonal, and restricts individual freedom and creativity. Science has become specialized and fragmented, and aspires to be value-free. Modern medicine sharply distinguishes between objectivity and subjectivity, evidence and interpretation. It can tell people what to do if they want to master health and disease but disregards the question whether it makes sense to do so and whether health is the most important value in life. Ethical discourse is problematic since moral values have moved from the public realm into the private sphere of personal relationships, and they have become the subject of empirical observation, measurement and testing. Social concerns are displaced into personal moral ones. In this context, it is difficult to address issues of meaning, emotion, connectedness with other people, affective human communication, social commitment, alternative lifestyles, and spirituality. Scientists cannot ask or answer questions of meaning and value; what they can provide are technical means and calculations [24].

During the pandemic disenchantment is observable in various ways. Certainly in the beginning many people are afraid but anxiety cannot be publicly addressed out of fear to create panic. In many countries, the risks are downplayed and underestimated. The emphasis on scientific evidence, though ambiguous and uncertain is presented as the most rational approach, not giving much consideration to emotions. Policy efforts to manage and control the pandemic are not completely reassuring but sometimes heightening fear and anxiety. Many patients avoid visits to physicians and hospitals in the hope not to be infected. The disease is first of all regarded as a medical problem. Quantitative data concerning infected, hospitalized and intensive care 'cases' dominate public discourse while personal patient stories are absent, at least initially. Epidemiologists apply mathematics to disease modelling which gains "quantitative authority" suggesting that uncertainty has been eliminated [25]. Objectivity rather than subjectivity is regarded as the hallmark of public health policies. Medical and scientific experts play a primary role in proposing policy recommendations and debating the necessity and effectiveness of measures. Dependency on expert advice may risk to undermine the capabilities of individual citizens to execute their own courses of action to cope with the viral threat. When consequences of policy measures are examined, most attention goes to the economic fallout while the human, social, and psychological effects receive less attention. Prevention of viral traffic and treatment of Covid patients have priority with lesser concern for care of vulnerable persons as well as affective and social relationships. Numerous ethical studies are published but many have a restricted scope focusing on a limited number of issues (primarily in connection to allocation of resources) with the chief purpose to assist the responsible utilization of medical interventions [26]. Control of the pandemic requires persistent distancing to each other and loss of mutual connectedness, while at the same time the main argument is that we have to do this out of concern for each other. Because we are embedded in the world, we have to abstract from it. Finally, the expectation that stringent measures are temporary and that life can return to normal as soon as the virus is under control with vaccination, is frustrated when new restrictions (notably vaccine certificates) are introduced for participating in social life.

Summarizing, efforts to control, manage, predict and calculate the spread of SARS-CoV-2 perfectly reflect the rationalization, bureaucratization and intellectualization of modern cultures but they simultaneously demonstrate the uncontrollability, uncertainty and unpredictability of the modern lifeworld. When the pandemic lasts longer than expected, and policy measures begin to oscillate and are less consistent, this uncontrollability becomes more apparent, and makes people aware what is lost when the focus is only on efforts to make the world controllable. This awareness calls for a broader and deeper ethical discourse.

9.4.2 Exceptionality

Another way of pre-structuring ethical concerns is performed with the notion of exceptionality. It can take two forms. Intrinsic exceptionality refers to the claim to be outside the general pattern, and thus especially privileged. Before Covid-19 some countries thought to be exceptional because they are well prepared for a global epidemic. After the outbreak of Covid-19, specific countries assume that they are less vulnerable and more resilient than others. During the pandemic, countries try to profile themselves as exceptional in their policy approaches, scientific contributions, or vaccination strategies. Special claims are made by the healthcare profession demanding priority in triage and vaccination because of the higher risks undertaken and their instrumental value for the healthcare system. From an ethical perspective, arguments in favor of intrinsic exceptionality may be true or false but what they do is to assign such value to a country or profession that it becomes difficult to criticize policy-makers, scientists or healthcare workers because they are special. The second form is extrinsic exceptionality, i.e. the argument that an emergency situation creates special conditions in which the usual standards and practices no longer apply. In this form, the ethical perspective itself is affected. It is argued that special circumstances justify actions that normally would not have been accepted, for example confining citizens to their homes, testing mandates, crisis standards of care, expediting of scientific research, or deprioritizing older patients for ventilatory interventions. Allegedly, as these examples illustrate, the ethical considerations that apply in normal circumstances can no longer be used but should be either bypassed or reversed into a utilitarian framework so that the individual interest of patients will be subordinated to the common interest of all. Previously, I have argued that rather than bypassing, reversing or shifting moral principles, the ethical framework guiding public health, clinical medicine and research should be broadened, so that more principles are taken into account. In the context of public health, extrinsic exceptionality shows itself in the safety standards that are used. The need for early release of new vaccines and their emergency use approval is associated with less rigorous surveillance of safety and effectiveness than usual. Even if the precedent has been set with pharmaceuticals, and the data of the vaccine trials has been persuasive, emergency approval is based on studies with limited groups of subjects, whereas the vaccines will be used for the entire world population. While in normal circumstances, all possibilities will be used to minimize potential risks, this has not been the case in public health policies, as evidenced in decisions to relax stringent measures and re-open the economy, not because the viral threat has diminished but because safety is balanced against other values such as economic recovery. This is also evident in early policy recommendations by the WHO and several governments not to use face masks. Even if the evidence for their effectiveness is not clear-cut, in ordinary conditions the precautionary principle would have led to the policy to advice their use. In ordinary life, safety first is a basic principle that has instigated many regulations for human traffic, industrial production, and occupational activities. That more risks are deemed acceptable in emergency conditions is illustrated in the discussion of adverse effects of some vaccines. In March 2021 serious side effects of the AstraZeneca vaccine are identified, particularly autoimmune reactions against blood platelets causing the death of several persons [27]. Many countries paused the deployment of this vaccine. Such accidents have happened before. The Pandemrix vaccine against the swine flu in 2009 caused narcolepsia in children. In 1976, vaccination against swine flu was stopped after 40 million Americans were vaccinated while a pandemic did not emerge but 450 people had developed Guillain-Barré syndrome, a rare autoimmune condition [28]. The pausing of vaccinations creates confusion. Initially, the adverse effect is played down as accidental but later recognized as causally connected to the specific vaccine. Experts respond critically to the pause and argue that vaccination should continue. Disease and death of many people can be prevented by the vaccine. The risks are so extraordinary low (1 in 100.000) that the probability to die from Covid is much greater than from the adverse effects of the vaccine. More people will die because of lack of vaccination than as a result of vaccination. Experts also point out that in daily life many other, and higher risks are accepted. For the general public, the safety issues magnify the idea that something is wrong with the AstraZeneca vaccine, and more and more people refuse this vaccine. Expert reassurances that side effects are rare, and much lower than the risk of serious illness due to infection are not convincing since a simple benefit-harm calculation at the population level will not suffice at the individual level. Vaccines are given to healthy people and protect against a disease that might affect them but not necessarily. Individuals do not compare the risk of side effects with the probability to die from Covid but with the probability to get infected. This last probability is in their own hands, and if they meticulously follow public health measures they assume that the risk of infection is extremely low. The side effect of the vaccine, though rare, is serious and might led to death, while it takes time before these effects are noticed so that the number may be growing. Minimizing risks by comparing them with risks in daily life, will not help either. It is true that many aspects of life are involving risks but they are often the result of things that happen to us (such as accidents or violence), not a consequence of a deliberate choice. Other risky behaviors (e.g. smoking) are chosen but not because they promote the public interest or because they prevent disease for others. Use of contraceptive pills is associated with risk of thrombosis but this cannot be compared with vaccination because of the personal advantage for individual users. The argument of exceptionality that more safety risks are acceptable during a pandemic applies to populations but does not work at the individual level where side effects are associated with the personal situation of people. The argument itself may have negative effects since it enhances the experience that individuals may be sacrificed for the greater good, and that the interests of individuals are disregarded since in war speed is more important than caution. Erosion of trust in public health measures may be the result. This concern is reflected in the justification of pausing vaccinations; it demonstrates that vaccination is carefully monitored and that serious consideration of risks may lead to adaptation of vaccination strategies [29]. Openness and transparency are the only way to restore confidence, showing that the conflict between the epidemiological perspective of experts (focused on the population level) and the individual perspective of citizens is taken seriously, and that the argument of exceptionality has limits. It is clear that the misfortunes of the AstraZeneca vaccine have enormous consequences for global health. This vaccine is the preferred option for many low-income countries because it is cheaper and easier to store and use than other vaccines. It is also a major component of Covax. But now the public impression is that it is a second-rate product that is no longer preferred in developed countries and that can be dumped on developing countries.

The debate on vaccine safety is an example of the downsides of exceptionality. It illustrates the impact of utilitarian thinking promoting a calculating, impersonal, abstract, and decontextualized approach, primarily focused on consequences, weighing benefits and harms, not for individuals but populations, and sidelining other ethical principles such as vulnerability and equity. It also is associated with a technocratic and paternalistic approach, giving experts (epidemiologists, virologist, and intensive care specialists) the first and last word in policy decision-making. This is highlighted in the development and application of triage systems, as discussed earlier, but also in vaccination strategies. Exceptionality is furthermore applied inconsistently. It is primarily used for individuals, not for more powerful agents such as pharmaceutical companies refusing to share data, patents, and property rights for the benefit of all. These agents may even take advantage of exceptional measures by arguing that expedited review of new medicines and vaccines should be maintained now that shorter review procedures have not impacted the reliability and safety of new products, assuming that the emergency conditions may be prolonged when the pandemic is over. The argument of exceptionality is also not applied to vulnerable people in nursing homes, prisons, and disadvantaged conditions who need special protection because they are exceptionally affected by Covid-19.

In mainstream bioethics, the basic principles of ethical discourse are respect for autonomy, beneficence, nonmaleficence, and justice. The principle of respect for autonomy is usually dominant, focusing on concrete individuals and interpreting vulnerability in an individualistic way. In the pandemic, the balance between principles changed. Public health and utilitarian ethics give priority to benefit and harm, focusing on abstract individuals as specimens of a collective, and ignoring issues of vulnerability. The ethical debate then shifts from individual to public interests but in both frameworks minor attention if given to the principle of justice. The notion of exceptionality defines the fundamental challenge as a conflict between individual and common good.

9.4.3 Binarity

Covid-19 has highlighted and aggravated existing disparities and contradictions within and between societies. While SARS-CoV-2 is a threat to everyone, not all people are 'in the same boat' since some are more heavily affected than others. This is especially true, as discussed in previous chapters, for persons who are already vulnerable and disadvantaged before the pandemic emerged. Morbidity and mortality rates are increased for marginalized communities with higher prevalence of underlying medical conditions and limited access to healthcare. Covid-19 exposes the existing health inequities and accentuates the significance of socio-economic determinants of health. At the same time, policy measures during the pandemic further exacerbate the impact of inequalities. People in low paid service jobs (such as retail, food services, childcare, and hospitality) must continue to work. In developed societies, people with lower socio-economic status (frequently immigrants with a various ethnic backgrounds) have to work in crowded conditions (e.g. in slaughterhouses), have to use public transportation, and often live in multigenerational households. Patients in intensive care units therefore do not reflect the average population. Especially, in developing countries people without jobs or with informal and irregular work are without income, support, and food if they stay home. Lockdowns, distancing, and self-isolation are measures that can be best carried out by wealthier citizens and those with better accommodation. But even in these circumstances, the burdens are not equally distributed: women are often more impacted than men when schools and day care centers are closed, and since they have more part-time employment which is more likely disrupted. The evidence that Covid-19 is worsening the existing inequalities in health and society, points to the need to pay special attention to notions of vulnerability, solidarity and equality to address disparities from a more encompassing ethical framework [30]. The structural causes of inequities will require a long-term effort and cannot be remediated during a pandemic emergency. The challenges of Covid-19 make bioethicists aware that their discipline could have done more to address them and could have presented a perspective beyond the emphasis on individual self-determination [31]. Policies could have better supported the basic needs of people which are varying within populations [32]. As discussed earlier, the utilitarian focus of triage systems for example proposes abstract categories of prioritization and is blind to structural healthcare disparities, not taking into account the social context and the variability of patient's needs and vulnerabilities. Guidelines usually do not include voices from marginalized groups [33].

Another disparity produced and intensified in the pandemic is intergenerational tension, putting the old against the young. Older people are the most vulnerable to serious consequences of infection. Younger persons are least affected but asked to stay at home, keep physical distance, while schools are closed. They experience the prevention paradox: they can disseminate the virus without being ill and at risk of serious effects but have to change behavior in order to protect more vulnerable citizens. Seniors may complain that the curve of the pandemic is not flattening due to irresponsible conduct of younger persons (who have corona parties, go on holiday, and gather in public parks without masks and physical distancing) while they themselves have to isolate and experience increasing loneliness. On the other hand, younger generations grumble that their social life is curtailed because of concerns with persons who are in the final stages of their lives, and that they have to wait longer to go back to normal since those persons are prioritized for vaccination. These tensions are magnified through some policies. An example is the use of age as a criterion of triage for ventilatory support. It is also reflected in the lack of attention to nursing and care homes where older residents with multiple comorbidities were often not transferred to hospitals in case of infection. The policy of herd immunity used in Sweden, and initially favored in the United Kingdom and the Netherlands has age as a determinant. Giving free reign to the virus and thus allowing people to continue social life but at the same time protecting the elderly and other vulnerable groups proved disastrous because transmission of the virus between generations could not be prevented. In May 2020, Sweden has the highest per-capita mortality rate in the world [34]. In order to avoid society's lockdown, in fact 'reverse quarantine' is applied, a form of 'vertical isolation' advocated by the president of Brazil, quarantining all people above a certain age [35]. Contrary to the intentions to protect the elderly in nursing homes, in most countries mortality rates in these homes are appalling. Despite these awful experiences, the debate on herd immunity resurfaced in October 2020 with the proclamation of the Great Barrington Declaration by scientists sponsored by a think tank dedicated to promoting free market principles [36]. In the Netherlands, a group of medical, economic and other experts launched a recovery plan to establish 'safe zones' for older and vulnerable citizens, insulating them from social life that can thus be re-opened [37]. Both proposals are clearly utilitarian: it is acceptable that more people succumb now, especially older ones, in order to reach immunity in the longer run, while excessive damage to current society and economy will be mitigated. The suggestion is that the lives of some people are expendable for the greater good which is usually interpretated as the free flow of the market and economic productivity [38].

One objection to the strategy of herd immunity is its impracticability [39]. Not only is it impossible to isolate older citizens from the rest of the population but it also embraces the illusion of controllability, assuming that the virus can be managed to spread in a controlled way so that it does not affect specific groups in the population. Following the war metaphor, it is expected that the disruption of society is temporary, that drastic measures are counterproductive, and that the viral threat will disappear over time. Moreover, a serious problem is that it will take a long time before 60–70%, or even 90% of the population will be immune, while it is not clear how long immunity will last. The example of the Brazilian city of Manaus, often used in this context, is illustrative. Six months after the first wave in April–May 2020, almost 70 percent of the inhabitants had antibodies to SARS-CoV-2. For protagonists of herd immunity Covid-19 is therefore neutralized. Nonetheless, the city was affected by a second wave with even more casualties than in the first wave [40]. The most important objection against the herd immunity strategy is ethical. It leads to catastrophic loss of lives particularly of the elderly, but also to serious illness and long term complaints in many younger people while these harms can be mitigated and prevented with traditional public health interventions. The resulting damage to individual people cannot be justified with reference to economic benefits. A fundamental principle of bioethics, as expressed in many international documents, is that the interest of the individual should have priority over the sole interest of science or society [41]. It is furthermore unacceptable because it promotes discrimination and stigmatization of the elderly population. It assumes that the life expectancy of human beings is an indicator of whether it is worth to preserve their lives [42]. This assumption is formulated in the fair-innings (or life-cycle) argument: each person should have an equal opportunity to live through the various phases of life. Human life has different stages: childhood, adulthood, middle age, and old age. Everybody is entitled to live through these stages. If choices have to be made, for example in providing intensive care treatment, younger patients have had fewer opportunities to experience these life stages so that they have a stronger claim to lifesaving treatment than older ones. This argument, advanced decades ago by utilitarian philosophers and health economists in the context of rationing healthcare resources, is now resuscitated in triage policy proposals. It is promoted because it emphasizes fairness and intergenerational solidarity, combining efficiency with equity. The argument postulates that everybody is entitled to a fair share of life (say 70 years). Usually fairness is limited to the length of life while other aspects of life are not taken into account (e.g. responsible health behavior, life time experiences, disabilities and genetic illnesses). Applying chronological age as a criterion to deny care violates the notion of fairness if the history and condition of the patient are not individually assessed. In clinical medicine it is known that chronological age is not a good prognostic indicator since there is enormous variability in health while benefits from treatment depend on comorbidities, underlying conditions, and impairments [43]. Rather than using abstract general categories to determine whether treatment is appropriate, individualized assessments of each patient are made in clinical practice [44].

The fair inning argument also assumes that the death of an older person is unfortunate while the death of a young person is tragic, and thus unjust since he has missed a reasonable share of life. But it is not clear why it is a matter of fairness and not of luck. Why is the death of a young joyrider in an accident more tragic than that of the older victim killed? Clearly, only the chronological age of a person is not decisive for a moral judgment [45]. Is it fair when the elderly who have most contributed to the health care system as it exists today are not receiving care when they need it? Moreover, not all people have had equal chances to flourish in life and to realize their plans. Since most societies offer inequal opportunities to their citizens, not all people had their fair innings at a specified age [46]. Furthermore, the idea that there is a 'natural' life-span is problematic. The number of years one might expect to live is dependent on socio-economic conditions and the state of medical science and technology.

The difficulties and problems with the fair innings argument make its use in triage proposals unjustified. Besides, it is rather artificial since the selection of Covid-19 patients for intensive care treatment is not a choice between young and old, but between old and older. Nonetheless, it is often used during the pandemic. This is associated, in my view, with some trends that already were visible before the coronavirus emerged. One is the anthropological vision of human beings as *homo economicus*. The human person is first of all a rational self-interested individual motivated by minimizing costs and maximizing gains for himself. This image emphasizes that contemporary citizens should do everything to responsibly govern their life and get the best out of life. In mainstream bioethics, the image is transformed into, and perfectly aligned with the corresponding image of the autonomous person. The terminology of 'innings' reiterates the same idea: life is a form of producing and collecting benefits. Human life is like a commodity, a resource that can be divided in parts and shares. It is not considered as a whole, in which all stages have a particular value and meaning. Another trend is the abstract approach in medical science. Fair innings is an attractive concept since it is quantitative. Rather than having an ambiguous and inconclusive debate about ethical principles, it can provide clear rules that can be consistently applied and evaluated because it quantifies benefits [47]. This approach regards 'the elderly' as a homogenous group, and an abstract category which is necessarily associated with vulnerability, frailty, dependency, and deterioration, rather than as individual people with distinct personal, clinical, and social characteristics, conveniently ignoring that the majority of people older than 60 are not weak, dependent or frail [48]. The result is an inflexible policy with blanket exclusions or inclusions for example in treatment and vaccination strategies. It is often combined with paternalism, manifested especially by intensive care physicians and policy-makers, as described in previous chapters. The terminology of 'intergenerational solidarity' in this regard is illustrative. The fact that a number of elderly people themselves decline to receive healthcare, or give preference to the young, does not imply that this can be imposed and applied to all people as a form of 'solidarity'. Finally, the reference to 'fair innings' during the pandemic accentuates a problem that existed before. Age discrimination that was often implicit, has now become explicit [49]. Covid-19 not only illustrates the divide between young and old but further articulates already prevailing ageism. There are many examples of stereotypes and prejudices concerning people because of their age, particularly following policies portraying Covid as a disease of older adults with age as the major lethal risk, and recommending to separate young and old people [50]. The initial emphasis on social rather than physical distancing severs intergenerational contacts for an extended period, leading to isolation and loneliness, especially for residents in nursing homes [51]. The idea of fair innings in particular stimulates negative views of ageing, regarding the last stage of life as less valuable than previous stages. The suggestion is that elderly people have in fact completed their life; they continue to exist in bonus time beyond the 'natural' lifespan. In the eyes of some, they are therefore expendable, 'dead wood' that will be weeded out by the virus [52].

To counter negative views of ageing, some organizations have taken a clear stance. The British Medical Association, for example, has explicitly stated that age should not be used to determine access to intensive treatment while the World Health Organization has launched a global campaign against ageism [53]. In other countries policies are clearly ageist. In the Netherlands, medical professional associations issued a triage protocol, supported by several ethicists, that prioritizes people younger than 60 years. The government initially rejected this and intended to prohibit age discrimination by law, arguing that all lives have equal value, but refrained from this after several, mostly liberal parties in Parliament opposed this point of view [54]. How the general population views the role of age in Covid care decision-making has not been systematically investigated. One study in England shows that the 'fair innings' principle is strongly rejected as arbitrary and unfair by members of the general public. They endorse the view that every life is of equal value, and that the utilitarian focus on efficiency "should be tempered with a concern for equality and vulnerability" [55].

9.5 A New Modus Operandi for Bioethics

Having examined how ethical thinking has been modelled and formatted during the public health emergency, the challenge is how to envision bioethics after Covid-19. The question also is to develop a synthetic view that goes beyond the disparities and dichotomies which have been prevalent during the pandemic and which have narrowed the ethical imagination. In order to do this, I propose to go back to the ecological perspective, discussed in Chap. 6 and earlier in this chapter. Connectedness, balance, and vulnerability are identified as key notions in this perspective, and they refer to the specific anthropological condition in which human beings are situated. What they underscore is the fundamental significance of relationality for bioethical discourse.

9.5.1 Relationality

Global bioethics articulates that human persons are essentially characterized by relationality. As integrated wholes of body and soul they are embedded within communities and they exist in a web of relationships with other beings and the environing world. This is why in previous chapters it is argued that the notion of 'sphere' is more appropriate than 'globe' and why global bioethics should primarily be regarded as social ethics. Relationality is a more fundamental characteristic than relatedness. A person is continuously engaging in relations but this is often conceived from the viewpoint of the individual. The notion of relationality expresses that individuals not merely connect and interact with each other but belong together and are mutually dependent, taking responsibility and shaping their lives together. The first experience of humans is that the world is shared with others. From this perspective, individual autonomy is redefined as 'relational autonomy'. A human person is constituted through encounters and dialogues with other beings. Authentic human being is being-together, in the words of Gabriel Marcel, being present and available to others [56]. Relationality and being situated in the world implies vulnerability since it exposes humans to other persons and the environing world. Relationality is not an option and we cannot make ourselves immune to the world.

It is evident that relationships and relatedness have become problematic in the pandemic. Other people are presented as a threat, and relations may have lethal outcomes since humans are the principal vector of the virus. The main objective of public health measures is to prevent connections and interactions. Distancing, masking, prohibition of visits, working remotely, and sheltering at home obstruct being too close together with other persons. Covid-19 affects therefore the anthropological condition of human beings. They risks to have their presence and availability reduced, and thus to lose what is specific for humanity. For many people the measures create significant problems, physical ones because they have problems in providing for their basic needs, and mental ones because they are lonely and depressed [57]. But all people face the same dilemma between being secluded or being open to the world since their fundamental relationality is not annulled. This renders the continuation of isolation policies untenable. It also explains why the term 'physical' distancing is considered inappropriate, 'bubbles' appeared in which closeness and intimacy with at least some others was allowed, and many other ways of interaction and communication emerged [58].

Relationality is discounted in the pandemic in ways that on the rebound demonstrate its crucial role. SARS-CoV-2 is an abstract, invisible entity that is a concern of policy-makers but for most people negligible and rather harmless since even when it infects them it does not necessarily produce any symptoms. The impact of the virus is measured and presented with numerical data, and while policy-makers often refer to the reproduction number, it does not reflect the experiences of people. Nonetheless, this abstract entity generates anxiety and fear, precisely because human beings are connected, even though they have to disconnect to mitigate its spread. The contamination frame that epidemiologists use to explain the pandemic, implying that everyone is equally vulnerable, is also abstract since some people are more affected than others, depending on a specific configuration of circumstances. The military metaphor enhances impersonal language. It categorizes people in abstract terms as victims and warriors. Categories of patients with non-Covid diseases are regarded as collateral damage, without paying attention to the distressing situation of individual patients. Abstract approaches are evident in the application of crisis standards of care and triage criteria, excluding categories such as 'the disabled', 'the demented', and 'the elderly' from live-saving treatment. The goal of saving lives is abstract since it is directed at the total number of lives rather than individual lives. Patients and their families are not involved in decision-making. Even paternalism is abstracted: the expert knows what is best for the individual patient, not because he understands this particular person but rather because he has expertise about this particular class of patients. At the same time, the implementation of these standards and criteria is different while many argue that assessments should be individualized and not based on blunt categories. Residents in nursing homes and dying patients are managed as a separate class of people and subjected to categorical policies (e.g., no referral to hospitals, and no visitors) with no room for personalized approaches. Finally, vaccination strategies prioritize groups of people, not allowing considered exceptions. Herd immunity for most people is an abstract term.

As argued earlier in this chapter, these tendencies towards abstraction are not new; they are related to the process of 'disenchantment of the world'. The problem that has become manifest during the pandemic is that ethical discourse itself has become an abstract system that provides principles and rules for conduct and decision-making. It does not allow an escape from what Max Weber has called the

Iron Cage, that is the rationalistic and bureaucratized impersonal system that dominates all spheres of contemporary life, that restricts freedom and creativity, and in which questions of meaning and value can no longer be answered [59]. However, during most of its existence, ethics has not been regarded simply as a theoretical discourse but as a way of life, as a crucial component of philosophy, not merely concerned with knowing and acting but first of all with being, attempting to interpret and critically reflect on what it means to be human and to live in a world with other beings [60]. In this perspective, ethics is a lifelong concern, rather than a rational exercise focused on successive discrete events that require decisions about what must be done, what should be allowed, and what can be justified. Moral sensibility assists the perception of situations as morally significant and relevant, and this precedes rational deliberation resulting in choices and decisions that can be justified. Such sensibility emerges from being touched by the particularity of situations and from recognizing moral features such as anxiety, distress, discomfort, and compassion. Moral reasoning can therefore not be separated from emotions and feelings that arise in the setting of encountering another person. By appealing to principles, norms or rules which are necessarily abstract, ethics in the context of healthcare may fail to realize the importance of concretely lived experiences of patients as well as care professionals [61]. Another feature of the broader view of ethics is that ethical life is never solitary but always shared with other people. Ethics is the manifestation as well as the fruit of relationality. It flows from interpersonal interactions, the tension of individual inclinations and preferences with the needs and demands of other persons, and it becomes apparent in a succession of perspectives: from first-person individual concerns, to the second-person stance of mutual engagement, to the third-person perspective of detached onlooker. A complete understanding of ethical life cannot be provided by the third-person point of view. A crucial feature of ethics is the capacity to move back and forth between perspectives [62]. Such orientations and changing perspectives are fundamentally what has been suggested in the ethical tradition as the moral point of view. Ethics exists because our sympathies are limited; we are therefore encouraged to take the point of view of other persons. Moral evaluation contributes to the amelioration of the human predicament. Since humans are vulnerable and dependent, their knowledge and resources limited while there is often competition with other people, ethical discourse encourages them to overcome limited sympathies and indifferences, and to seek cooperation and to care about each other [63]. This view relates to what is called the expanding circle of moral concern. Over time, the scope of ethics has been widening and more beings are taken into account as morally relevant.

9.5.2 Individual Versus Common Interests

The recognition that ethical life is socially shared is helpful to re-interpret another opposition which is often surfacing in debates during the Covid-19 pandemic, that is the antagonism between individual and common interests. In mainstream bioethics, respect for personal autonomy is one of the most - perhaps the most - important ethical principles. In healthcare, individual responsibility is therefore a basic good. On the other hand, in public health ethics the health of the population is the primary good, and maximizing societal benefit has priority over respect for individual autonomy. In an emergency such as the Covid-19 pandemic, saving as many lives as possible takes precedence over the wishes, desires, and needs of individual patients. If people do not follow policy measures such as distancing, masking and self-isolating, they will be mandated for the sake of the collective of all citizens, even if some individuals are reluctant to do so. When medical facilities are no longer able to provide the usual level of care, there is a need for rationing so that at least as many patients will benefit as much as possible, even if some individuals will die who would have survived in normal conditions. The debate on herd immunity has the same implications: for the benefit of the community it is better to let the virus rage, knowing that a substantial number of individuals will die.

However, this opposition between individual and common interests is false because it ignores the fundamental relationality of human beings. Individuals are not isolated, abstract entities but social beings. This point of view is not accepted in the ideology of individualism, prevailing especially in the West, according to which human beings are independent and self-reliant, the masters of their own life, and choosing their own values, and thus as unique individuals separated and demarcated from other beings. The normative implication of this view is that respect for individual autonomy means non-interference: individual decisions and actions should be respected as long as they do not harm other human beings. Individualism is often associated with the view of the human person as *homo economicus*. The assumption is that the individual is a rational self-regarding actor. Only self-interest is guiding and motivating human behavior, making individuals into self-managers driven by calculations to maximize expected utility for themselves. In healthcare, this government of life should be encouraged by treating patients as responsible consumers who actively seek information and produce health as the outcome of their choices. What is needed is correct information and proper education because health is primarily a matter of individual responsibility. In this ideology of individualism there is no distinction between individual and collective interests since they coincide. The common good is in fact the aggregate of individual interests. If each individual can pursue his or her self-interest, and can take responsibility for his or her action, this will work for the benefit of all [64]. In this perspective, the opposition between individual and common interests disappears because the last type of interests are annulled and reduced to individual interests which are the only interests that are really relevant.

Taking into account the crucial role of relationality, the assumptions of the notion of *homo economicus* and individualism must be rejected. Regarding the human being as *homo economicus* presents a reductive view since many studies show that individual behavior is not merely focused on outcomes in terms of goods and services obtained, possessed, and consumed; it is also concerned with how people treat each other. Rather than being self-regarding, human agents care about the wellbeing of other persons; they frequently cooperate, and appreciate reciprocity and fairness.

The image of *homo economicus* is fictional because in practice people are motivated by other things than material self-interest and greed. For most of them, morality is a stronger motive to act than self-interest [65]. Besides these empirical studies, the assumptions of a reductive view of human beings are objectionable from the perspective of global ethics. This view ignores that humans are social beings and that ethics is a social activity. Societies are not mere collections of individuals and social interactions are not simply exchanges and communications between autonomous individuals. On the contrary, individuals are social products. They are not abstract beings but become concrete and individual within a communal and social context. The reductive view furthermore narrows vulnerability to primarily an individual affair, and neglect dependencies and inequalities between human beings. The interconnectedness of human life is lost whereas it is in fact the interaction with other people that makes us into autonomous individuals. Also lost is the significance of networking and cooperation, as well as the human relationships with the environing world whereas human beings are not only embedded in society and culture but also in the natural world of animals and plants. To surmount the inadequacies of the reductive view an ethical perspective is required that transcends the dichotomies between humans, nature, and culture, and that widens the usual scope of bioethics from individual to social, cultural, economic, and political concerns. Relationality involves the sharing and exchanging of perspectives, and is therefore a precondition for moral agency.

In the perspective of global bioethics, the opposition between individual and common interests is false because the first type of interests must be reinterpreted, while the last type is taken seriously. Several arguments contribute to this position. One argument is that personal autonomy is a relational notion. Not only has it originated and been nurtured within a context of dependency but it is also always exercised in interaction with other people, dependent on social and cultural conditions [66]. Another argument is that preferences, values, and beliefs are not merely individual but conditioned by the social context. Societies transmit values across generations because norms are internalized. The human capacity to internalize norms means that human preferences are socially 'programmable' and human behavior is guided by the moral values of social life. Because human agents are socially entangled and networked, their conduct cannot be explained by self-regarding rationality directed at maximizing self-interests but by social rationality, that is taking into account the well-being of other people and the needs of larger society [67]. A further argument, especially expressed in global bioethics documents is that autonomy is intrinsically connected to responsibility. Individual actions and decisions have social consequences, so individual autonomy and social responsibility cannot be opposed. Personal autonomy is not abstract and decontextualized but has impacts on concrete other people [68].

The Covid pandemic clearly illustrates that individual behavior affects the well-being of the community. Widespread use of face masks will protect not only the individual but also other people against possible infection. Testing will identify whether someone is infected, but it is a warning signal that others may be at risk. The aim of vaccination is not only to protect individuals but society as a whole. In a public health emergency, appeals to self-interest cannot be separated from concerns with the interests of others. Individual decisions whether or not to adhere to public health measures have an inherently social dimension. Appeals to individual responsibility will therefore not be sufficient without articulating social responsibility, and without creating the social, political and economic conditions for the exercise of responsible autonomy [69]. That individualistic policies fail without this dimension of social responsibility is evident in debates concerning quarantine, isolation, lockdown and distancing where it is argued that human dignity and human rights are violated. In these debates, dignity and rights are frequently regarded as notions that apply strictly to individuals. Human dignity is considered as a theoretical and abstract construct, an intrinsic quality that applies equally to every human being. It does not depend on human characteristics or conditions such as age, gender, or disease. It cannot be diminished or taken away by any authority or political system, or disregarded in emergency conditions. Dignity does not depend on whether it is recognized or respected since it continues to exists even in the most dismal or cruel circumstances. However, this is only part of the story of human dignity. It is also a practical experiential phenomenon, a lived experience. It refers to how humans behave and are treated; human dignity is thus a relational quality. In this perspective, dignity can be disrespected, lost or destroyed. In certain situations and practices, humans experience threats to their dignity and are confronted with undignified conditions [70]. Human dignity is an important notion in bioethics since it emphasizes that human beings are not like things which are objects that can be used for various purposes and exchanged in transactions. Human beings are different because they are subjects, "creatures with needs, tendencies and directions of their own" [71]. The notion of dignity is used to protect subjects against objectification, exploitation, and degradation. The implication is that dignity should be respected and protected because it is a collective value, not dependent on individual decisions and interpretations. The need for protection arises from the fact that as subjects, human beings are necessarily embedded in society and nature, and therefore vulnerable and dependent. Against this background, the two perspectives on human dignity (as intrinsic and relational quality) both derive from the awareness that all humans share basic needs and vulnerabilities. Both have similar implications. Human dignity, for example, has motivated the search for shared humanity, and therefore human rights [72]. It is the basis for mutual respect in decent societies across the world, linking bioethics to civilization. In requesting respect and recognition of all human beings, the concept of human dignity furthermore provides a bridge to overcome usual dichotomies between individual and social, abstract and concrete, rational and emotional. Dignity is not merely an abstract concept but shows itself in experiences of humiliation, disrespect and injustice. In other words, for a human being, dignity is a "form of conduct and interaction between people, a way of being with others and with herself" [73]. Human rights are the focus of similar discussions. Sometimes they are interpreted as rights of individuals, especially emphasizing non-interference to protect individuals against the state, and thus regarding civil and political rights as more important than social and economic rights. In this view, moral individualism is at the core of human rights language as the discourse of individual empowerment. However, all human rights are interdependent. Civil and political rights cannot be exercised if basic conditions for human existence are not provided, as expressed by social and economic rights. Individual persons can only be empowered within a relational context with others. The right to health illustrate that the individual dimension of human rights is connected to a dimension of solidarity and collective good: if appropriate conditions such as access to health care and quality health services do not exists, individuals cannot enjoy their right to health. Like human dignity, human rights are based on the recognition that human beings share fundamental needs and vulnerabilities [74].

9.5.3 National Versus Global Approaches

The above arguments against antagonism of individual and common interests apply to another opposition which is intensified during the pandemic: nationalism versus globalism. In emergency conditions, national interests dominate the approach to Covid-19. Countries are first concerned for their own citizens, and try to seize as much protective equipment, masks, and testing materials as possible, often in competition with each other, and without consideration for the needs of other, less powerful and economically weaker nations. Vaccines are pre-ordered and purchased in enormous quantities by high-income countries, leaving other countries at the end of the queue. The World Health Organization and international actors have argued multiple times that national policies will not be sufficient to control the pandemic as long as a global approach is missing [75]. The arguments are familiar: nationalism will hurt everybody and is self-defeating because all people are connected. The virus does not recognize borders and nations but affects the global population and requires global solutions. Even if the virus can be eliminated in one country, trade and travel will remain affected, economies will not recover, and stability and prosperity will not return as long as the virus is rampaging across the world. It is therefore in the interest of each nation to engage in global efforts to address Covid-19 [76]. This has been the lesson from global health thus far. Rather than making a distinction between more and less developed countries, it should be acknowledged that all countries are confronted with health challenges and health inequities. An example of global solidarity and cooperation has been the field of vaccine diplomacy, but not as a tactical move as in the current pandemic, but as an opportunity between nations, based on intensive scientific and diplomatic cooperation. The two most successful vaccination campaigns (polio and smallpox) have been the result of a global mutual effort, acknowledging a common rather than national problem. For instance, during the Cold War (in the 1950s and 1960s) the American Albert Sabin (the developer of the first oral vaccine against polio) cooperated with Soviet scientists to jointly produce and test vaccines to eradicate polio. Scientists from countries with different ideologies can overcome political and nationalistic tendencies to work for the common good [77].

9.5.4 *The Framework of Global Bioethics*

Previously, several principles of global bioethics have been discussed: vulnerability, human dignity, justice and equity, non-discrimination (particularly ageism), social responsibility, benefit sharing (for example, the pooling of vaccine patents), and protection of the environment and biodiversity. These principles present a broader framework for discussion of relevant ethical considerations during the pandemic. One principle of global bioethics has not been examined so far: solidarity and cooperation.

Numerous times it is argued that global solidarity is necessary for an adequate response to the global threat of the pandemic. The World Health Organization has set up the Solidarity Trial, an international clinical study involving thousands of patients from many countries to find and test effective treatments for Covid-19. In May 2020, it launched the Covid-19 Technology Access Pool, following a Solidarity Call to Action, aimed at sharing intellectual property, know-how and data in order to accelerate the manufacturing of critical equipment, treatments and vaccines. A further initiative is Covax with the purpose to provide fair and equitable access to vaccines in all countries of the world. Though these initiatives are important as vehicles for global solidarity, their implementation is slow and limited. As of 25 October May 2021, Covax has rolled-out 406 million vaccine doses to 144 participating countries (compared to 102 million administered in the United Kingdom alone, and 415 million in the United States) [78]. The Technology Access Pool so far has not received any contributions from the pharmaceutical industry [79]. The discrepancy between idealistic intentions and practical application makes solidarity one of the biggest challenges in the confrontation with the coronavirus.

In the context of public health, solidarity has since long been endorsed as a key ethical value. Because health systems are interdependent, and disease threats are global, collaboration between healthcare institutions is necessary at national, regional and global levels, requiring open communication, sharing of information, and coordination of policy responses [80]. In the Covid-19 pandemic, international bodies have repeatedly emphasized solidarity as a core concept. The ethical committees of UNESCO call it "an ethical duty to build solidarity and cooperation," while the ethics advisors of the European Union refer to solidarity as "a social vaccine" against indifference and exclusion [81]. Remarkably, the WHO's Working Group on Ethics and Covid-19 lists solidarity as the first ethical principle to apply [82]. The Vatican Covid-19 Commission states that the principle of solidarity must be the basis of any specific and concrete intervention in response to the pandemic, which implies that vaccines must be available and accessible to all [83].

Defining solidarity as "the practice of standing up together and acting in common" is attractive since it highlights that solidarity is the expression of moral relationship as well as principle of action [84]. Like the concept of human dignity, solidarity is both a theoretical construct and practical relational experience. Solidarity is often explained with references to the same grounds as the notion of relationality: it is based on the mutual recognition that human beings share the same needs, that their destiny is interconnected, that vulnerabilities are crucial human features but not equally experienced, and that the well-being of all citizens of the world should be the primary concern of global policies. Yet, what is typical for solidarity are not just these explanations, but its practical implications. Rather than a feeling of connectedness, and intentions to act, solidarity shows itself in supporting a specific cause and in common action [85]. It requires that understanding interdependency and willingness to assist others translates in public action, demonstrating that one's own interests are subordinated to those of others. Such action can be motivated by mutual self-interest, especially in a pandemic where it is everybody's interests to reduce and eliminate infections but the core of solidarity is moral concern for others, selfless commitment to the other rather than the expectation of unilateral benefit. It is not manifested because other people are a threat to our health, but because our health is connected and interdependent. Solidarity differs from charity, aid, and generosity: it signifies mutuality, a symmetrical relation between equals, and implies therefore inclusion and cooperation [86].

Solidarity is reflected in many aspects of society and healthcare, operating at various levels. At the interpersonal level, people care for each other because they share the same situation. In the pandemic, solidarity is manifested in many ways: students delivering food to isolated older people, citizens sharing face masks when they are scarce, and family and friends interacting with quarantined nursing home residents. In numerous countries, the majority of people modified their behavior and accepted the burdens of lockdown out of concern for vulnerable persons at higher risks. At the institutional level, solidarity means that equal access to healthcare is available to all citizens regardless of health risks and wealth. Healthcare is regarded as a collective responsibility so those with lower risks and higher incomes make the highest contributions to sustain health services [87]. While over the last few decades public financing of social arrangements (particularly in public health) has been cut back, and services privatized, joined with appeals to individual responsibility to health, during the Covid-19 pandemic massive government interventions supported healthcare and prevention. In many countries, tests, treatments and vaccinations are freely offered to every citizen while in some nations economic support is provided to those who lost their jobs and income. Impressive solidarity is shown in science with researchers networking and sharing knowledge and study results. Also healthcare professionals share a disproportional burden in caring for patients. However, at the global level, examples of solidarity are more difficult to find. Some countries provided intensive care to patients from other countries. India, severely affected by a wave of Covid in Spring 2021, received assistance (for example, oxygen supplies) but this show of solidarity came too late. In fact, global solidarity was absent when the world community faced shortages of protective equipment, diagnostic tests, and medication, while national self-interest dominated the production and distribution of vaccines.

The German national ethics committee points out that a crisis like Covid-19 is a test of the solidarity structures in societies [88]. The same is true for solidarity between countries. That lack of solidarity and cooperation has deleterious consequences for all countries is demonstrated in Latin America as the world region most affected by the pandemic. It has the highest death rates in the world, even when some countries (e.g. Chile) have efficient vaccination campaigns. Part of the explanation is the surge of a new, more contagious virus variant, but the catastrophe is also the result of structural problems such as poverty and underfunded health systems. A major factor is the lack of regional cooperation. While during the 2009 swine flu pandemic effective policy coordination existed and common strategies were employed in the Union of South American Nations under the leadership of Brazil, such spirit of solidarity and cooperation is now absent [89]. In Europe, while the pandemic crisis could have been an opportunity for increased cooperation, each country responded in its own way.

The failure of solidarity at the global level is startling since the need for it is endorsed numerous times. Every head of governments knows that without solidarity and cooperation, relief from the virus in some areas will only be temporary. Nonetheless, it is not surprising because the conditions for solidarity have been eroded. Global policies and international cooperation have primarily focused on economic interests. For example, in the European Union, protection of human health has not received priority since the organization and delivery of health services and medical care is the primary responsibility of individual member states. Global institutions such as the World Health Organization have been systematically weakened by budget cuts and attempts to delegitimize its work [90]. In most countries, public health infrastructure has been reduced, and health is first of all regarded as an individual rather than collective responsibility. The main driving force for cooperation is the neoliberal ideology of the free market, emphasizing competition, free trade, and commercialization of all aspects of human life. In this ideology, government interference must be reduced as much as possible, and deregulation, privatization, reduction of taxes and public expenditures encouraged. In this philosophy of rational egoism, societies are mere collections of individuals, and solidarity is rejected or regarded as a superfluous value [91]. The same processes have undermined solidarity within societies. The dominance of individualism and the view of the human person as *homo economicus* have diminished the experience of human beings that they are embedded within communities, cultures and environments, and that their destiny is connected to distant others as citizens of the world. Since solidarity cannot be imposed unilaterally or top-down, it will not emerge in these conditions [92].

This diagnosis clarifies that the failure of global solidarity is the result of policies which advance specific values at the expense of others. Bioethical debate has usually further articulated these values and employed them in the context of healthcare. The Covid-19 pandemic demonstrates the inadequacies of neoliberal policies as well as mainstream bioethical discourse. Public health infrastructures in most countries prove incapable of coping with the virus. Appeals to individual responsibility alone do not manage to control viral transmission. Massive government interventions are necessary to support the healthcare system, the economy, and all sectors of social life. The free market is not able to produce sufficient quantities of protective equipment, medication, and vaccines without substantial public support. Mainstream bioethics, relying on the language of autonomy, interests, utility, efficiency, and negative rights presents a myopic view of relevant ethical concerns. Starting from the point of view of the autonomous individual, it cannot recognize the connectedness of human beings, and the global dimensions of the pandemic, and thus the need for global responses [93]. After Covid-19 bioethics can no longer assume that autonomy is the dominant ethical principle; it must recognize that taking human relationality seriously implies enhancing and embracing social and structural conditions that make solidarity possible.

9.6 The World After Corona

Previous pandemics have affected and transformed human civilizations. It is expected that the experience with the contagious and lethal global threat of Covid-19 will influence societies and cultures. It is the first time in history that the majority of human beings have been confined to their homes for a rather extended time. The awareness that human interaction can be dangerous, is changing communications and relationships with other people. At the same time, the pandemic has impacted most sectors of public life, first of all the economy. Many businesses are seriously affected, so that the livelihood of people, particularly in countries where social safety nets are missing or deficient, is jeopardized. Education is also threatened. Schools and universities have been closed for a long time which has important consequences when alternative approaches (such as online teaching) are impossible. Furthermore, most social, cultural and religious activities are impeded. The global system of travel and tourism is disrupted. The pandemic has aggravated pre-existing political tensions between democratic and authoritarian states, and between the United States and China [94].

These large-scale effects of the pandemic have motivated the question how the world will be after the corona pandemic. New ways of interacting and communicating have been instituted (e.g. working from home, virtual education, zoom meetings, digital funerals and church services) and will they persist in the near future? Will life resume as it used to be when the global threat has disappeared? How will social, cultural and economic systems be rebuilt? Will the opportunity be taken to effectuate significant changes in our way of life and social behavior? Reflecting on such questions is important for two reasons. One is the expectation (and scientific prediction) that Covid-19 will not be last pandemic, and that humans should prepare for future, perhaps worse global diseases. The other reason is that an even more serious global threat is at the immediate horizon. Climate change is accelerating and the window to address it is rapidly closing. The pandemic experience can therefore be an incentive to move towards a green economy and a more sustainable world so that the planet is healed not only from infectious diseases but from annihilation.

Like other crises, the coronavirus pandemic has instigated a wealth of reflections and predictions. The pandemic experience is transformational in that it prompts rethinking of the usual patterns of living, acting and communicating. Being exposed to and having survived a global threat is a liminal experience, signifying the transition from an old to a new phase, reimaging human history as before and after Covid. This need for rethinking is called for by its comprehensive consequences, illustrating that the disease is not just a medical crisis, but also a social, economic, political, and moral upheaval. Some regard the pandemic primarily as a warning. The Covid pandemic has confronted humanity with the inescapable materiality of nature. Physical reality and the ubiquitous proximity of micro-organisms demand to use the pandemic to prepare for more disruptive times ahead, especially due to climate change [95]. Like the Great Depression in the 1930s initiated the creation of new infrastructures, the current crisis should be the occasion to remake our energy systems, to divest in fossil fuel, and to deploy new ways of transportation. There is no return to normality because the global capitalist system is over, and a radical change towards an alternate society based on global solidarity and cooperation is imperative [96]. Other commentators are more positive, though not less prophetic. Covid-19 is the occasion to "reset" global thinking and acting in order to create a world that is more resilient, inclusive and sustainable [97]. The virus exposes deficiencies in social life and is therefore an occasion to ameliorate them and make human existence more tolerable for everyone [98]. The expectation is that the coronavirus pandemic will change societies (making them more resilient and sustainable), governments (finding better means for leadership and coordination to solve society's problem, rather than relying on markets), publics (expecting stronger systems of social protection, and regarding disease as a pathology of society), medicine (focusing on planetary health and One Health, and paying more attention to vulnerable populations), and science (committing to equity and sharing of benefits) [99]. Most of all, the pandemic is interpreted as a "moral provocation": it provides an opportunity "to rethink the ethical basis of our society" [100]. It is a catalyst for changes in ethical reflection since it has intensified the experience that we are linked to each other. Rather than revealing hidden messages and meanings, the pandemic experience demonstrates the fundamental relationality of human beings, and may therefore inspire new ways of living [101]. Perhaps the experience will produce "an extension of empathic consciousness" because all of us have gone through the same ordeal although it may also increase divisiveness (old versus young, healthy versus vulnerable) with scapegoating and stigmatization [102]. Whatever the outcome, Covid-19 presents the opportunity for potential change, to ask questions about the current and future situation of humanity, and is thus an incentive to ethically reassess the human condition [103].

Of course, we have to wait and see whether such predications will materialize when the pandemic is over. Many assume, or perhaps hope that within a few years everything has returned to how it was before. Nonetheless, in the midst of the confusion and perplexity of the pandemic experience several themes for reconsideration have recurred frequently: policies, prevention and public health, global governance, and globalization. Rethinking each of these themes can be supported with the framework of global bioethics.

9.6.1 Evaluation of Policies

Without any doubt national, regional, and global public health policies in the Covid-19 pandemic will be intensely evaluated. Not only the short-term effects of policies on the transmission and health consequences of the virus will be taken into account, but also the longer-term repercussions for the economy, education, culture, social life and well-being, religious activities, travel and tourism. It is obvious that policies failed in many aspects. Preliminary studies after the first wave point out lack of preparedness, shortages of protective materials, and insufficient testing and tracing regimes while most governments were incapable to remediate these inadequacies because public health systems and institutions have been weakened by years of austerity measures [104]. While formal inquiries are postponed until 2022, hoping that the pandemic has receded by that time, evaluation studies are increasingly published. In the United Kingdom, governmental policy is ranked as one of the most important public health failures that the country has ever experienced. The slow and fatalistic machinery of government has led to many excess deaths. This was the result of deliberate policy despite the fact that the country has some of the best experts and universities in the world [105]. Similar devastating conclusions are drawn in an interim-report in the Netherlands. The country has more Covid death than many other countries (24,000 excess mortality while countries such as Norway and Denmark hardly had any excess mortality). The reason is that policies were made on the basis of the number of patients in the ICUs rather than of the number of infections, so that public health measures usually came too late [106]. The Dutch policy has therefore been called a form of 'necro-politics,' accepting the surge of infections until a national emergency with overflowing hospitals has occurred, and allowing for the death of thousands of people [107].

Evaluation of policies needs to examine the factors determining how countries 'managed' the pandemic. Weak leadership, ineffective bureaucracies, corruption, lack of social trust and no involvement of civil society are associated with poor responses to the pandemic. Fast and consistent interventions have been the most effective [108]. Responses were also better in societies with cultural values that articulate less individualism and more communal well-being, with long-term orientation, and less power disparities [109]. A critical issue is the relationship between science and policy-making. Scientific knowledge of the virus expanded rapidly and was generally available, while it was also known since a long time which public health measures could interrupt viral transmission. Despite this shared knowledge, preventive strategies were very diverse among nations. In some countries, experts played a major role (e.g. Sweden) but they were not more successful than other nations. In some other countries, experts were sidelined by populist leaders (e.g. USA, UK, Brazil and India), often with catastrophic results [110]. In both groups of countries, a sentiment of exceptionalism impeded taking drastic measures, using the argument of herd immunity as justification. Particularly Western countries mistakenly believed that they were well organized and prepared to cope with a crisis, while Covid-19 ruthlessly exposed massive disorganization and incompetence. Public health measures were delayed until there was really no other way, so in fact they were too late. Clearly disastrous decisions were made such as discharging thousands of untested patients from hospitals to nursing homes, while older and vulnerable patients were no longer referred to hospitals and workers in care homes did not receive protective materials and testing. Health professionals who spoke out were censored and disciplined. Policy-makers mistakenly assumed that the pandemic would be over soon, and rapidly relaxed measures as soon as a wave receded. Reliable evidence about the virus was certainly at hand when infections surged in a new wave, whereas many governments again dithered and delayed stringent measures. They also did not use the time to expand testing capacities, and track and trace infrastructure. No wonder that in many countries, leaders came under fire for their blunders and botched responses. It also raises the question whether politicians should be held accountable for mismanaging the pandemic [111].

Evaluation will take into consideration abundant empirical data but needs to investigate the normative dimensions of policies, too. Incompetent and inadequate policies manifested numerous moral failures. Disregard for existing inequities led to neglect of disadvantaged populations that were not only disproportionately affected by the virus but also by the public health measures. The focus on hospital care was linked to a failure to protect the vulnerable [112]. Inattention to prevention caused moral perplexities, for example for healthcare workers with insufficient protective materials, and for intensivists required to triage patients. Applying the precautionary principle could have saved thousands of lives [113]. Discounting the global dimension of the pandemic produced not only incoherent and nationalistic policies but increased the disparities between well- and less-resourced countries. Such negative implications of policies were not simply the result of uncertainty and paucity of scientific evidence, or the product of lack of sense of urgency and denial, but at least additionally the outcome of normative policy choices, often motivated by the anxiety to avoid the economic consequences of public health measures, and by appealing to individual responsibility, delaying drastic measures and fostering premature reopening of the economy. While the source of the virus is still unknown, and Chinese authorities suppressed any information about the outbreak, the genome sequence was communicated across the world in early January 2020 so that PCR tests could be developed and the first vaccines developed [114]. A fortnight later, the first clinical study of Covid cases was published in *The Lancet* [115]. Troubling information was available early but not taken into account.

9.6.2 The Imperative of Prevention

The Covid-19 pandemic presents multiple ethical challenges: healthcare professionals requested to care for infected patients without proper protective materials, older and disabled persons being denied access or referral to hospital treatment, people with symptoms going to work because testing capacity is limited, workers in crowded places disadvised to wear face masks since these are scarce, cancer patients needing surgery but waiting for months since care is postponed. Bioethical scholars have dutifully analyzed these challenges but the point is that they could have been avoided with appropriate preparedness and preventive action. One of the lessons of the pandemic is that the focus of ethical debate should be anticipatory and proactive rather than reactive. It has been argued many times that a pandemic could be expected. Most countries wrongly assumed that they are well-prepared, while some have downscaled their preparedness infrastructures because major global threats did not occur. Without an immediate menace, preventive activities are postponed in the belief that disruptive infections are rare. The same attitude is expressed when it is argued that the Covid-19 experience is a once in a century phenomenon like the nightmare of the Spanish flu. It forgets that over the last hundred years humanity is confronted with several pandemics which could have affected the entire world. The experience with the current pandemic and its global devastation must be a turning point. One reason is that Covid-19 may not be the most disastrous infection that can affect humanity; its fatality rate is low compared to other virus infections [116]. The second reason is that there are millions of pathogens with the potential to develop into serious human diseases so the risk for future pandemics is real and imminent. The Covid experience should therefore be a lesson to take global health seriously, and to leave behind complacency and indifference. In May 2021, a WHO commissioned Independent Panel concludes that the Covid-19 pandemic was a preventable disaster [117]. Lessons have not been learned from the past; pandemic prevention, preparedness and response were characterized by weaknesses, failures, gaps, and incoherencies. If the world populations want to be safe and healthy, immediate action will be necessary [118].

Manifold proposals and recommendations are now circulating to strengthen public health, for example the creation of a Global Health Threats Council and the adoption of a Pandemic Framework Convention [119]. The aim is to improve global governance and international cooperation. However, public health measures have not only failed because of inconsistent and uncoordinated application but because the concept of public health itself has become impoverished and deflated. Since decades, budgets for public health in many countries have been declining. Public health programs and agencies which already received only a small share of total health expenditures have been closed, particularly due to austerity policies since 2007 [120]. The number of employees in public health services has substantially decreased. In developing countries with good public health systems neoliberal policies of globalization made these systems collapse in favor of private contractors and NGOs [121]. The same policies transformed the concept of public health: health is regarded as an individual asset for which each individual is responsible. The emphasis is on healthy lifestyles and self-management, shifting the burden of prevention to individuals; they should show self-restraint because they have internalized the norms of public health and should demonstrate individual responsibility in choosing healthy behavior and voluntarily implementing these norms in getting vaccinated, adapting their diets, and regularly exercising [122]. In fact, the 'public' in public health is redefined, relying on individual autonomy rather than attending to social determinants of health and considering health as common good.

Against this background, it is explainable why existing public health infrastructures have been unable to cope with the Covid-19 pandemic. The lesson is that the concept of public health must be revisited [123]. In view of a global threat, public health should be holistic, proactive, and horizontal. Its focus is to approach the individual as embedded in social, economic, and political contexts. It considers diseases as connected to environmental circumstances taking into account that pandemic risks can be reduced by influencing human activities that exploit biodiversity. With such holistic view it counterbalances the biological-technological approach that regards the disease outbreak first of all as a medical, specifically intensive care phenomenon. It directs the public and policy discussion toward prevention of emerging diseases and preservation of health. Public health should furthermore be proactive, rather than reactive. COVID-19 is a reminder that human health is linked to the health of the planet. Coronaviruses cause zoonotic diseases, which means that they are transmitted between animals and humans. These diseases represent about 75% of all emerging infectious diseases. Instead of waiting for the next pandemic, future outbreaks should be prevented by addressing threats to ecosystems and wildlife, including habitat loss, illegal trade, pollution and climate change. Many current proposals suggest improved global surveillance of 'hotspots' to early identify pathogens with pandemic potential and to estimate the risks of newly discovered viruses [124].

Finally, public health should be horizontal. Rather than vertical interventions directed against specific diseases (such as the WHO campaign to eradicate poliomyelitis), horizontal interventions aim at strengthening the healthcare system so that a public health infrastructure is created that can cope with the determinants of health in the long run. Policies will not be effective if the socio-economic contexts of health problems and health inequalities are not addressed. More than disease management, a focus on the systemic dimensions of health and healthcare requires intersectoral action to overcome departmentalism and governmental silos, as well as transdisciplinary cooperation. An example of the last is the One Health movement, arguing that human, animal and environmental health are linked. Approximately 60% of all diseases recognized in humans are due to pathogens that move across species lines [125]. That closer surveillance of animal diseases is crucial to prevent the emergence of human diseases, has been clear in the 2009 outbreaks of avian influenza in poultry. Cooperation at national, regional, and global levels will be required between agencies and departments that usually are separated [126]. The horizontal orientation of public health furthermore implies a configuration perspective. More is needed than identifying possible contamination and emerging viral threats because diseases arise within specific socio-economic conditions in association with human practices [127]. The central role of human behavior for transmission of micro-organisms and for the effectiveness of public health interventions should be recognized. The implication is that teams and councils advising governments about public health measures not only include physicians, virologists, and epidemiologists, but represent a broader range of expertise.

9.6.3 Transforming Global Governance

Since long it is recognized that global challenges such as climate change, poverty, terrorism, migration but also health threats like antibiotic resistance and infectious diseases cannot be faced by each country alone but require global cooperation and governance. The Covid-19 challenge demonstrates the deficiencies of global governance. While the WHO appeals to solidarity and started several initiatives, many high-income countries are primarily interested in furthering their own interests. Fragmentation and lack of coordination in global health is not a new phenomenon [128]. Countries wish to preserve their national sovereignty and are not interested in a stronger role of the WHO. Evaluations of previous pandemics (SARS, Ebola, and Zika) have pointed out exactly the same issues that come to light during Covid-19 but these have not been taken into account to improve preparation. The hope is that this time will be different and that global health governance will be reframed. In contrast to earlier pandemics, SARS-CoV-2 affects not a limited number of countries but devastates all, and particularly wealthy countries. Across the world, focusing on national interests is insufficient to solve the global threat. Now also high-income countries experience that health for their own populations can only be secured through promoting global health. Domestic approaches provide a false sense of security and ignore that vulnerability is global. Despite this global connectedness, disparate maneuvers have been undertaken from the perspective of national concerns. Chinese authorities did not share early information during the first outbreak in December 2019. The declaration of a Public Health Emergency of International Concern by the WHO on 30 January could have been made earlier, but was not followed by emergency responses in most countries. Recommendations to use traditional public health measures (test, trace, and isolate) when it was still possible to interrupt the spread of the virus, were ignored. Contrary to the WHO advice, countries closed borders and imposed export bans, creating shortages of health supplies (particularly oxygen), protective materials, tests and equipment. Short-term nationalistic interests led countries competing in buying up medication, testing materials, and vaccines [129]. It is clear that self-centered approaches impede international cooperation, and that despite the rhetoric of global solidarity, a proper balance between nationalism and globalism in global governance is missing [130].

To bring global bioethics discourse beyond wishful thinking, concrete steps are necessary to enhance global governance. Some leaders have proposed the adoption of a global pandemic treaty but this seems to shift solutions to the future. Countries could have decided to cooperate since the outbreak and do not need a new treaty to do that. For example, a decision can be made to waive Covid vaccine patents so that as many manufacturers as possible can produce vaccines quickly in order to overcome current vaccine inequity and nationalism. There is no lack of agreements and proposals, and it should be examined why they were not working. Even when there would be a new treaty, after long and intense negotiations, there are no guarantees that it will be implemented in a future crisis [131]. The European Union that is promoting a new treaty could show global solidarity and fairness simply by no longer resisting the sharing of vaccine patents [132].

Countries should acknowledge that they have weakened the only existing international organization potentially capable to deal with global health challenges. Mandatory contributions to the WHO have been reduced over the years, with voluntary and earmarked financing increasing, making it difficult for the organization to respond to emergencies [133]. It is argued that the WHO lacks authority but it has as much authority as member states allow it to have. It relies for example on national authorities for information. While it was known on social media and through whistleblowers that something was going on, the WHO was dependent on information that China wished to share, which came late and was inaccurate. Countries cannot be forced to give information so that the same secretiveness in China was repeated that marked the SARS outbreak in 2002. Recommendations now emphasize that the role and authority of the WHO should be strengthened [134].

Stronger and independent global governance will be necessary not merely for future pandemics but also for the other major global risk: climate change and its deleterious effects on health. It is estimated that 250,000 deaths each year will be the result of extreme weather, air pollution, conflicts over natural resources, rising sea levels, reduced food production, and wildfires [135]. The pandemic is a surface phenomenon of deeper changes in the human relations with the conditions for life on earth. It is therefore essential to zoom out from Covid and look at the broader context in which the disaster unfolds [136]. Rather than being fixated on the virus, it is important to consider the human ways of life which gave it the opportunity to emerge. The connection between the spread of SARS-CoV-2 and exploitation of the planet is an incentive to other manners to inhabit the world, for example to change our ways of transport, systems of production and consumption of food. Some argue that Covid-19 has been a blessing in disguise: the lockdowns have significantly reduced environmental pollution [137]. The shared experience of catastrophe has not only created a new awareness of our sense of belonging but also showed the possibilities to protect the common good of humanity and implement a green agenda. It is time "to imagine and implement a project of human coexistence that allows a better future for each and every one" [138].

9.6.4 Redirecting Globalization

The Covid-19 pandemic has revealed the shortcomings of globalization. The concern is not so much that interconnectedness greatly enhanced the global dissemination of the virus, but rather that the virus particularly affected those populations that already were disadvantaged and marginalized within the global system. Decades of neoliberal policies have increased general vulnerability and inequalities so that poor populations are disproportionally affected by the infectious disease. These policies have also weakened public health care systems in many countries through reducing budgets and personnel, and privatizing services. The major challenge to post-pandemic bioethical debate therefore is to assess the positive and negative dimensions of globalization, and to identify how more humane, equal and resilient forms of globalization can be effectuated, recognizing the fundamental values of health, security, and human dignity in addition to a primary focus on the economy.

The shortcomings of globalization have been known for some time but Covid-19 makes perfectly clear that the neoliberal ideology with its basic values of self-interest, competition, and private ownership, and its policies of free market economy and efficiency do not work to overcome a global health emergency. The idea that the market is self-regulating and the most fair and efficient form of social organization proves not only deceptive but outright dangerous. While criticism of globalization has been growing over the past years, many commentators nowadays argue that a redirection of global processes is necessary [139].

This is first of all noticeable in a re-evaluation of the role of the state and the public sector. During the pandemic, citizens are asked to subordinate their individual interests to the public interest, the common good of public health. Decades of policies have amplified individualism and reduced the role of the state and the community, regarding government as a business to manage efficiently and frugally the interactions between citizens, treating them as responsible clients and consumers of services, facilitated by public officials who as process managers do not need extensive knowledge of their service portfolio. Institutional systems are very lean, without surplus capacity and resilience in the face of setbacks and uncertainties, often decentralized with insufficient budgets, and pushed to work as efficiently as possible. The public sector is tightly designed with shortages and high workloads already common before the pandemic in healthcare, education and law enforcement. During the pandemic, healthcare infrastructures have become unstable due to increasing demand for care and decline in the supply of healthcare because many healthcare workers became ill or even died as a result of lack of personal protective equipment. Inventories of PPE were inadequate at various levels. Emphasis on efficiency and cost minimization did not encourage hospital systems to maintain stockpiles of PPE. Private systems in particular regarded care as business and stockpiles as waste. Multinational for-profit companies owning care and elderly homes have failed dramatically to protect residents [140]. Without governments intervening, the burden was on healthcare workers without awareness that it was a shared interest of employers, healthcare workers and patients to provide safe care. The global supply chain of PPE was seriously disrupted while corrective actions (such as increased domestic production) came late. These examples illustrate that the market and its profit motive is not an adequate mechanism to address shortages of resources necessary for health [141]. A similar example is the initial shortfall of diagnostic tests, and the scarcity of laboratory equipment, machines and ingredients which could only be remediated with rigorous interventions of central governments. Another example is the development and productions of vaccines that have not been left to the market but were heavily supported by governmental and philanthropic funding. The irony is that while most of us are self-isolating and avoiding close contacts, the pandemic reinvigorated the sense of belonging and common purpose, requesting a more active role of the state and the abandonment of neoliberal policies of individualism [142].

Secondly, the role of the pharmaceutical industry is questioned. While the production of medicines is largely outsourced, and drug shortages, particularly for generic drugs, are common before the pandemic, Covid-19 highlights that the production of drugs is concentrated in only a few countries (especially India and China). Most key components (80% of active pharmaceutical ingredients) are produced in these two countries with only small stocks of essential drugs in other countries. Disruption of supply chains and export bans immediately affect numerous countries. Outsourcing and off-shoring drug production is a cheap business strategy but again a disadvantage in times of emergency which market ideology cannot solve [143]. That market incentives are insufficient is amply demonstrated in the case of Covid vaccines. Without massive injections of public money, they could not have been developed so quickly whereas most of the benefits go to the companies which are in control of pricing, production and distribution. Despite public investment, companies refuse to share patents and manufacturing knowledge which could bring the global spread of SARS-CoV-2 under control. Some pharmaceutical companies are undoubtedly the big winners in the pandemic. The announcement of company Merck that they will share the patent on molnupivar, as soon as this first oral antiviral Covid drug is approved by the FDA and EMA, so that it can be manufactured and sold cheaply in developing countries, should be an incentive to other, especially vaccine producing, companies to take their social responsibility more serious [144]. Pharmaceutical companies have special obligations in a global health emergence, and fair distribution of vaccines, with a priority for countries in great need, is one of them [145]. Amnesty International has pointed out that the companies that developed vaccines do not meet this obligation, and violate their own human rights commitments [146].

Thirdly, more than before, the pandemic makes global connectedness visible and palpable. It reveals itself not as usually an opportunity for trade and travel but as systemic vulnerability. Disconnections due to public health measures have severely impacted various countries. This has enhanced the awareness that at least some crucial services should not be outsourced at the global market, to reduce the dependency on a few production sites in Asia. The Serum Institute of India is the world's largest vaccine manufacturer but when the country prioritizes vaccination of its population, many countries were affected, and contributions to Covax seriously diminished. The vulnerability of global production chains has led to rethinking neoliberal policies. Until recently, the Netherlands for example, had its own institute to produce vaccines for governmental vaccination programs. In 2012 it was privatized and sold to the Serum Institute of India. Germany closed in 2017 one of last European factories for components of penicillin. Outsourcing and off-shoring of medical products is also associated with quality issues. Without a global framework it is difficult to assure the quality of diagnostic tests, drugs, and vaccines. Especially in times of a pandemic, substandard and falsified products will flood the market which has been experienced with inadequate face masks, tests, and hand sanitizer. Inspections are usually suspended because of public health measures [147]. Reversing the trend to outsourcing and stimulating domestic manufacturing of at least essential medical products again indicates that global trade cannot be left to the free market. Severely affected by Covid and confronted with chaotic and incoherent national policies, the European Union, once a bastion of neoliberal austerity, can no longer accept that healthcare and public health are a member state responsibility and has proposed to develop a union-wide public health policy, funding and setting up a common capacity to test for diseases, building stockpiles of key resources, a common disaster response agency, and a coordinating committee for health security [148].

Finally, global dependencies clarify that the idea of a dichotomy between health and economy is false [149]. Some policy-makers have argued that letting the virus spread in the population is the best way to preserve economic activities. Most policy-makers have struggled with the balance between health safety and reopening society to enable the economy to resume. In retrospect, it is clear that this choice between health and economy is incorrect. The most effective way to restore economic life is to curb the spread of the virus as fast and drastic as possible, Only when the infection is 'controlled', economic life will revive. Reopening society too soon will lead to more loss of life and protracted lockdowns so that it takes longer for the economy to recover [150].

Global bioethics can play a significant role in redirecting globalization. Since its beginning it has argued that many contemporary moral problems are related to processes of globalization and that a critical analysis of these processes is essential for the explanation and elucidation of moral problems. In particular, it has provided a broader scope, focusing attention to the social, political and economic context of ethical problems, and arguing for a reversal of priorities in policy and society. The economic and financial considerations that used to drive globalization should serve principles of social justice and human dignity rather than being regarded as ends in themselves. Policies should include strategies for social inclusion and institutional support for vulnerable people, engaging collective capacities to act, not relying on individual responsibility for public health measures. They should emphasize that public health is a common good, inspiring solidarity and a sharing ethos.

9.7 Conclusion

This chapter has started with the notion of global health, pointing out that whereas health and disease are localized experiences of individuals, they essentially connect all people across the world. This is especially true for infectious diseases as demonstrated in the Covid-19 pandemic. When a new virus affects human beings somewhere in the world, it is a potential threat to everyone since connectedness among humans can never be completely stopped. While policy responses follow the patterns of previous pandemics such as closing borders, interrupting travel, and quarantining and isolating people, viruses do not recognize domestic boundaries, making nationalistic policies ineffectual. Regarding health as a global concern has three implications. First is the need for global governance. The threat of an infectious disease can never be eliminated at a domestic level, perhaps only temporarily, but in the longer run, it continuous to be threatening as long as the pathogen is circulating among a few populations. If people in some countries are fully vaccinated, they still risk to be infected as long as the majority of the world population is not protected, while the uncontrolled circulation of the virus in other parts of the world increases the possibility for mutations which can jeopardize the advantages of fast vaccination in a limited number of countries. The second implication is that global health demands an ecological perspective. The origin as well as the transmission of emerging infectious diseases is fundamentally associated with human behavior. Humans are not isolated individuals but embedded in spheres that include other beings, not merely other humans but also animals. They are furthermore dependent for their survival and health on healthy ecosystems and a thriving biodiversity. Pandemics are therefore more than medical or virological events but social and political phenomena. Rather than an approach that focuses on contamination, a configuration view is necessary to understand how viral emergence and transmission are related to social and economic contexts. This highlights the third implication: global health is not the same for everyone. Socio-economic determinants of health differ within and between countries. That health disparities prevail in all countries has become clear in the Covid-19 pandemic: even in resource-rich countries with excellent healthcare systems, disadvantaged populations are disproportionately affected. Although everybody is vulnerable to SARS-CoV-2, it is evident that some people are more vulnerable because of age, comorbidities, poverty, race, ethnicity, and socio-economic circumstances. The pandemic exacerbates vulnerabilities and inequalities within and between countries, and it is an occasion to address the inequities that are illuminated.

The significance of global health accentuated by Covid-19 underscores the need for a more encompassing discourse of global bioethics. This chapter investigated how the scope of ethical debate used to be limited and focused on specific concerns. It is often orientated towards disease management, technocratic approaches, and individual treatment rather than attention to conditions in which diseases arise and expand. These tendencies are reinforced during the emergency of the pandemic, emphasizing controllability, exceptionality and binarity, thus structuring and formatting ethical considerations in a specific and narrow way, relegating concerns with vulnerability, human dignity, inequity, cooperation, and solidarity to a lower level of urgency and interest. In this chapter it is argued that the Covid-19 pandemic illustrates that another way of thinking and working is helpful to clarify the ethical dimensions of present-day life. The starting point of the argument is the basic relationality of human beings. This is not just the anthropological finding that as integrated wholes of body and soul, human persons are connected to other beings and the environing world but the philosophical view that being human means being-together. Human persons can only exist and flourish because they share the world, belong together, are present and available to each other. Basic relationality has ethical significance. Being situated within relationships and engaging with other beings means vulnerability. Openness to the world and mutual dependency is necessary to make human persons grow, develop and flourish but also exposes them to possible harm and injury. Humans cannot immunize themselves to the world since they would then lose what characterizes them as human beings. Ethical discourse is one way to mitigate and remediate vulnerability. It encourages us to change perspectives, and to imagine ourselves in the position of other people who have the same needs, desires and feelings as we have. It expresses that when we share the world, we must recognize others who have lives to live, and we must treat their interests as equal to our own.

Covid-19 has revived the collective memories of the past, especially of the influenza pandemic of one century ago. Humans now realize that they live in a pandemic era that begun in 1918 and that the idea that infectious diseases can be controlled is false. More than other disasters, Covid-19 has affected all dimensions of everyday life for all people across the globe. The spread of SARS-CoV-2 makes visible and tangible to everyone that human beings are connected, illustrating that globalization is a phenomenon of health and disease, and not simply of trade, travel, and finance. Globalization no longer is an abstract set of processes but an experience of mutual and personal vulnerability. Everybody is confronted with the same threat, and scientific knowledge of the virus is the same for everyone and rapidly shared across the globe. Nonetheless, responses to the pandemic are diverse and heterogenous. Some countries have managed the impact of the virus rapidly and efficiently, when in fact numerous others have bungled, delayed, and vacillated in applying public health measures. One reason why global strategies in the face of the pandemic differ has to do with values. The experience of contagion and responses to it necessitate choices about basic values: individual vs social responsibility; voluntary compliance and self-control vs state enforcement and external control; individual liberties vs solidarity, collective action and security; persuasive vs mandatory policies. That Covid-19 has ethical relevancy is furthermore manifested in the social inequities that it has revealed and aggravated. It exposes socio-economic and racial disparities in health and healthcare, as well as the privileges of people who have homes to shelter, and work that can be done remotely. Trends towards discrimination of elderly and disabled people are magnified, and stigmatization and scapegoating are not past. The pandemic also discloses the lack of preparedness of most countries and the insufficiency of public health infrastructures. Furthermore it clarifies that the economic order promoted by the neoliberal policies of globalization over the last few decades have led to the moral impoverishment of the social life-world and to multiplication of experiences of injustice, especially of humiliation, disrespect, and inequality.

For these reasons, the pandemic is an opportunity to rethink global governance, public health, and globalization with a new appreciation of the common good and the role of governments in protecting citizens, with more emphasis on resilience rather than efficiency. If bioethics as a social and global endeavor mobilizes the moral imagination in order to expand the scope of moral concern by applying the human capacity to empathize, it crucially contributes to enhancing social life and civilization.

References

1. Howard-Jones, N. 1975. *The scientific background of the international sanitary conferences, 1851–1938*. Geneva: World Health Organization.
2. Brown, T.M., M. Cueto, and E. Fee. 2006. The World Health Organization and the transition from "international" to "global" public health. *American Journal of Public Health* 96 (1): 62–72.
3. A PubMed search with the key-word 'global health' shows 564 publications in 1990, 2,071 in 2000, and 45,624 in 2020. See also: Holst, J. 2020. Global health – Emergence, hegemonic trends and biomedical reductionism. *Globalization and Health* 16, 42.
4. Ten Have, H.A.M.J. 2019. *Wounded planet. How declining biodiversity endangers health and how bioethics can help*. Baltimore: Johns Hopkins University Press, 50 ff.
5. Harrison, M. 2015. A global perspective: Reframing the history of health, medicine, and disease. *Bulletin of the History of Medicine* 89: 639–689.
6. For example, Koplan et al. define global health as "an area for study, research and practice that places a priority on improving health and achieving equity in health for all people worldwide." Koplan, J.P., T. C. Bond, M. H. Merson, et al. 2009. Towards a common definition of global health. *Lancet* 373, 1995. The notion of global health differs from public health in its geographical reach (public health is focused on countries, communities and even cities), its level of cooperation (public health usually does not require global cooperation), and objectives (public health mainly focuses on prevention activities for populations).
7. Planetary health is defined as "the health of human civilisation and the state of the natural systems on which it depends." Whitmee, S., A. Haines, C. Beyrer, et al., 2015. Safeguarding human health in the Anthropocene epoch. Report of the Rockefeller Foundation-*Lancet* Commission on planetary health. *Lancet* 386, 1978. See also, Horton, R., R. Beaglehole, R. Bonita, et al. 2014. From public to planetary health: A manifesto. *Lancet* 383, 847.
8. World Health Organization. 2007. *World health report 2007 – A safer future: Global public health security in the 21st century*. Geneva: WHO.
9. Ten Have, H. 2014. Vulnerability as the antidote to neoliberalism in bioethics. *Revista Redbioetica/UNESCO* 5 (1; no.9): 87–92; Ten Have, H. 2015. Respect for human vulnerability: The emergence of a new principle in bioethics. *Journal of Bioethical Inquiry* 12 (3): 395–408.
10. In 1995 the journal *Emerging Infectious Diseases* was launched. A PubMed search with the keyword 'emerging diseases' shows 1,651 publications in 1992, and 40,356 in 2020.
11. "Although it is impossible to predict their individual emergence in time and place, we can be confident that new microbial diseases will emerge." Institute of Medicine. 1992. *Emerging infections. Microbial threats to health in the United States*. Washington: National Academy Press, 23.
12. "In fact, environmental changes probably account for most emerging diseases." Institute of Medicine, *Emerging infections*, 42.
13. Pandemics "… almost always begin with the transmission of an animal microbe to a human…" Wolfe, N. 2011. *The viral storm. The dawn of a new pandemic age*. New York: St. Martin's Griffin, 3.
14. Wolfe, *The viral storm*, 161. See also, Wolfe, N. D., C. P. Dunavan, and J. Diamond. 2007. Origins of major infectious diseases. *Nature* 447: 279–283.
15. Wolfe, *The viral storm*, 242.
16. We are confronted with an "endless dance between microbes and humans." Khan, A.S. 2020. *The next pandemic. On the frontlines against humankind's gravest dangers*. New York: PublicAffairs, 255.
17. Mackenzie, D. 2020. *Covid-19. The pandemic that never should have happened, and how to stop it*. London: The Bridge Street Press, 233 ff.
18. Rosa, H. 2020. *The uncontrollability of the world*. Cambridge: Polity Press, viii.
19. Rosa, *The uncontrollability of the world*, 4.

20. Rosa, *The uncontrollability of the world*, 5 ff.
21. See for example, Macip, S. 2020. *Modern epidemics. From the Spanish flu to Covid-19.* Cambridge: Polity Press.
22. Khan, *The next pandemic*; Wolfe, *The viral storm.*
23. Weber, M. 1958. Science as a vocation. *Daedalus* 87 (1): 111–134.
24. Ten Have, H., and R. Pegoraro. 2021. *Bioethics, healthcare and the soul.* London: Routledge.
25. Mansnerus, E. 2013. Using model-based evidence in the governance of pandemics. In *Pandemics and emerging infectious diseases: The sociological agenda*, edited by R. Dingwall, L. M. Hoffman, and K. Staniland. Chichester: Wiley, 111.
26. Exceptions are, for example, Woesler, M., and H.-M. Sass, eds. 2020. *Medicine and ethics in times of corona.* Zürich: LIT Verlag; Child, B.H., and L. Vearrier, eds. 2021. Special issue: Covid 2020: Historical and ethical perspectives. *HEC Forum* 33 (1–2): 1–164.
27. One month later, in April 2021, similar adverse effects are observed with the Johnson & Johnson vaccine, although in more rare cases (1 in 1 million). Wallis, C. 2021. Few would fear Covid vaccines if policy makers explained their risks better. *Scientific American*, April 30.
28. Coutinho, R. 2021. *Vaxx. Hoe vaccinaties onze wereld beter hebben gemaakt.* Amsterdam: Ambo/Anthos, 152–153; Marsa, L. 2021. Can AstraZeneca dispel doubts about its shots? *National Geographic*, April 2.
29. Ledford, H. 2021. Covid vaccines and blood clots: Five key questions. *Nature* 592 (7855): 495–496.
30. Horton, R. 2020. *The Covid-19 catastrophe. What's gone wrong and how to stop it happening again.* Cambridge: Polity.
31. Fins, J.J. 2020. Covid-19 makes clear that bioethics must confront health disparities. *The Hastings Center*, July 9; Martins, A.A. 2021. Global bioethics in a pandemic: A dialogical approach. *Health Care Ethics USA.*
32. Berkowitz, S.A., C.W. Cené, and A. Chatterjee. 2020. Covid-19 and health equity – Time to think big. *New England Journal of Medicine* 383: e76(1)–e76(3).
33. Childs, B.H., and I. Vearrier. 2021. A journal of the Covid-19 (plague) year. *HEC Forum* 33: 1–6.
34. Baldwin, P. 2021. *Fighting the first wave. Why the coronavirus was tackled so differently across the globe*, 79. Cambridge: Cambridge University Press.
35. Baldwin, *Fighting the first wave*, 36, 112.
36. Burki, T.K. 2021. Herd immunity for Covid-19. *Lancet. Respiratory Medicine* 9 (2): 135–136.
37. Herstel-NL. 2021. *Het plan van herstel-NL.*
38. Symons, X. 2020. Should we sacrifice older people to save the economy? *BioEdge*, March 28.
39. Aschwanden, C. 2020. The false promise of heard immunity for Covid-19. *Nature* 587: 26–28.
40. Sabino, E.C., L.F. Buss, M.P.S. Carvalho, et al. 2021. Resurgence of Covid-19 in Manaus, Brazil, despite high seroprevalence. *Lancet* 397: 452–455; Taylor, L. 2021. Covid 19: Is Manaus the final nail in the coffin for natural herd immunity? *British Medical Journal* 372: m394.
41. See, for example Article 3 of the UNESCO *Universal Declaration on Bioethics and Human Rights*, 2005.
42. Symons, X. 2020. Is seeking 'herd immunity' ethical? *BioEdge*, May 9.
43. "Chronological age is a poor marker of vitality and ability to benefit from treatment." Ebrahim, S. 2002. The medicalization of old age. *British Medical Journal* 324: 862.
44. Fisher, A. 2013. Fair innings? Against healthcare rationing in favour of the young over the elderly. *Studies in Christian Ethics* 26 (4): 439; Hunt, R. W. 1993. A critique of using age to ration health care. *Journal of Medical Ethics* 19: 19–23.
45. Rivlin, M.M. 2000. Why the fair innings argument is not persuasive. *BMC Medical Ethics* 1: 1.
46. Wareham gives the example of South Africa where the vast majority of the elderly population has been deprived, discriminated and exploited during most of their lives. Wareham, C.S. 2015. Youngest first? Why it is wrong to discriminate against the elderly in healthcare. *South African Journal of Bioethics and Law* 8 (1): 37–39.

47. Williams, A. 1997. Intergenerational equity: An exploration of the 'fair innings' argument. *Health Economics* 6: 117–132.
48. Fisher, Fair innings?, 446; Bramstedt, K.A. 2001. Age-based health care allocation as a wedge separating the person from the patient and commodifying medicine. *Reviews in Clinical Gerontology* 11: 185–188; De Medeiros, K. 2020. A Covid-19 side effect: Virulent resurgence of ageism. *The Hastings Center*, May 14.
49. Peisah, C., A. Byrnes, I. Doron, et al. 2020. Advocacy for the human rights of older people in the Covid pandemic and beyond: A call to mental health professionals. *International Psychogeriatrics* 32 (10):1199–1204; Shortt, S. 2001. Venerable or vulnerable? Ageism in health care. *Journal of Health Services Research & Policy* 6 (1):1–2; De Medeiros, A Covid-19 side effect.
50. Ayalon, L. 2020. There is nothing new under the sun: ageism and intergenerational tension in the age of the Covid-19 outbreak. *International Psychogeriatrics* 32 (10): 1221–1224; Fraser, S., M. Lagacé, B. Bongué, et al. 2020. Ageism and Covid-19: What does our society's response say about us? *Age and Ageing* 49 (5): 692–695.
51. Brooke, J., and D. Jackson. 2020. Older people and Covid-19: Isolation, risk and ageism. *Journal of Clinical Nursing* 29: 2044–2046.
52. DutchNews.nl. 2020. *Dispensable? Dead wood? Coronavirus dilemma sparks social media debate.* August 10.
53. British Medical Association (BMA). 2021. *Statement/ briefing about the use of age and/ or disability in our guidance*; World Health Organization. 2021. *Global report on ageism.* Geneva: WHO.
54. Den Exter, A. 2020. The Dutch critical care triage guidelines on Covid-19: Not necessarily discriminatory. *European Journal of Health Law* 27 (5): 495–498.
55. Kuylen, M.N.I., S.Y. Kim, A.R. Keene, and G.S. Owen. 2021. Should age matter in Covid-19 triage? A deliberative study. *Journal of Medical Ethics* 47: 291–295.
56. See chapter 6.
57. De Castro, L., and J. Yasol-Naval. 2020. Sustainable Covid-19 response measures: An ethical imperative for enhancing core human capabilities. In *Medicine and ethics in times of corona*, edited by Woesler, M., and H-M. Sass. Zürich: LIT Verlag, 283–299.
58. It can be argued that physical distancing is also dependent on social conditions. Keeping physical distance to others is not merely controlled by individual decision-making but dependent on what other people do and what is appropriate in social interactions as a form of nonverbal communication. Kaminsky, C. 2020. Normality "ex post": social conditions of moral responsibility. In *Medicine and ethics in times of corona*, edited by Woesler, M., and H-M. Sass. Zürich: LIT Verlag, 63–74.
59. Weber, Science as vocation, 133.
60. Hadot, P. 1995. *Philosophy as a way of life. Spiritual exercises from Socrates to Foucault.* Malden: Blackwell.
61. Ten Have, and Pegoraro, *Bioethics, healthcare and the soul.*
62. Webb, K. 2016. *Ethical life. Its natural and social histories.* Princeton and Oxford: Princeton University Press.
63. Warnock, G.J. 1971. *The object of morality*. London: Methuen & Co.
64. Wilson, D., and W. Dixon. 2012. *A history of Homo Economicus. The nature of the moral in economic theory*. London and New York: Routledge; Hayek, F. 2005. *The road to serfdom.* Chicago and London: University of Chicago Press.
65. Bowles, S. 2016. *The moral economy. Why good incentives are no substitute for good citizens*. New Haven and London: Yale University Press; Granovetter, M. 2017. *Society and economy. Framework and principles*. Cambridge and London: The Belknap Press of Harvard University Press; Christakis, N.A. 2019. *Blueprint. The evolutionary origins of a good society*. New York, Boston, London: Little, Brown Spark.
66. Mackenzie, C., and N. Stoljar. 2000. *Relational autonomy: Feminist perspectives on autonomy, agency, and the social self.* New York: Oxford University Press.

67. Gintis, H. 2017. *Individuality and entanglement: The moral and material bases of social life*. Princeton and Oxford: Princeton University Press.
68. World Emergency Covid-19 Pandemic Ethics (WeCope) Committee. 2020. Statement on individual autonomy and social responsibility within a public health emergency. In *Medicine and ethics in times of corona*, edited by Woesler, M., and H-M. Sass. Zürich: LIT Verlag, 419–425.
69. World Emergency Covid-19 Pandemic Ethics (WeCope) Committee, Statement on individual autonomy and social responsibility within a public health emergency, 421; Macer, D. 2020. The foundation and functioning of the world emergency Covid-19 pandemic ethics committee. In *Medicine and ethics in times of corona*, edited by Woesler, M., and H-M. Sass. Zürich: LIT Verlag, 115–125.
70. Bieri, P. 2017. *Human dignity: A way of living*. Cambridge: Polity Press.
71. Midgley, M. 2014. *Are you an illusion?* 62. London and New York: Routledge.
72. Horton, R. 2004. Rediscovering human dignity. *The Lancet* 364: 1081–1085.
73. Sayer, A. 2011. *Why things matter to people. Social science, values and ethical life*, 193. Cambridge: Cambridge University Press.
74. Ten Have, H. 2016. *Global bioethics. An introduction*. London and New York: Routledge, 113 ff.
75. World Health Organization. 2021. *WHO Director-General's opening remarks at the media briefing on Covid-19* – 8 January 2021.
76. Eurasia Group. 2020. *Ending the Covid-19 pandemic: The need for a global approach*. November 25.
77. Hotez, P.J. 2021. *Preventing the next pandemic. Vaccine diplomacy in a time of anti-science*. Baltimore: Johns Hopkins University Press.
78. GAVI. 2021. *Covax Facility*; Our World in Data. 2021. *Coronavirus (Covid-19) vaccinations.*.
79. Safi, M. 2021. WHO platform for pharmaceutical firms unused since pandemic began. *The Guardian*, January 22; Billette de Villemeur, E., V. Dequiedt, and B. Versaevel. 2021. Pool patents to get Covid-19 vaccines and drugs to all. *Nature* 591: 529.
80. Thompson, A.K., K. Faith, J.L. Gibson, and R.E.G. Upshur. 2006. Pandemic influenza preparedness. An ethical framework to guide decision-making. *BMC Medical Ethics* 7: 12.
81. UNESCO. 2020. *Statement on Covid-19: Ethical considerations from a global perspective*. Paris, March 26; European Group on Ethics in Science and New Technologies. 2020. *Statement on European solidarity and the protection of fundamental rights in the Covid-19 pandemic*, April 2.
82. Dawson, A., E.J. Emanuel, M. Parker, M.J. Smith, and T.C. Voo. 2020. Key ethical concepts and their application to Covid-19 research. *Public Health Ethics* 13 (2): 127–132.
83. Vatican Covid-19 Commission/Pontifical Academy for Life. 2020. *Vaccine for all. 20 points for a fairer and healthier world*. December 29; Sahm, S. 2021. "Vaccine for all" – Prüfstein globaler Bioethik. *Zeitschrift für Medizinische Ethik* 67: 201–206.
84. This is the definition of solidarity provided by Dawson, Emanuel, Parker, Smith, and Voo, Key ethical concepts and their application to Covid-19 research, 128.
85. West-Oram, P.G.N., and A. Buyx. 2017. Global health solidarity. *Public Health Ethics* 10 (2): 212–224.
86. Ten Have, *Global bioethics*, 216–218; Jennings, B., and A. Dawson. 2015. Solidarity in the moral imagination of bioethics. *Hastings Center Report* 45 (5): 31–38.
87. Ten Have, H. 1993. Physicians' priorities – Patients' expectations. In: *Solidarity, justice and health care priorities,* edited by Szawarski, Z., and D. Evans. Linköping: Linköping University, 42–52.
88. Deutscher Ethikrat. 2020. *Solidarity and responsibility during the coronavirus crisis*. Berlin, 27 March, 5.
89. Phillips, T. 2021. Political chaos and poverty leave a continent at virus's mercy. *The Guardian*, May 2; Pagliarini, A. 2021. Latin America's lack of a united front on Covid has had disastrous

consequences. *The Guardian*, April 23; See also: Litewska, S.G., and J.D. Moreno. 2021. Covid-19 in Argentina and the abuse of bioethics. *The Hastings Center*, May 20.
90. Brown, G., and D. Susskind. 2020. International cooperation during the Covid-19 pandemic. *Oxford Review of Economic Policy* 36 (S1): S64–S76.
91. Jennings, and Dawson, Solidarity in the moral imagination of bioethics, 31–32.
92. West-Oram, and Buyx, Global health solidarity, 213.
93. Ho, A., and I. Dascalu. 2020. Global disparity and solidarity in a pandemic. *Hastings Center Report* 50 (3): 65–67; Venkatapuram, S. 2020. Covid-19 and the global ethics freefall. *The Hastings Center,* March 19; Ravitsky, V. 2020. Post-Covid bioethics. *The Hastings Center*, May 20; Heilinger, J-C., S. Venkatapuram, M. Voss, and V. Wild. 2020. Bringing ethics into the global coronavirus response. *The Hastings Center*, June 22.
94. Brands, H., and F.J. Gavin, eds. 2020. *Covid-19 and world order. The future of conflict, competition, and cooperation*. Baltimore: Johns Hopkins University Press.
95. "There is going to be nothing normal anywhere about the rest of this century… we can look forward to a continual, and accelerating series of crisis that will knock us off balance again and again." McKibben, B. 2020. The end of the world as we know it. *Times Literary Supplement* July 31: 4.
96. Žižek, S. 2020. *Pandemic! Covid-19 shakes the world*. New York: Polity Press.
97. Schwab, K., and T. Malleret. 2020. *Covid-19: The great reset*. Cologny/Geneva: World Economic Forum.
98. "…calamity can be an occasion for making intolerable social conditions visible – and for reforming them." Witt, J.F. 2020. *American contagions. Epidemics and the law from smallpox to Covid-19*. New Haven and London: Yale University Press, 140.
99. Horton, *The Covid-19 catastrophe*, 118 ff.
100. Horton, *The Covid-19 catastrophe*, 126.
101. "… we will have to learn to live in a viral world, a new way of living will have to be painfully reconstructed." Žižek, *Pandemic!*, 118.
102. "Troubled times could also lead to an extension of empathic consciousness – 'we're all in this together' – as we heighten our sensitivity to each other's common plight." Rifkin, J. 2009. *The empathic civilization. The race to global consciousness in a world in crisis*. Cambridge: Polity Press, 590.
103. Cipoletta, S., and M. C. Ortu. 2020. Covid-19: Common constructions of the pandemic and their implications. *Journal of Constructivist Psychology* 29 (4): 340–356; Komesaroff, P.A. 2020. Not all bad: Sparks of hope in a global disaster. *Journal of Bioethical Inquiry* 17: 515–518.
104. Baldwin, *Fighting the first wave*, 101.
105. Sample, I., and P. Walker. 2021. Covid response 'one of the UK's worst ever public health failures.' *The Guardian*, October 12.
106. KPMG. 2021. Dit zijn de lessen van 1,5 jaar coronacrisis. *KPMG Health*, October 14.
107. Schinkel, W. 2021. Het Nederlandse coronabeleid is een vorm van necropolitiek. *NRC*, July 26.
108. Christakis, N.A. 2020. *Apollo's arrow. The profound and enduring impact of coronavirus on the way we life*. New York, Boston and London: Little, Brown Spark; Calvert, J., and G. Arbuthnott. 2021. *Failures of state. The inside story of Britain's battle with coronavirus*. London: Mudlark; Baldwin, *Fighting the first wave*; Wright, L. 2021. *The Plague year. America in the time of Covid*. London: Allen Lane.
109. Windsor, L.C., G.Y. Reinhardt, A.J. Windsor, et al. 2020. Gender in the time of Covid-19: Evaluating national leadership and Covid-19 fatalities. *PLoS One* 15 (12): e0244531.
110. See, for example, Calvert, and Arbuthnott, *Failures of state*.
111. The indifference of ruling powers to protect human lives and their failures to respond adequately to the pandemic has been called "social murder." Abbasi, K. 2021. Covid-19: Social murder, they wrote – elected, unaccountable, and unrepentant. *British Medical Journal* 372: n314; Rosenthal, M.S., and A. Caplan. 2021. Why we need a Covid-19 commission. *The*

Hastings Center, March 3; Rosenthal, M.S., and A. Caplan. 2021. How to make it right: Covid reparations. *The Hastings Center*, March 15.
112. Baldwin, *Fighting the first wave*, 116.
113. Universal usage of face masks could have saved nearly 100,000 lives in the United States alone. Peeples, L. 2020. What the data say about wearing face masks. *Nature* 586: 186–189.
114. The genome sequence had already been identified by a Chinese laboratory at the beginning of January but Chinese authorities prohibited its publication. Voormolen, S. 2020. De eerste noodkreten kwamen rond Kerst. Die gingen in de doofpot. *NRC*, 19–20 December; Calvert, and Arbuthnott, *Failures of state*, 30 ff.
115. Horton, *The Covid-19 catastrophe*, 41 ff.
116. Davey, M. 2020. WHO warns Covid-19 pandemic is 'not necessarily the big one.' *The Guardian*, December 29.
117. Independent Panel for Pandemic Preparedness & Response. 2021. *Covid-19: Make it the last pandemic.*
118. "The world needs a new international system for pandemic preparedness and response, and it needs one fast, to stop future infectious disease outbreaks from becoming catastrophic pandemics." The Independent Panel for Pandemic Preparedness & Response. 2021. *Covid-19: Make it the last pandemic. A summary*, 1.
119. The Independent Panel for Pandemic Preparedness & Response, *Covid-19: Make it the last pandemic.*
120. Rechel, B. 2019. Funding for public health in Europe in decline? *Health Policy* 123 (1): 21–26.
121. Couto, M. 2020. A message from the pangolins. *Times Literary Supplement*, August 7: 11; Pfeiffer, J., and R.R. Chapman. 2019. NGOs, austerity, and universal health coverage in Mozambique. *Globalization and Health* 15, 0.
122. Jambroes, M., T. Nederland, M. Kaljouw, et al. 2016. Implications of health as 'the ability to adapt and self-manage' for public health policy: A qualitative study. *European Journal of Public Health* 26 (3): 412–416; Baldwin, *Fighting the first wave*, 170 ff; Witt, *American contagions.*
123. Jennings, B. 2020. Beyond the Covid crisis – A new social contract with public health. *The Hastings Center.*
124. The Independent Panel for Pandemic Preparedness & Response, *Covid-19: Make it the last pandemic;* Grange, Z.L., T. Goldstein, C.K. Johnson, et al. 2021. Ranking the risk of animal-to-human spillover for newly discovered viruses. *Proceedings of the National Academy of Sciences of the United States of America* 118 (15): e2002324118.
125. Hinchliffe, S. 2015. More than one world, more than one health: Re-configuring interspecies health. *Social Science & Medicine* 129: 29.
126. At the global level, it implies cooperation between the World Health Organization, the World Organization for Animal Health, and the Food and Agricultural Organization. Jerolmack, C. 2013. Who's worried about turkeys? How 'organisational silos' impede zoonotic disease surveillance. In *Pandemics and emerging infectious diseases. The sociological agenda*, edited by Dingwal, R., L. M. Hoffman, and K. Staniland. Chichester: Wiley-Blackwell, 33–45; Chien, Y-J. 2013. How did international agencies perceive the avian influenza problem? The adoption and manufacture of the 'One World, One Health' framework. In *Pandemics and emerging infectious diseases. The sociological agenda*, edited by Dingwal, R., L.M. Hoffman, and K. Staniland. Chichester: Wiley-Blackwell, 46–58; Thoradeniya, T., and S. Jayasinghe. 2021. Covid-19 and future pandemics: A global systems approach and relevance to SDGs. *Globalization and Health* 17: 59.
127. For this point, see Hinchliffe, More than one world, more than one health.
128. Spicer, N., I. Agyepong, I. Ottersen, A. Jahn, and G. Ooms. 2020. "It's far too complicated": Why fragmentation persists in global health. *Globalization and Health* 16 (1): 60.
129. The Independent Panel for Pandemic Preparedness & Response, *Covid-19.*
130. Gostin, L.O., S. Moon, and B.M. Meier. 2020. Reimagining global health governance in the age of Covid-19. *American Journal of Public Health* 110 (11): 1615–1619.

131. Campbell, L. 2021. Global treaty needed to protect states from pandemics, say world leaders. *The Guardian*, March 30; Editorial. 2021. Learn from Covid before diving into a pandemic treaty. *Nature* 592; 165–166.
132. Venkatapuram, S., and A.C. Zielinska. 2021. Covid vaccine patent waivers are for health sovereignty. *The Hastings Center*, June 1.
133. Brown, G., and D. Susskind. 2020. International cooperation during the Covid-19 pandemic. *Oxford Review of Economic Policy* 36: S64–S76; Rutkow, L. 2020. Origins of the Covid-19 pandemic and the path forward. A global public health policy perspective. In *Covid-19 and world order. The future of conflict, competition, and cooperation*, edited by H. Brand, and F.J. Gavin. Baltimore: Johns Hopkins University Press, 93–113.
134. Wenham, C. 2021. What went wrong in the global governance of covid-19? *British Medical Journal* 372: n303: Wilson, K., S. Halabi, and L.O. Gosting. 2020. The International Health Regulations (2005), the threat of populism and the Covid-19 pandemic. *Globalization and Health* 16: 70; Maxmen, A. 2021. Why did the world's pandemic warning system fail when Covid hit? *Nature* 589: 499–500.
135. World Health Organization. 2009. *Global health risks. Mortality and burden of disease attributable to selected major risks*, 24. Geneva: WHO Press.
136. Zwart, H. 2020. Emerging viral threats and the simultaneity of the non-simultaneous: Zooming out in times of Corona. *Medicine, Health Care and Philosophy* 23: 589–602.
137. Sulaman, M., X. Long, and M. Salman. 2020. COVID-19 pandemic and environmental pollution: A blessing in disguise? *Science of the Total Environment* 728: 138820.
138. Pontifical Academy for Life. 2020. *Humana communitas in the age of pandemic: Untimely meditations on life's rebirth.* Vatican City, July 22, 8.
139. Brands, and Gavin, eds., *Covid-19 and world order.*
140. Grabowski, D.C. 2021. The future of long-term care requires investment in both facility- and home-based services. *Nature Aging* 1: 10–11.
141. Cohen, J., and Y. van der Meulen Rodgers. 2020. Contributing factors to personal protective equipment shortages during the Covid-19 pandemic. *Preventive Medicine* 141: 106263.
142. Žižek, S. 2020. *Pandemic! Covid-19 shakes the world.* New York: Polity Press, 42 ff.
143. Gustafsson, L. 2020. Covid-19 highlights problems with our generic supply chain. *The Commonwealth Fund*, May 7.
144. Nolen, S. 2021. Merck will share formula of its Covid pill with poor countries. *New York Times*, October 27.
145. Emanuel, E.J., A. Buchanan, S.Y. Chan, et al. 2021. What are the obligations of pharmaceutical companies in a global health emergency? *Lancet* 398: 1015–1020.
146. Johnson, S. 2021. Big pharma fuelling human rights crisis over Covid vaccine inequity – Amnesty. *The Guardian*, September 22.
147. Newton, P.N., K.C. Bond, et al. 2020. Covid-19 and risks to the supply and quality of tests, drugs, and vaccines. *The Lancet Global Health* 8 (6): e754–e755.
148. Editorial. 2020. Europe must think more globally in crafting its pandemic response. *Nature* 587: 329; Clemens, T., and H. Brand. 2020. Will Covid-19 lead to a major change of the EU Public Health mandate? A renewed approach to EU's role is needed. *The European Journal of Public Health* 30 (4): 625–626.
149. Inglesby, T. 2020. Make pandemics lose their power. In *Covid-19 and world order. The future of conflict, competition, and cooperation*, edited by H. Brands, and F.J. Gavin. Baltimore: Johns Hopkins University Press, 75–92.
150. Calvert, and Arbuthnott, *Failures of state*, 9, 279, 309, 340.

Bibliography

Abbas, M.Z. 2020. Treatment of the novel Covid-19: Why Costa Rica's proposal for the creation of a global pooling mechanism deserves serious consideration. *Journal of Law and the Biosciences* 7 (1): 1–10.

Abbasi, J. 2021. Study suggests lasting immunity after Covid-19, with a big boost from vaccination. *JAMA* 326 (5): 376–377.

Abbasi, K. 2021. Covid-19: Social murder, they wrote – Elected, unaccountable, and unrepentant. *British Medical Journal* 372: n314.

Abbott, J., D. Johnson, and M. Wynia. 2020. Ensuring adequate palliative and hospice care during Covid-19 surges. *JAMA* 324 (14): 1393–1394.

Abraham, T. 2007. *Twenty-first century plague. The story of SARS*. Baltimore: The Johns Hopkins University Press.

Access to Medicine Foundation. 2021. *Access to Medicine Index*, 25 January. https://accesstomedicinefoundation.org/news/now-online-2021-access-to-medicine-index#:~:text=Amsterdam%2C%20the%20Netherlands%2C%2026%20January,to%20Medicine%20Index%2C%20published%20Tuesday.

Adam, M. 2020. An enemy to fight or someone to live with, how Covid-19 is metaphorically described in Indonesian media discourse. *Conference paper*, July, 2nd National Webinar on English Linguistics and Literature.

Adams, J.G., and R.M. Walls. 2020. Supporting the health care workforce during the Covid-19 global epidemic. *JAMA* 323 (15): 1439–1450.

Agence France-Presse. 2021. US joins calls for transparent, science-based investigation into Covid origins. *The Guardian*, May 26. https://www.theguardian.com/world/2021/may/26/us-joins-calls-for-transparent-science-based-investigation-into-covid-origins.

Agence France-Press. 2021. 'Last chance': WHO reveals new team to investigate Covid origins. *The Guardian*, October 14. https://www.theguardian.com/world/2021/oct/14/last-chance-who-reveals-new-team-to-investigate-covid-origins.

Agoramoorthy, G., M.J. Hsu, and P. Shieh. 2020. Queries on the Covid-19 quick publishing ethics. *Bioethics* 34 (6): 633–634.

Akinwotu, E. 2021. Tanzania's missing president is in Kenya with Covid, says opposition leader. *The Guardian*, March 10. https://www.theguardian.com/world/2021/mar/10/tanzania-missing-president-kenya-covid-says-opposition-leader.

Aksoy, C.G., M. Ganslmeier, and P. Poutvaara. 2020. *Public attention and policy responses to Covid-19 pandemic*. Bonn: IZA Institute of Labor Economics. http://ftp.iza.org/dp13427.pdf.

Alicandro, G., G. Remuzzi, and C. La Vecchia. 2020. Italy's first wave of the Covid-19 pandemic has ended: No excess mortality in May, 2020. *Lancet* 396 (10253): e27–e28.

H. ten Have, *The Covid-19 Pandemic and Global Bioethics*, Advancing Global Bioethics 18, https://doi.org/10.1007/978-3-030-91491-2

Al Jazeera Staff. 2021. Young people first: Indonesia's Covid vaccine strategy questioned. *Aljazeera*, January 13. https://www.aljazeera.com/news/2021/1/13/young-people-first-indonesias-covid-vaccine-strategy-questioned.

Allen, A. 2020. For billion-dollar Covid vaccines, basic government-funded science laid the groundwork. *Scientific American*, November 18. https://www.scientificamerican.com/article/for-billion-dollar-covid-vaccines-basic-government-funded-science-laid-the-groundwork/.

Allen, T., K.A. Murray, C. Zambrana-Torrelio, S.S. Morse, C. Rondini, M. Di Marco, N. Breit, K.J. Olival, and P. Daszak. 2017. Global hotspots and correlates of emerging zoonotic diseases. *Nature Communications* 8 (1): 1–10.

Almond, J.W. 2007. Vaccine renaissance. *Nature Reviews Microbiology* 5: 478–481.

American Red Cross. 2003. *Fact sheet on shelter-in-place*. https://www.redcross.org/content/dam/redcross/atg/PDF_s/Preparedness___Disaster_Recovery/Disaster_Preparedness/Terrorism/shelterinplace.pdf.

Andersen, K.G., A. Rambaut, W.I. Lipkin, E.C. Holmes, and R.F. Garry. 2020. The proximal origin of SARS-CoV-2. *Nature Medicine* 26: 450–452.

Anderson, T. 2021. Covid-19 vaccines: Resolving deployment challenges. *Bulletin of the World Health Organization* 99: 174–175.

Anonymous. 2012. Challenges in critical care in Africa: Perspectives and solutions. *ICU Management & Practice* 12 (4). https://healthmanagement.org/c/icu/issuearticle/challenges-in-critical-care-in-africa-perspectives-and-solutions.

Arora, A., and A. Arora. 2020. Ethics in the age of Covid-19. *Internal and Emergency Medicine* 15: 889–890.

Artenstein, A.W. 2012. The discovery of viruses: Advancing science and medicine by challenging dogma. *International Journal of Infectious Diseases* 16: e470–e473.

Arya, A., S. Buchman, B. Gagnon, and J. Downar. 2020. Pandemic palliative care: Beyond ventilators and saving lives. *Canadian Medical Association Journal* 192: E400–E404.

Aschwanden, C. 2020. The false promise of herd immunity. *Nature* 587: 26–28.

Asghari, F., A. Parsapour, and E.S. Gooshk. 2020. Priority Setting of Ventilators in the COVID-19 Pandemic from the public's perspective. *medRxiv preprint*; https://doi.org/10.1101/2020.06.10.2012779.

Åslund, A. 2020. Responses to the Covid-19 crisis in Russia, Ukraine, and Belarus. *Eurasian Geography and Economics* 61 (4-5): 1–14.

Associated Press. 2021. 'Wildly unfair': UN says 130 countries have not received a single Covid vaccine dose. *The Guardian*, February 18. https://www.theguardian.com/world/2021/feb/18/wildly-unfair-un-says-130-countries-have-not-received-a-single-covid-vaccine-dose.

———. 2021. Canada suspends use of AstraZeneca Covid vaccine for those under 55. *The Guardian*, March 30. https://www.theguardian.com/world/2021/mar/30/canada-suspends-use-of-astrazeneca-covid-vaccine-for-those-under-55.

Audureau, W., G. Dagom, A. Maad, and J. Parienté. 2020. Covid-19: 54 scientifiques évaluent la stratégie sanitaire. *Le Monde*, September 30. https://www.lemonde.fr/les-decodeurs/article/2020/09/30/masques-mesures-communication-comment-les-scientifiques-jugent-la-strategie-sanitaire-face-au-covid-19_6054250_4355770.html.

Auriemma, C.L., M.O. Harhay, K.J. Haines, et al. 2021. What matters to patients and their families during and after critical illness: A qualitative study. *American Journal of Critical Care* 38 (1): 11–20.

Ayalon, L. 2020. There is nothing new under the sun: Ageism and intergenerational tension in the age of the Covid-19 outbreak. *International Psychogeriatrics* 32 (10): 1221–1224.

Bachelard, G. 2014. *The poetics of space*. New York: Penguin Books.

Baden, L.R., H.M. El Sahly, B. Essink, et al. 2021. Efficacy and safety of the mRNA-1273 SARS-CoV-2 vaccine. *New England Journal of Medicine* 384 (5): 403–416.

Badre, J. 2021. How can we deal with 'pandemic fatigue' *Scientific American*, January 24. https://www.scientificamerican.com/article/how-we-can-deal-with-pandemic-fatigue/.

Badshah, N. 2021. British Medical Association says 'time is now' for Covid plan B. *The Guardian*, October 20. https://www.theguardian.com/society/2021/oct/20/british-medical-association-says-time-is-now-for-covid-plan-b.

Baker, M.G., and N. Wilson. 2020. Successful elimination of Covid-19 transmission in New Zealand. *New England Journal of Medicine* 383 (8): e56.

Back, A., J.A. Tulsky, and R.M. Arnold. 2020. Communication skills in the age of Covid-19. *Annals of Internal Medicine* 172 (11): 759–760.

Baldwin, P. 2021. *Fighting the first wave. Why the coronavirus was tackled so differently across the globe*. Cambridge: Cambridge University Press.

Ball, P. 2021. What the lighting-fast quest for Covid vaccines means for other diseases. *Nature* 589: 16–18.

Banerjee, R., J. Bhattacharya, and P. Majumbar. 2020. *Exponential-growth prediction bias and compliance with safety measures in the times of COVID-19*. Bonn: IZA Institute of Labor Economics. http://ftp.iza.org/dp13257.pdf.

Bargain, O., and U. Aminjonov. 2020. Trust and compliance to public health policies in times of Covid-19. *Journal of Public Economics* 192: 104316. https://doi.org/10.1016/j.jpubeco.2020.104316.

Barilan, Y.M., and M. Brusa. 2016. Triage. In *Encyclopedia of global bioethics*, ed. H. ten Have, vol. 3, 2839–2847. Cham: Springer.

Barratt, R.L., R. Shaban, and W. Moyle. 2011. Patient experience of source isolation: Lessons for clinical practice. *Contemporary Nurse* 39 (2): 180–193.

Barry, J.M. 2005. 1918 revisited: Lessons and suggestions for further inquiry. In *The threat of pandemic influenza: Are we ready? Workshop summary*, ed. S.L. Knobler, A. Mack, A. Mahmoud, et al., 58–68. Washington: Institute of Medicine (US) Forum on microbial threats.

———. 2018. *The great influenza. The story of the deadliest pandemic in history*. New York: Random House. (original 2004).

Bates, B.R. 2020. The (in)appropriateness of the WAR metaphor in response to SARS-CoV-2: A rapid analysis of Donald J. Trump's rhetoric. *Frontiers in Communication* 5: 50. https://doi.org/10.3389/fcomm.2020.00050.

BBC News, 7 October. https://www.bbc.com/news/world-africa-54418613.

Beale, R. 2020. Short cuts. *London Review of Books*, 21 May.

Beaumont, T. 2020. 'Landmark moment': 156 countries agree to Covid vaccine allocation deal. *The Guardian*, September 21. https://www.theguardian.com/global-development/2020/sep/21/landmark-moment-156-countries-agree-to-covid-vaccine-allocation-deal.

Beaumont, P. 2021. UK and US criticize WHO's Covid report and accuse China of withholding data. *The Guardian*, March 30. https://www.theguardian.com/world/2021/mar/30/who-criticises-chinas-data-sharing-as-it-releases-covid-origins-report.

———. 2021. Did Covid come from a Wuhan lab? What we know so far. *The Guardian*, May 27. https://www.theguardian.com/world/2021/may/27/did-covid-come-from-a-wuhan-lab-what-we-know-so-far

Beduschi, A. 2020. *Digital health passports for Covid-19: Data privacy and human rights law*. University of Exeter. https://socialsciences.exeter.ac.uk/media/universityofexeter/collegeofsocialsciencesandinternationalstudies/lawimages/research/Policy_brief_-_Digital_Health_Passports_COVID-19_-_Beduschi.pdf.

Bedyński, W. 2020. Liminality: Black Death 700 years later. What lessons are for us from the medieval pandemic? *Society Register* 4 (3): 129–144.

Belayneh, A. 2020. Off-Label Use of Chloroquine and Hydroxychloroquine for COVID-19 Treatment in Africa Against WHO Recommendation. *Research and Reports in Tropical Medicine* 11: 61–72.

Bell, V. 2020. French care home where staff locked themselves in with patients for 47 days avoids coronavirus. *Yahoo News* UK, May 4. https://uk.news.yahoo.com/french-care-home-where-staff-locked-themselves-in-with-patients-for-47-days-avoids-coronavirus-120331650.html?guccounter=1&guce_referrer=aHR0cHM6Ly93d3cuZ29vZ2x

lLmNvbS8&guce_referrer_sig=AQAAAIgkeDv22NZtEZ25X2Qa229ZMSgSXDgrz3wFoxJhZSQJoxObmnYLIUpD9ltW_Rs_WspfJqt052gjx4nmxSRq9qhfnb08NVu68elTgxVjEasp_ezyqN3EpJvpJnCzuvgFTWriAqL-YKsSuoqtQcjfq3TlD_ke8rhzwfaqOmYVyjFn.

Bell, A., and T. Gift. 2020. 200,00 losses later. Some lessons for the 'wartime president' *Newsweek*, September 22. https://www.newsweek.com/trump-wartime-president-coronavirus-losses-1533573.

Bellagio Initiative on the Global Virome Project. 2016. https://static1.squarespace.com/static/581a4a856b8f5bc98311fb03/t/582120e4ff7c5080cc611fd6/1478566120350/GVP+Bellagio+Initiative.pdf.

Bellieni, C. 2021. Ethical drawbacks of treating doctors as heroes during the COVID pandemic. *Academia Letters*, Article 2700; https://doi.org/10.20935/AL2700.

Benedictow, O.J. 2004. *The Black Death 1346–1353. The complete history*. Woodbridge: The Boydell Press.

Benezra, A., J. DeStefano, and J.I. Gordon. 2012. Anthropology of Microbes. *Proceedings of the National Academy of Sciences of the United States of America* 109 (17): 6378–6381.

Berkessel, J.B., T. Ebert, J.E. Gebauer, T. Jonsson, and S. Oishi. 2021. Pandemics initially spread among people of higher (not lower) social status: Evidence from Covid-19 and the Spanish fly. *Social Psychological and Personality Science* 11: 20098. https://doi.org/10.1038/s41598-021-99060-y.

Berkowitz, S.A., C.W. Cené, and A. Chatterjee. 2020. Covid-19 and health equity – Time to think big. *New England Journal of Medicine* 383: e76(1)–e76(3).

Berlinger, N., M. Wynia, T. Powell, et al. 2020. *Ethical framework for health care institutions responding to novel coronavirus SARS-CoV-2 (Covid-19)*. The Hastings Center, March 16. https://www.thehastingscenter.org/ethicalframeworkcovid19/.

———. 2021. *Ethical challenges in the middle tier of Covid-19 vaccine allocation: Guidance for organizational decision-making*. The Hastings Center, January 15. https://www.thehastingscenter.org/wp-content/uploads/COVID-guidelines-supplement-vaccines-2.pdf.

Berlinger, N., J. Abbott, A. Milliken, et al. 2020. *Access to therapeutic and palliative drugs in the context of Covid-19*. The Hastings Center, July 14. https://www.sicp.it/wp-content/uploads/2020/03/The-Hastings-Center-Covid-Ethical-Framework-Supplement-Drugs-14-Jul-20.pdf.

Bernal, J. L., N. Andrews, C. Gower, et al. 2021. Early effectiveness of Covid-19 vaccination with BNT162b2 mRNA vaccine and ChAdOx1 adenovirus vector vaccine on symptomatic disease, hospitalization and mortality in older adults in England. *MedRxiv preprint*, March 2; https://doi.org/10.1101/2021.03.01.21252652.

Bertherat, E. 2019. Plague around the world in 2019. *Weekly Epidemiological Record* 25: 289–292.

Bhatia, N. 2020. We need to talk about rationing: The need to normalize discussion about healthcare rationing in a post Covid-19 era. *Journal of Bioethical Inquiry* 17: 731–735.

Bieri, P. 2017. *Human dignity: A way of living*. Cambridge: Polity Press.

Blastland, M., A.L.J. Freeman, S. van der Linden, et al. 2020. Five rules for evidence communication. *Nature* 587: 362–364.

BMA Survey. 2020. 14 September. https://www.bma.org.uk/media/3233/bma_second_peak_survey_england_september_2020.pdf.

Bierer, B.E., S.A. White, J.M. Barnes, and L. Gelinas. 2020. Ethical challenges in clinical research during the Covid-19 pandemic. *Journal of Bioethical Inquiry* 17: 717–722.

Billette de Villemeur, E., V. Dequiedt, and B. Versaevel. 2021. Pool patents to get Covid-19 vaccines and drugs to all. *Nature* 591: 529.

Blinderman, C.D., R. Adelman, D. Kumaraiah, et al. 2021. A comprehensive approach to palliative care during the coronavirus pandemic. *Journal of Palliative Medicine* 24 (7): 1017–1022.

Boccaccio, G. 1353. *The Decameron*. https://docs.google.com/viewer?a=v&pid=sites&srcid=ZGVmYXVsdGRvbWFpbnxjaWFzY3VuZGVsbGFicmlnYXRhfGd4OmVmZGZiNTYyZTBhMzczZg.

Boccia, S., W. Ricciardi, and J.P.A. Ioannidis. 2020. What other countries can learn from Italy during the Covid-19 pandemic. *JAMA Internal Medicine* 180 (7): 927–928.

Boffey, D. 2021. Vaccine row: EU has exported 34 m doses – Including 9m to the UK. *The Guardian*, March 10. https://www.theguardian.com/world/2021/mar/10/britain-has-no-ban-on-covid-vaccine-exports-eu-concedes.

Boffey, D., and J. Elgot. 2021. EU to widen criteria for possible Covid vaccine export bans. *The Guardian*, March 23. https://www.theguardian.com/society/2021/mar/23/eu-expand-criteria-used-decide-block-covid-vaccine-shipments.

Boin, A., W. Overdijk, C. van der Ham, J. Hendriks, and D. Sloof. 2020. *Covid-19. Een analyse van de nationale crisisresponse*. Leiden: The Crisis University Press.

Bolsen, T., R. Palm, and J.T. Kingsland. 2020. Framing the origins of Covid-19. *Science Communication* 42 (5): 562–585.

Boseley, S. 2020. US secures world stock of key Covid-19 drug remdesivir. *The Guardian*, 30 June. https://www.theguardian.com/us-news/2020/jun/30/us-buys-up-world-stock-of-key-covid-19-drug.

———. 2020. US and UK 'lead push against global patent pool for Covid-19 drugs.' *The Guardian*, May 7. https://www.theguardian.com/world/2020/may/17/us-and-uk-lead-push-against-global-patent-pool-for-covid-19-drugs.

———. 2021. WHO chief: Waive Covid vaccine patents to put the world on 'war footing.' *The Guardian*, March 5. https://www.theguardian.com/world/2021/mar/05/covid-vaccines-who-chief-backs-patent-waiver-to-boost-production.

———. 2021. Covid: AstraZeneca vaccine 79% effective with no increased blood clot risk – US trial. *The Guardian*, March 22. https://www.theguardian.com/society/2021/mar/22/astrazeneca-covid-vaccine-79-effective-with-no-increased-blood-clot-risk-us-trial.

———. 2021. US agency questions AstraZeneca's Covid vaccine trial data. *The Guardian*, March 23. https://www.theguardian.com/business/2021/mar/23/us-health-agency-astrazeneca-covid-vaccine-trial-data.

Boshart, M. 2016. *De blauwe dood. Cholera in Nederland*. Soesterberg: Uitgeverij Aspekt.

Bowen, L. 2021. Whose underlying conditions count for priority in getting the vaccine? *Scientific American*, February 6. https://www.scientificamerican.com/article/whose-underlying-conditions-count-for-priority-in-getting-the-vaccine/.

Bowen, R.A.R. 2021. Ethical and organizational considerations for mandatory Covid-19 vaccination of health care workers: A clinical laboratorian's perspective. *Clinica Chimica Acta* 510: 421–522.

Bowles, S. 2016. *The moral economy. Why good incentives are no substitute for good citizens*. New Haven/London: Yale University Press.

Bramstedt, K.A. 2001. Age-based health care allocation as a wedge separating the person from the patient and commodifying medicine. *Reviews in Clinical Gerontology* 11: 185–188.

———. 2020. Antibodies as Currency: COVID-19's Golden Passport. *Bioethical Inquiry* 17: 687–689.

———. 2020. The carnage of substandard research during the Covid-19 pandemic: A call for quality. *Journal of Medical Ethics* 46: 803–807.

Brands, H., and F.J. Gavin, eds. 2020. *Covid-19 and world order. The future of conflict, competition, and cooperation*. Baltimore: Johns Hopkins University Press.

Brauner, J.M., S. Mindermann, M. Sharma, et al. 2020. Inferring the effectiveness of government interventions against Covid-19. *Science* 371 (6531): eabdg338.

British Medical Association. 2020. *Covid-19: Refusing to treat where PPE is inadequate*. https://www.bma.org.uk/advice-and-support/covid-19/ppe/covid-19-refusing-to-treat-where-ppe-is-inadequate.

———. 2020. *Covid-19 – Ethical issues. A guidance note*. https://www.bma.org.uk/media/2226/bma-covid-19-ethics-guidance.pdf.

British Medical Association (BMA). 2021. *Statement/ briefing about the use of age and/or disability in our guidance*. https://www.bma.org.uk/media/2358/bma-statement-about-ethics-guidance-and-disability-april-2020.pdf.

Broadbent, A. 2019. *Philosophy of medicine*. New York: Oxford University Press.
Brody, H., and E.N. Avery. 2009. Medicine's duty to treat pandemic illness: Solidarity and vulnerability. *Hastings Center Report* 39 (1): 40–48.
Brooke, J., and D. Jackson. 2020. Older people and Covid-19: Isolation, risk and ageism. *Journal of Clinical Nursing* 29: 2044–2046.
Brown, B. 2020. *Ethics of emergency use authorization during the pandemic*. The Hastings Center, October 30. https://www.thehastingscenter.org/ethics-of-emergency-use-authorization-during-the-pandemic/.
Brown, J. 2018. *Influenza. The hundred-year hunt to cure the deadliest disease in history*. New York: Simon & Schuster.
Brown, T.M., M. Cueto, and E. Fee. 2006. The World Health Organization and the transition from "international" to "global" public health. *American Journal of Public Health* 96 (1): 62–72.
Brown, G., and D. Susskind. 2020. International cooperation during the Covid-19 pandemic. *Oxford Review of Economic Policy* 36 (S1): S64–S76.
Bruni, L. 2012. *The wound and the blessing. Economics, relationships, and happiness*. Hyde Park/New York: New City Press.
Bruno, B., and S. Rose. 2020. Patients left behind: Ethical challenges in caring for indirect victims of the Covid-19 pandemic. *Hastings Center Report* 50 (4): 19–23.
Bubar, K.M., K. Reinholt, S.M. Kissler, et al. 2021. Model-informed Covid-19 vaccine prioritization strategies by age and serostatus. *Science* 371 (6532): 916–921.
Buchanan, K., L.B. Aknin, S. Lotun, and G.M. Sandstrom. 2021. Brief exposure to social media during the Covid-19 pandemic: Doom-scrolling has negative emotional consequences, but kindness-scrolling does not. *PLoS ONE* 16 (10): e0257728. https://doi.org/10.1371/journal.pone.0257728.
Buchholz, K. 2021. Where coronavirus vaccines will be produced. *Statista*, January 8. https://www.statista.com/chart/23885/coronavirus-vaccine-production-capabilities-by-country/.
Buckwalter, W., and A. Peterson. 2020. Public attitudes toward allocating scarce resources in the COVID-19 pandemic. *PLoS ONE* 15 (11): e0240651. https://doi.org/10.1371/journal.pone.0240651.
Bundgaard, H., J.S. Bundgaard, D.E.T. Raaschau-Pedersen, et al. 2021. Effectiveness of adding a mask recommendation to other public health measures to prevent SARS-CoV-2 infection in Danish mask wearers. *Annals of Internal Medicine* 174 93): 335–343.
Burgess, A., and M. Horii. 2012. Risk, ritual and health responsibilisation: Japan's 'safety blanket' of surgical face mask-wearing. *Sociology of Health & Illness* 34 (8): 1184–1198.
Burki, T. 2020. The origin of SARS-CoV-2. *The Lancet Infectious Diseases* 20 (9): 1018–1019.
———. 2021. Understanding variants of SARS-CoV-2. *Lancet* 397: 462.
Burki, T.K. 2021. Herd immunity for Covid-19. *Lancet. Respiratory Medicine* 9 (2): 135–136.
Burton, J.K., G. Bayne, C. Evans, et al. 2020. Evolution and effects of Covid-19 outbreaks in care homes: A population analysis of 189 care homes in one geographical region of the UK. *Lancet Healthy Longevity* 1: e21–e31.
Buruma, I. 2020. Virus as metaphor. *New York Times*, March 28. https://www.nytimes.com/2020/03/28/opinion/coronavirus-racism-covid.html.
Cacciapaglia, G., C. Cot, and F. Sannino. 2020. Second wave Covid-19 pandemics in Europe: A temporal playbook. *Scientific Reports* 10: 15514. https://doi.org/10.1038/s41598-020-72611-5.
Callaway, E. 2020. Making sense of coronavirus mutations. *Nature* 585: 174–177.
———. 2021. Multitude of coronavirus variants found in the US – But the threat is unclear. *Nature* 591: 190.
Callaway, E., and H. Ledford. 2021. How to redesign Covid vaccines so they protect against variants. *Nature* 590: 15–16.
Calvert, J., and G. Arbuthnott. 2021. *Failures of state. The inside story of Britain's battle with coronavirus*. London: Mudlark.
Campbell, L. 2021. Global treaty needed to protect states from pandemics, say world leaders. *The Guardian,* March 30. https://www.theguardian.com/world/2021/mar/30/global-treaty-needed-to-protect-states-from-pandemics-say-world-leaders.

Cantor, N.F. 2001. *In the wake of the plague. The Black Death and the world it made*. New York: Simon & Schuster.

Capano, G. 2020. Policy design and state capacity in the Covid-19 emergency in Italy; if you are not prepared for the (un)expected, you can be only what you already are. *Policy and Society* 39 (3): 326–344.

Caplan, A., and D. Reiss. 2020. Fair compensation for rare vaccine harms. *The Hastings Center*, September 9. https://www.thehastingscenter.org/fair-compensation-for-rare-vaccine-harms/.

Cardona, M., M. Anstey, E.T. Lewis, et al. 2020. Appropriateness of intensive care treatments near the end of life during the Covid-19 pandemic. *Breathe* 16 (2): 1–9.

Carrington, D. 2021. World leaders 'ignoring' role of destruction of nature in causing pandemics. *The Guardian*, June 4. https://www.theguardian.com/world/2021/jun/04/end-destruction-of-nature-to-stop-future-pandemics-say-scientists.

Carvalho, A.C.C., and A. Kritski. 2020. Learning from the Italian experience in coping with Covid-19. *Journal of the Brazilian Society of Tropical Medicine* 53: e20200199.

Cash, R., and V. Patel. 2020. Has Covid-19 subverted global health? *Lancet* 395 (10238): 1687–1688.

Castells, M.C., and E.J. Phillips. 2021. Maintaining safety with SARS-CoV-2 vaccines. *New England Journal of Medicine* 384 (7): 643–649.

Centers for Disease Control and Prevention. 1994. *Addressing emerging infectious disease threats: A prevention strategy for the United States* (Executive Summary). MMWR 43 (No. RR-5): 1–23.

Centers for Disease Control and Prevention. 2020. *Interim clinical guidance for management of patients with confirmed coronavirus disease (Covid-19)*. December 8. https://www.cdc.gov/coronavirus/2019-ncov/hcp/clinical-guidance-management-patients.html.

Centers for Disease Control and Prevention. 2020. *CDC's Covid-19 vaccine rollout recommendations*. Updated March 25, 2021. https://www.cdc.gov/coronavirus/2019-ncov/vaccines/recommendations.html.

———. 2020. *Covid-19 in children and teens*. Centers for Disease Control and Prevention, September 17. https://www.cdc.gov/coronavirus/2019-ncov/daily-life-coping/children/symptoms.html.

———. 2020. *Health equity considerations and racial and ethnic minority groups*. July 24. https://www.cdc.gov/coronavirus/2019-ncov/community/health-equity/race-ethnicity.html.

Centers for Disease Control and Prevention. 2020. *Covid-19 and animals*, December 4. https://www.cdc.gov/coronavirus/2019-ncov/daily-life-coping/animals.html.

———. 2020. *If you have pets*. September 9. https://www.cdc.gov/coronavirus/2019-ncov/daily-life-coping/pets.html.

Centers for Disease Control and Prevention. 2021. *Post-Covid conditions*, April 8. https://www.cdc.gov/coronavirus/2019-ncov/long-term-effects.html,

Cevik, M., M. Tate, O. Lloyd, et al. 2021. SARS-Co-V-2, SARS-CoV, and MERS-CoV viral load dynamics, duration of viral shedding, and infectiousness: A systematic review and meta-analysis. *The Lancet Microbe* 2: e13–e22.

Cha, I. 2012. *The mundialization of home in the age of globalization. Towards a transcultural ethics*. Műnchen: LIT Verlag.

Chan, M.S., C.R. Jones, K.H. Jamieson, and D. Albarracin. 2017. Debunking: A meta-analysis of the psychological efficacy of messages countering misinformation. *Psychological Science* 28 (11): 1531–1546.

Chan, T.K. 2020. Universal masking for Covid-19: Evidence, ethics and recommendations. *BMJ Global Health* 5: e002819. https://doi.org/10.1136/bmjgh-2020-002819.

Changeux, J.P., Z. Amoura, F.A. Rey, and M. Miyara. 2020. A nicotine hypothesis for Covid-10 with preventive and therapeutic implications. *Comtes Rendu Biologies* 343 (1): 33–39.

Chapman, C.M., and D.S. Miller. 2020. From metaphor to militarized response: The social implications of 'we are at war with Covid-19' – Crisis, disaster, and pandemics yet to come. *International Journal of Sociology and Social Policy* 40 (9/10): 1107–1124.

Chaudhry, R., G. Dranitsaris, T. Mubashir, J. Bartoszko, and S. Riazi. 2020. A country level analysis measuring the impact of government actions, country preparedness and socioeconomic factors on Covid-19 mortality and related health outcomes. *EClinicalMedicine* 25: 100464.

Cheng, V.C.C., S.K.P. Lau, P.C.Y. Woo, and K.Y. Yuen. 2007. Severe acute respiratory syndrome coronavirus as an agent of emerging and reemerging infection. *Clinical Microbiology Reviews* 20 (4): 660–694.

Cheng, K.K., T. H. Lam, and C. C. Leung. 2020. Wearing face masks in the community during the Covid-19 pandemic: Altruism and solidarity. *Lancet*, April 16. https://doi.org/10.1016/S0140-6736(20)30918-1.

Chien, Y.-J. 2013. How did international agencies perceive the avian influenza problem? The adoption and manufacture of the 'One World, One Health' framework. In *Pandemics and emerging infectious diseases. The sociological agenda*, ed. R. Dingwal, L.M. Hoffman, and K. Staniland, 46–58. Chichester: Wiley-Blackwell.

Childs, B.H., and l. Vearrier. 2021. A journal of the Covid-19 (plague) year. *HEC Forum* 33: 1–6.

Child, B.H., and L. Vearrier, eds. 2021. Special issue: Covid 2020: Historical and ethical perspectives. *HEC Forum* 33 (1–2): 1–164.

Chin-A-Fo, H., and C. van de Wiel. 2021. Heeft de EU zich door farmaceuten laten bespelen? *NRC*, 30-31 January. https://www.nrc.nl/nieuws/2021/01/29/heeft-de-eu-zich-door-de-farmaceuten-laten-bespelen-a4029788.

Christakis, N.A. 2019. *Blueprint. The evolutionary origins of a good society*. New York/Boston/London: Little, Brown Spark.

Christakis, N. A. 2020. *Apollo's arrow. The profound and enduring impact of coronavirus on the way we life*. New York/Boston/London: Little, Brown Spark.

Chu, D.K., E.A. Akl, S. Duda, et al. 2020. Physical distancing, face masks, and eye protection to prevent person-to-person transmission of SARS-CoV-2 and Covid-19: A systematic review and meta-analysis. *Lancet* 395: 1973–1987.

Cipoletta, S., and M.C. Ortu. 2020. Covid-19: Common constructions of the pandemic and their implications. *Journal of Constructivist Psychology* 29 (4): 340–356.

Clemens, T., and H. Brand. 2020. Will Covid-19 lead to a major change of the EU Public Health mandate? A renewed approach to EU's role is needed. *The European Journal of Public Health* 30 (4): 625–626.

Cleveland Manchanda, E.C., C. Sanky, and J.M. Appel. 2020. Crisis standards of care in the USA: A systematic review and implications for equity amidst Covid-19. *Journal of Racial and Ethnic Health Disparities*: 1–13. https://doi.org/10.1007/s40615-020-00840-5.

Cleveland Manchanda, E., C. Couillard, and K. Sivashanker. 2020. Inequity in crisis standards of care. *New England Journal of Medicine* 383 (4): e16. https://doi.org/10.1056/NEJMp2011359.

Cockerell, T.D.A. 1916. The Black Death, and its lessons for to-day. *The Scientific Monthly* 3 (1): 81–86.

Cohen, J., and Y. Van der Meulen Rodgers. 2020. Contributing factors to personal protective equipment shortages during the Covid-19 pandemic. *Preventive Medicine* 141: 106263. https://doi.org/10.1016/j.ypmed.2020.106263.

Cohn, J. 2021. Delay second doses? A guide to the latest Covid-19 vaccine debate. *HuffPost*, February 6. https://www.huffpost.com/entry/covid-19-delay-second-dose-pfizer-moderna_n_601d8bf2c5b618b31988384d.

COMEST. 2005. *The precautionary principle*. Paris: United Nations Educational, Scientific and Cultural Organization.

Comité consultatif national d'éthique. 2020. *Covid-19. Contribution of the French National Consultative Ethics Committee; Ethical issues in the face of a pandemic*. March 13. https://www.ccne-ethique.fr/sites/default/files/publications/ccne_contribution_march_13_2020.pdf.

Committee on Guidance for Establishing Crisis Standards of Care for Use in Disaster Situations; Institute of Medicine. 2012. *Crisis Standards of Care: A Systems Framework for Catastrophic Disaster Response*, 1–78. Washington, DC: National Academies Press.

Congregation for the Doctrine of the Faith. 2020. *Note on the morality of using some anti-Covid-19 vaccines*. https://www.vatican.va/roman_curia/congregations/cfaith/documents/rc_con_cfaith_doc_20201221_nota-vaccini-anticovid_en.html.

Conti, C., L. Fontanesi, R. Lanzara, I. Rosa, and P. Porcelli. 2020. Fragile heroes. The psychological impact of the Covid-19 pandemic on health-care workers in Italy. *PLoS ONE* 15 (11): e0242538. https://doi.org/10.1371/journal.pone.0242538.

Cook, M. 2020. A ghastly incident in Spain shows what could happen. *BioEdge*, March 29. https://www.bioedge.org/bioethics/a-ghastly-incident-in-spain-shows-what-could-happen/13381.

———. 2021. Germans shocked by fake vaccinations. *BioEdge*, August 15. https://www.bioedge.org/bioethics/germans-shocked-by-fake-vaccinations/13873.

Cori, L., F. Bianchi, E. Cadum, and C. Anthonj. 2020. Risk perception and Covid-19. *International Journal of Environmental Research and Public Health* 17 (9): 3114. https://doi.org/10.3390/ijerph17093114.

Corman, V.M., O. Landt, M. Kaiser, et al. 2020. Detection of 2019 novel coronavirus (2019-nCoV) by real-time RT-PCR. *Euro Surveillance* 25 (3): pii=2000045. https://doi.org/10.2807/1560-7917.ES.2020.25.3.2000045.

Cousings, E., K. de Vries, and K.H. Dening. 2020. Ethical care during Covid-19 for care home residents. *Nursing Ethics*: 1–12; https://doi.org/10.1177/0969733020976194.

Coutinho, R. 2021. *Vaxx. Hoe vaccinaties onze wereld beter hebben gemaakt*. Amsterdam: Ambo/Anthos.

Couto, M. 2020. A message from the pangolins. *Times Literary Supplement*, August 7: 11.

Covid-19 vaccine tracker. https://www.smh.com.au/interactive/2021/coronavirus/vaccine-tracker/index.html?resizable=true.

Cowling, B.J., and W.W. Lim. 2020. They've contained the coronavirus. Here's how. *New York Times*, March 13. https://www.nytimes.com/2020/03/13/opinion/coronavirus-best-response.html.

Cox, C., and M. Dixon-Woods. 2020. Need for ethical framework to guide mass testing for asymptomatic covid-19. *British Medical Journal* 371: m4567. https://doi.org/10.1136/bmj.m4567.

Cox, C.L. 2020. 'Healthcare heroes': Problems with media focus on heroism from healthcare workers during the Covid-19 pandemic. *Journal of Medical Ethics* 46: 510–513.

Crager, S.E. 2014. Improving global access to new vaccines: Intellectual property, technology transfer, and regulatory pathways. *American Journal of Public Health* 104 (11): e85–e91.

Craig, D. 2020. Pandemic and its metaphors: Sontag revisited in the Covid-19 era. *European Journal of Cultural Studies* 23 (6): 1025–1032.

Craxi, L., M. Vergano, J. Savulescu, and D. Wilkinson. 2020. Rationing in a pandemic: Lessons from Italy. *Asian Bioethics Review* 12: 325–330.

Cunha, B.A. 2004. The cause of the plague of Athens: Plague, typhoid, typhus, smallpox, or measles? *Infectious Disease Clinics of North America* 18: 29–43.

Curley, M.A.Q., E.G. Broden, and E.C. Meyer. 2020. Alone, the hardest part. *Intensive Care Medicine* 46: 1974–1976.

Cylus, J., D. Panteli, and E. van Ginneken. 2021. Who should be vaccinated first? Comparing vaccine prioritization strategies in Israel and European countries using the Covid-19 Health System Response Monitor. *Israel Journal of Health Policy Research* 10: 16. https://doi.org/10.1186/s13584-021-00453-1.

Cyranoski, D. 2020. How to stop restaurants seeding Covid infections. *Nature* 587: 344.

Dagan, N., N. Barda, E. Kepten, et al. 2021. BNT162b2 mRNA Covid-19 Vaccine in a Nationwide Mass Vaccination Setting. *New England Journal of Medicine* 384: 1412–1423.

Daher, M., G. Rouhana, N. Souaiby, et al. 2020. Ethical consideration in response to the Covid-19 pandemic. *Lebanese Medical Journal* 68 (1-2): 99–104.

Daily Sabah, 2021. *Patient left for dead in Netherlands brought to Turkey*, February 11. https://www.dailysabah.com/turkey/diaspora/patient-left-for-dead-in-netherlands-brought-to-turkey.

Daoust, J.-F. 2020. Elderly people and responses to Covid-19 in 27 countries. *PLoS ONE* 15 (7): e0235590. https://doi.org/10.1371/journal.pone.0235590.

Daszak, P. 2020. We are entering an era of pandemics – It will end only when we protect the rainforest. *The Guardian,* 28 July. https://www.theguardian.com/commentisfree/2020/jul/28/pandemic-era-rainforest-deforestation-exploitation-wildlife-disease.

Davey, M. 2020. WHO warns Covid-19 pandemic is 'not necessarily the big one.' *The Guardian*, December 29. https://www.theguardian.com/world/2020/dec/29/who-warns-covid-19-pandemic-is-not-necessarily-the-big-one.

———. 2021. Scientific paper claiming smokers less likely to acquire Covid retracted over tobacco industry links. *The Guardian*, April 22. https://www.theguardian.com/science/2021/apr/22/scientific-paper-claiming-smokers-less-likely-to-acquire-covid-retracted-over-tobacco-industry-links.

Davidson, H. 2020. Wuhan Covid citizen journalist jailed for four years in China crack-down. *The Guardian*, December 28. https://www.theguardian.com/world/2020/dec/28/wuhan-citizen-journalist-jailed-for-four-years-in-chinas-christmas-crackdown.

Davies, M., and R. Furneaux. 2021. World is on course for a coronavirus vaccine 'apartheid', experts warn. *Independent*, February 6. https://www.independent.co.uk/news/health/coronavirus-crisis-vaccine-apartheid-b1798326.html.

Davis, N. 2020. How has a Covid vaccine been developed so quickly? *The Guardian*, December 8. https://www.theguardian.com/society/2020/dec/08/how-has-a-covid-vaccine-been-developed-so-quickly.

Dawson, A., D. Isaacs, M. Jansen, et al. 2020. An ethics framework for making resource allocation decisions with clinical care: Responding to Covid-19. *Journal of Bioethical Inquiry* 17: 749–755.

Dawson, A., E.J. Emanuel, M. Parker, M.J. Smith, and T.C. Voo. 2020. Key ethical concepts and their application to Covid-19 research. *Public Health Ethics* 13 (2): 127–132.

Dearden, N. 2021. System change, not charity, will end the vaccine apartheid. *Al Jazeera*, February 26. https://www.aljazeera.com/opinions/2021/2/26/system-change-not-charity-will-end-the-vaccine-apartheid.

De Beaumont Foundation. 2020. *Poll: New national conversation about Covid-19 urgently needed to overcome partisan divide and save lives*. November 30. https://www.globenewswire.com/fr/news-release/2020/11/30/2136371/0/en/Poll-New-National-Conversation-About-COVID-19-Urgently-Needed-to-Overcome-Partisan-Divide-and-Save-Lives.html

DeBruin, D., and J.P. Leider. 2020. Covid-19: The shift from clinical to public health ethics. *Journal of Public Health Management and Practice* 26 (4): 306–309.

De Bruin, W. 2010. Samuel Sarphati (1813–1866). Schepper van een nieuwe stad, *Historisch Nieuwsblad* 8. https://www.historischnieuwsblad.nl/samuel-sarphati-1813-1866-schepper-van-een-nieuwe-stad/.

De Castro, L., and J. Yasol-Naval. 2020. Sustainable Covid-19 response measures: An ethical imperative for enhancing core human capabilities. In *Medicine and ethics in times of corona*, ed. M. Woesler and H.-M. Sass, 283–299. Zürich: LIT Verlag.

Deb, P., D. Furceri, J.D. Ostry, and N. Tawk. 2020. *The effect of containment measures on the Covid-19 pandemic*. IMF Working paper.

Deen, K. 2020. Dit kan Nederland leren van de Duitse aanpak van de coronacrisis. *Trouw*, 22 September.

Dekkers, W. 2011. Dwelling, house and home: Towards a home-led perspective on dementia care. *Medicine Health Care and Philosophy* 14: 291–300.

De Medeiros, K. 2020. *A Covid-19 side effect: Virulent resurgence of ageism*. The Hastings Center, May 14. https://www.thehastingscenter.org/a-covid-19-side-effect-virulent-resurgence-of-ageism/.

Deming, M.E., N.L. Michael, M. Robb, et al. 2020. Accelerating development of SARS-CoV-2 vaccines – The role for controlled human infection models. *New England Journal of Medicine* 383: e63. https://doi.org/10.1056/NEJMp2020076.

Den Exter, A. 2020. The Dutch critical care triage guidelines on Covid-19: Not necessarily discriminatory. *European Journal of Health Law* 27 (5): 495–498.

Department of Health and Human Services (HHS). 2020. *Crimson Contagion 2019 Functional Exercise After-Action Report*. Washington. https://www.governmentattic.org/38docs/HHSaarCrimsonContAAR_2020.pdf.
Desson, Z., E. Weller, P. McMeekin, and M. Ammi. 2020. An analysis of the policy responses to the Covid-19 pandemic in France, Belgium, and Canada. *Health Policy and Technology* 9 (4): 430–446.
Deutscher Ethikrat. 2020. *Solidarity and responsibility during the coronavirus crisis*. Berlin, 27 March. https://www.ethikrat.org/fileadmin/Publikationen/Ad-hoc-Empfehlungen/englisch/recommendation-coronavirus-crisis.pdf, 5.
DeVita, M.A., and L.S. Parker. 2021. *Ethics supports seeking population immunity, not immunizing priority groups*. Hastings Center, January 26. https://www.thehastingscenter.org/ethics-supports-seeking-population-immunity-not-immunizing-priority-groups/.
De Waal, A. 2021. *New pandemics, old politics. Two hundred years of war on disease and its alternatives*. Cambridge: Polity Press, 180 ff.
Dinnes, J., J.J. Deeks, S. Berhane, et al. 2021. Cochrane COVID-19 Diagnostic Test Accuracy Group. Rapid, point-of-care antigen and molecular-based tests for diagnosis of SARS-CoV-2 infection. *Cochrane Database of Systematic Reviews* 2021, Issue 3. Art. No.: CD013705. https://doi.org/10.1002/14651858.CD013705.pub2.
Dishman, L., and V. Schroeder. 2020. A Covid-19 patient's experience: Engagement in disease management, interactions with care teams and implications on health policies and managerial practices. *Patient Experience Journal* 7 (2): 10.35680/2372-0247.1487.
Dolgin, E. 2021. Covid vaccine immunity is waning – How much does it matter? *Nature* 597: 606–607.
Downar, J., and D. Seccareccia. 2010. Palliating a pandemic: "All patients must be cared for". *Journal of Pain and Symptom Management* 39 (2): 291–295.
Dryhurst, S., C.R. Schneider, J. Kerr, et al. 2020. Risk perception of Covid-19 around the world. *Journal of Risk Research* 23 (7-8): 994–1006.
Dudzinski, D.M., B.Y. Hoisington, and C.E. Brown. 2020. Ethics lessons from Seattle's early experience with Covid-19. *American Journal of Bioethics* 20 (7): 67–74.
DutchNews.nl. 2020. *Dispensable? Dead wood? Coronavirus dilemma sparks social media debate*. August 10. https://www.dutchnews.nl/news/2020/08/dispensable-dead-wood-coronavirus-dilemma-sparks-social-media-debate/.
Earnest, M. 2020. On becoming a plague doctor. *New England Journal of Medicine* 383: e64. https://doi.org/10.1056/NEJMp2011418.
Ebrahim, S. 2002. The medicalization of old age. *British Medical Journal* 324: 861–863.
Economist Intelligence Unit, 2021. *Q1 global forecast 2021 Coronavirus vaccines: Expect delays*. https://www.eiu.com/n/campaigns/q1-global-forecast-2021/.
Editorial. 2020. Global collaboration for health: Rhetoric versus reality. *Lancet* 396: 735.
———. 2020. Communication, collaboration and cooperation can stop the 2019 coronavirus. *Nature Medicine* 26: 151.
———. 2020. Everyone wins when patents are pooled. *Nature* 581: 240.
———. 2020. Covid vaccine confidence requires radical transparency. *Nature* 586: 8.
———. 2020. Europe must think more globally in crafting its pandemic response. *Nature* 587: 329.
———. 2021. Learn from Covid before diving into a pandemic treaty. *Nature* 592: 165–166.
Edridge, A.W.D., J. Kaczorowska, A.C.R. Hoste, et al. 2020. Seasonal coronavirus protective immunity is short-lasting. *Nature Medicine* 26: 1691–1693.
Ehni, H.-J., U. Wiesing, and R. Ranisch. 2021. Saving the most lives – A comparison of European triage guidelines in the context of the Covid-19 pandemic. *Bioethics* 35: 125–134.
Elliott, C. 2020. An ethical path to a Covid vaccine. *New York Review of Books*, July 2.
Editorial. 2020. The Covid-19 testing debacle. *Nature Biotechnology* 38 (653). https://doi.org/10.1038/s41587-020-0575-3.
Emanuel, E.J., G. Persad, R. Upshur, et al. 2020. Fair allocation of scarce medical resources in the time of Covid-19. *New England Journal of Medicine* 382 (21): 2049–2055.

Emanuel, E.J., G. Persad, A. Kern, et al. 2020. An ethical framework for global vaccine allocation. *Science* 369 (6509): 1309–1312.

Emanuel, E.J., A. Buchanan, S.Y. Chan, et al. 2021. What are the obligations of pharmaceutical companies in a global health emergency? *Lancet* 398: 1015–1020.

Eng, K.F. 2020. 5 things that scientists now know about Covid-19 – And 5 things they're still figuring out. *Ideas.Ted.com*, September 16. https://ideas.ted.com/5-things-that-scientists-now-know-about-covid-19-and-5-things-theyre-still-figuring-out/.

Ercetin, G., and H. Raba. 2021. Turkije haalt coronapatiënten in Nederland en andere landen op. *NOS Nieuws*, February 11. https://nos.nl/artikel/2368254-turkije-haalt-coronapatienten-in-nederland-en-andere-landen-op.html.

Erdema, H., and D.R. Lucey. 2020. Healthcare worker infections and deaths due to Covid-19: A survey from 37 nations and a call for WHO to post national data on their website. *International Journal of Infectious Diseases* 102: 239–241.

Eyal, N., P.G. Lipsitch, and P.G. Smith. 2020. Human challenge studies to accelerate coronavirus licensure. *Journal of Infectious Diseases* 221: 1752–1756.

Eyal, N. 2020. Why challenge trials of SARS-CoV-2 vaccines could be ethical despite risk of severe adverse events. *Ethics & Human Research* 42: 24–34.

Eurasia Group. 2020. *Ending the Covid-19 pandemic: The need for a global approach*. November 25. https://www.who.int/docs/default-source/coronaviruse/act-accelerator/2020-summary-analysis-of-ten-donor-countries-11_26_2020-v2.pdf.

Euronews. 2020. We need a 'war economy' to deal with COVID-19 crisis, UN chief Antonio Guterres tells Euronews. *Euronews Online*. https://www.euronews.com/2020/03/25/coronavirusantonio-guterres-speaks-to-euronews-about-un-s-covid-19-response.

European Center for Disease Prevention and Control. 2021. *Considerations on the use of self-tests for Covid-19 in the EU/EEA*. Stockholm: ECDC. https://www.ecdc.europa.eu/sites/default/files/documents/Considerations-for-the-use-of-self-tests-for-COVID-19-in-the-EU-EEA_0.pdf.

European Group on Ethics in Science and New Technologies. 2020. *Statement on European solidarity and the protection of fundamental rights in the Covid-19 pandemic*, April 2. https://ec.europa.eu/info/sites/default/files/research_and_innovation/ege/ec_rtd_ege-statement-covid-19.pdf.

Evanega, S., M. Lynas, J. Adams, and K. Smolenyak. 2020. *Coronavirus misinformation: Quantifying sources and themes in the Covid-19 'infodemic;'* https://allianceforscience.cornell.edu/wp-content/uploads/2020/09/Evanega-et-al-Coronavirus-misinformationFINAL.pdf.

Fairbanks, E. 2020. A pandemic is not a war. *HuffPost*, April 14. https://www.huffpost.com/entry/war-pandemic-coronavirus-covid-19_n_5e90d449c5b672672149e1fa.

Fee, E., and T.M. Brown. 2001. Preemptive biopreparedness: Can we learn anything from history? *American Journal of Public Health* 91 (5): 721–726.

Feiring, E., R. Førde, S. Holm, O.F. Norheim, B. Solberg, C.T. Solberg, and G. Wester. 2020. *Advice on priority groups for coronavirus vaccination in Norway*. In *Report 2020*. Oslo: Norwegian Institute of Public Health.

Fenn, E.A. 2001. *Pox Americana. The great smallpox epidemic of 1775–82*, 5. New York: Hill and Wang.

Feral-Pierssens, A.-L., P.-G. Claret, and T. Chouihed. 2020. Collateral damage of the Covid-19 outbreak: Expression of concern. *European Journal of Emergency Medicine* 27 (4): 233–234.

Fernandez, R., and T.J. Klinge. 2020. *The financialization of Big Pharma*. Amsterdam: SOMO. https://www.somo.nl/wp-content/uploads/2020/04/Rapport-The-financialisation-of-Big-Pharma-def.pdf.

Fidler, D.P. 2010. Negotiating Equitable Access to Influenza Vaccines: Global Health Diplomacy and the Controversies Surrounding Avian Influenza H5N1 and Pandemic Influenza H1N1. *PLoS Med* 7 (5): e1000247. https://doi.org/10.1371/journal.pmed.1000247.

Fink, S. 2020. The hardest questions doctors may face: Who will be saved? Who won't? *The New York Times*, March 21. https://www.nytimes.com/2020/03/21/us/coronavirus-medical-rationing.html.

Fins, J.J. 2020. Pandemics, protocols, and the plague of Athens: Insights from Thucydides. *Hastings Center Report* 50 (3): 50–53.

Fins, J. J. 2020. *Covid-19 makes clear that bioethics must confront health disparities*. The Hastings Center, July 9. https://www.thehastingscenter.org/covid-19-makes-clear-that-bioethics-must-confront-health-disparities/.

Fisher, A. 2013. Fair innings? Against healthcare rationing in favour of the young over the elderly. *Studies in Christian Ethics* 26 (4): 431–450.

Flaxman, S., S. Mishra, A. Gandy, et al. 2020. Estimating the number of infections and the impact of non-pharmaceutical interventions on Covid-19 in 11 European countries. Imperial College Covid-19 Response Team. *Nature* 584: 257–261.

———. 2020. *Estimating the number of infections and the impact of non-pharmaceutical interventions on COVID-19 in 11 European countries*. Imperial College London (30-03-2020). https://doi.org/10.25561/77731.

Fleck, L. 1935. *Entstehung und Entwicklung einer wissenschaftliche Tatsache: Einführung in die Lehre vom Denkstil und Denkkollektiv*. Frankfurt am Main: Suhrkamp Verlag.

Flood, C.M.B., Thomas, and K. Wilson. 2021. Mandatory vaccination for health care workers: An analysis of law and policy. *Canadian Medical Association Journal* 193 (6): E217–E220.

Folegatu, P.M., K.J. Ewer, P.K. Aley, et al. 2020. Safety and immunogenicity of the ChAdOx1 nCoV-19 vaccine against SARS-CoV-2: A preliminary report of a phase ½, single-blind, randomized controlled trial. *Lancet* 396 (10249): 467–478.

Fraser, M.R. 2020. Leading in the Covid-19 crisis: Challenges and solutions for state health leaders. *Journal of Public Health Management and Practice* 26 (4): 380–383.

Fraser, S., M. Lagacé, B. Bongué, et al. 2020. Ageism and Covid-19: What does our society's response say about us? *Age and Ageing* 49 (5): 692–695.

Frey, C.B., C. Chen, and G. Presidente. 2020. Democracy, culture, and contagion: Political regimes and countries' responsiveness to Covid-19. *Covid Economics* 18: 222–238.

Friedman, T. 2020. Finding the 'common good' in a pandemic. *New York Times*, March 24.

Friedrich, A. 2021. *Fear of doing too much too soon or too little too late: Research on Covid-19*. The Hastings Center, August 20. https://www.thehastingscenter.org/fear-of-doing-too-much-too-soon-or-too-little-too-late-research-on-covid-19/.

Frontera, J. A., S. Sabadia, R. Lalchan et al. 2020. A prospective study of neurological disorders in hospitalized Covid-19 patients in New York City. *Neurology*, October 5. https://n.neurology.org/content/neurology/early/2020/10/05/WNL.0000000000010979.full.pdf.

Gandhi, M., C. Beyrer, and E. Goosby. 2020. Masks do more than protect others during Covid-19: Reducing the inoculum of SARS-CoV-2 to protect the wearer. *Journal of General Internal Medicine* 35 (10): 3063–3066.

Gandhi, M., D.S. Yokoe, and D.V. Havlir. 2020. Asymptomatic transmission, the Achilles' heel of current strategies to control Covid-19. *New England Journal of Medicine* 382 (22): 2158–2160.

Gatto, M., E. Bertuzzo, L. Mari, et al. 2020. Spread and dynamics of the Covid-19 epidemic in Italy: Effects of emergency containment measures. *Proceedings of the National Academy of Sciences of the United States of America* 117 (19): 10484–10491.

GAVI. 2021. *Covax Facility*. https://www.gavi.org/covax-facility.

Geddes, L. 2021. *Five things we know about the Delta variant (and two things we don't*. GAVI, June 15. https://www.gavi.org/vaccineswork/five-things-we-know-about-delta-coronavirus-variant-and-two-things-we-still-need.

General Medical Council. 2019. *Good medical practice*. https://www.gmc-uk.org/ethical-guidance/ethical-guidance-for-doctors/good-medical-practice.

———. 2020. *Working safely*. https://www.gmc-uk.org/ethical-guidance/ethical-hub/covid-19-questions-and-answers#Working-safely.

George, D.R., E.R. Whitehouse, and P.J. Whitehouse. 2016. Asking more of our metaphors: Narrative strategies to end the 'war on Alzheimer's' and humanize cognitive aging. *American Journal of Bioethics* 16 (10): 22–24.

Gerberding, J.L., and B.F. Haynes. 2021. Vaccine innovations – Past and future. *New England Journal of Medicine* 384 (5): 393–396.

German Ethics Council. 2020. *Solidarity and responsibility during the coronavirus crisis. Ad hoc recommendation.* March 27. https://www.ethikrat.org/fileadmin/Publikationen/Ad-hoc-Empfehlungen/englisch/recommendation-coronavirus-crisis.pdf.

Gervais, A. 1972. À propos de la 'Peste' d'Athènes: Thucydide et la littérature de l'épidémie. *Bulletin de l'Association Guillaume Budé: Lettres d'humanité* 31: 395–429. https://doi.org/10.3406/bude.1972.3490.

Gibney, E. 2020. Whose coronavirus strategy worked best? Scientists hunt most effective policies. *Nature* 581 (7806): 15–16.

Gintis, H. 2017. *Individuality and entanglement: The moral and material bases of social life.* Princeton/Oxford: Princeton University Press.

Glenza, J. 2021. 'Oversupplied' US faces pressure to send Covid vaccine doses to less wealthy countries. *The Guardian*, March 13. https://www.theguardian.com/world/2021/mar/13/us-faces-pressure-send-covid-vaccine-doses-less-wealthy-countries-biden.

Glenza, J., and M. Pengelly. 2021. Catholics in New Orleans and St Louis told to avoid Johnson & Johnson vaccine. *The Guardian*, March 2. https://www.theguardian.com/world/2021/mar/02/new-orleans-archdiocese-catholics-avoid-johnson-johnson-vaccine.

Glenza, J. 2021. The Delta variant is spreading. What does it mean for the US? *The Guardian*, June 16. https://www.theguardian.com/world/2021/jun/16/delta-variant-coronavirus-us.

Global Health Security Index. 2019. *Building collective action and accountability.* https://www.ghsindex.org/wp-content/uploads/2019/10/2019-Global-Health-Security-Index.pdf.

Goalkeepers Report. 2020, September. *Covid-19. A global perspective.* Bill & Melinda Gates Foundation. https://www.gatesfoundation.org/goalkeepers/downloads/2020-report/report_a4_en.pdf.

Goenka, A. 2021. Israel's vaccine rollout has been fast, so why is it controversial and what can other countries earn? *The Conversation*, January 27. https://theconversation.com/israels-vaccine-rollout-has-been-fast-so-why-is-it-controversial-and-what-can-other-countries-learn-153687.

Goldie, S., A. Hill, D. Eagles, and T.W. Drew. 2020. The effect of temperature on persistence of SARS-CoV-2 on common surfaces. *Virology Journal* 17: 145. https://doi.org/10.1186/s12985-020-01418-7.

Goldman, E. 2020. Exaggerated risk of transmission of Covid-19 by fomites. *Lancet Infectious Diseases* 20: 892–893.

———. 2021. SARS wars: The fomites strike back. *Applied and Environmental Microbiology* 87 (13): e00653–e00621.

Goldstein, R.H., and R.P. Walensky. 2020. The challenges ahead with monoclonal antibodies. From authorization to access. *JAMA* 324 (21): 2151–2152.

Goldstein, J.R., T. Cassidy, and K.W. Wachter. 2021. Vaccinating the oldest against Covid-19 saves both the most lives and most years of life. *Proceedings of the National Academy of Science of the United States of America* 118 (11): e2026322118. https://doi.org/10.1073/pnas.2026322118.

Goodman, J.R. 2020. Welcome to the virosphere. *New Scientist* 245 (3264): 40–43.

Gopichandran, V. 2020. Clinical ethics during the Covid-19 pandemic: Missing the trees for the forest. *Indian Journal of Medical Ethics* 5 (3). https://doi.org/10.20529/IJME.2020.053.

Gordon, A.L., C. Goodman, W. Achterberg, et al. 2020. Commentary: Covid in care homes – Challenges and dilemmas in healthcare delivery. *Age Ageing* 49 (5): 701–705.

Gostin, L.O., S. Moon, and B.M. Meier. 2020. Reimagining global health governance in the age of Covid-19. *American Journal of Public Health* 110 (11): 1615–1619.

Gostin, L.O., D.A. Salmon, and H.J. Larson. 2021. Mandating Covid-19 vaccines. *JAMA* 325 (6): 532–533.

Gouglas, D., T.T. Le, K. Henderson, et al. 2018. Estimating the costs of vaccine development against epidemic infectious diseases: A cost minimization study. *Lancet Global Health* 6: e1386–e1396.

Goulabchand, R., P.-G. Claret, and B. Lattuca. 2020. What if the worst consequences of Covid-19 concerned non-Covid patients? *Journal of Infection and Public Health* 13: 1237–1239.

Grabowski, D.C. 2021. The future of long-term care requires investment in both facility- and home-based services. *Nature Aging* 1: 10–11.

Grady, C. 2004. Ethics of vaccine research. *Nature Immunology* 5 (5): 465–468.

Graham, B.S. 2020. Rapid Covid-19 vaccine development. *Science* 368: 945–946.

Grange, Z.L., T. Goldstein, C.K. Johnson, et al. 2021. Ranking the risk of animal-to-human spillover for newly discovered viruses. *Proceedings of the National Academy of Sciences of the United States of America* 118 (15): e2002324118. https://doi.org/10.1073/pnas.2002324118.

Granovetter, M. 2017. *Society and economy. Framework and principles*. Cambridge/London: The Belknap Press of Harvard University Press.

Green, M.H. 2004. Taking 'pandemic' seriously: Making the Black Death global. *The Medieval Globe* 1: 27–61.

Greenhalgh, T. 2020. Face coverings for the public; laying straw men to rest. *Journal of Evaluation in Clinical Practice* 26: 1070–1077.

Greenhalgh, T., M.B. Schmid, T. Czypionka, et al. 2020. Face masks for the public during the Covid-19 crisis. *British Medical Journal* 369: m1435. https://doi.org/10.1136/bmj.m1435.

Greshko, M. 2021. Covid-19 will likely be with us forever. Here's how we'll live with it. *National Geographic*, January 22. https://www.nationalgeographic.co.uk/science-and-technology/2021/01/covid-19-will-likely-be-with-us-forever-heres-how-well-live-with-it.

Greve, J. E. 2021. Joe Biden orders US intelligence to intensify efforts to study Covid's origins. *The Guardian*, May 27. https://www.theguardian.com/us-news/2021/may/26/joe-biden-us-intelligence-community-covid-19-origins-china.

Grover, N. 2021. Covid: New vaccines needed globally within a year, say scientists. *The Guardian*, March 30. https://www.theguardian.com/world/2021/mar/30/new-covid-vaccines-needed-within-year-say-scientists.

———. 2021. Delta variant renders herd immunity from Covid 'mythical.' *The Guardian*, August 10. https://www.theguardian.com/world/2021/aug/10/delta-variant-renders-herd-immunity-from-covid-mythical.

Gur-Arie, R., E. Jamrozik, and P. Kingori. 2021. No jab, no job? Ethical issues in mandatory Covid-19 vaccination of healthcare personnel. *BMJ Global Health* 6: e004877.

Gustafsson, L. 2020. *Covid-19 highlights problems with our generic supply chain*. The Commonwealth Fund, May 7. https://www.commonwealthfund.org/blog/2020/covid-19-highlights-problems-our-generic-supply-chain.

Hadot, P. 1995. *Philosophy as a way of life. Spiritual exercises from Socrates to Foucault*. Malden: Blackwell.

Haim-Boukobza, S., B. Roquebert, S. Trombert-Paolantoni, et al. 2021. Detecting rapid spread of SARS-CoV-2 variants, France, January 26 – February 16, 2021. *Emerging Infectious Diseases* 27 (5): 1496–1499.

Halpern, J., and D.J. Opel. 2020. Sustaining clinical empathy during the pandemic. *The Hastings Center Bioethics Forum*. https://www.thehastingscenter.org/sustaining-clinical-empathy-during-the-pandemic/.

Hameleers, M., T.G.L.A. van der Meer, and A. Brosius. 2020. Feeling 'disinformed' lowered compliance with Covid-19 guidelines: Evidence from the US, UK, Netherlands and Germany. *The Harvard Kennedy School Misinformation Review* 1. https://doi.org/10.37016/mr-2020-023.

Hamilton, G. 2008. Welcome to the virosphere. *New Scientist* 199 (2671): 38–41.

Hansen, C.H., D. Michlmayr, S.M. Gubbels, et al. 2021. Assessment of protection against reinfection with SARS-CoV-2 among 4 million PCR-tested individuals in Denmark in 2020: A population-level observational study. *Lancet* 397: 1204–1212.

Hantel, A., J. M. Marron, M. Casey, et al. 2020. US state government crisis standards of care guidelines. Implications for patients with cancer. *JAMA Oncology*, December 3; https://doi.org/10.1001/jamaoncol.2020.6159.

Haque, A., and A.B. Pant. 2020. Efforts at Covid-19 vaccine development: Challenges and successes. *Vaccines* 8 (4): 739. https://doi.org/10.3390/vaccines8040739.

Harding, L. 2020. 'Weird as hell': The Covid-19 patients who have symptoms for months. *The Guardian*, May 15. https://www.theguardian.com/world/2020/may/15/weird-hell-professor-advent-calendar-covid-19-symptoms-paul-garner.

Harrison, M. 2015. A global perspective: Reframing the history of health, medicine, and disease. *Bulletin of the History of Medicine* 89: 639–689.

Haseltine, W. A. 2020. The risks of rushing a Covid-19 vaccine. *Scientific American*, June 22. https://www.scientificamerican.com/article/the-risks-of-rushing-a-covid-19-vaccine/.

Hassan, B., and T. Arawi. 2020. The care of non-Covid-19 patients: A matter of choice or moral obligation? *Frontiers in Medicine* 7: 564038. https://doi.org/10.3389/fmed.2020.564038.

Hassoun, N. 2020. What is COVAX and why does it matter for getting vaccines to developing nations? *The Conversation*, October 2. https://theconversation.com/what-is-covax-and-why-does-it-matter-for-getting-vaccines-to-developing-nations-146284.

———. 2020. How to distribute a Covid-19 vaccine ethically. *Scientific American*, September 25. https://www.scientificamerican.com/article/how-to-distribute-a-covid-19-vaccine-ethically/.

———. 2021. How to make 'immunity passports' more ethical. *Scientific American*, February 24. https://www.scientificamerican.com/article/how-to-make-immunity-passports-more-ethical/.

Haug, N., L. Geyrhofer, A. Londei, et al. 2020. Ranking the effectiveness of worldwide Covid-19 government interventions. *Nature Human Behavior* 4: 1303–1312.

Hawkins, D. 2020. Differential occupational risk for COVID-19 and other infection exposure according to race and ethnicity. *American Journal of Industrial Medicine* 63 (9): 817–820.

Hayek, F. 2005. *The road to serfdom*. Chicago/London: University of Chicago Press.

Haynes, B.F. 2021. A new vaccine to battle Covid-19. *New England Journal of Medicine* 384 (5): 470–471.

Health Council of the Netherlands. 2020. *Covid-19 vaccination strategies*. The Hague: Health Council of the Netherlands.

Heilinger, J.-C., S. Venkatapuram, M. Voss, and V. Wild. 2020. *Bringing ethics into the global coronavirus response*. The Hastings Center, June 22. https://www.thehastingscenter.org/bringing-ethics-into-the-global-coronavirus-response/.

Henley, J. 2021. Covid vaccine acceptance rising across Europe but falling in parts of Asia. *The Guardian*, January 22. https://www.theguardian.com/world/2021/jan/22/covid-vaccine-acceptance-rising-across-europe-but-falling-in-parts-of-asia.

———. 2021. A quarter of people in France, Germany and the US may refuse Covid vaccine. *The Guardian*, February 4. https://www.theguardian.com/world/2021/feb/04/covid-vaccine-refuse-france-germany-us-quarter.

———. 2021. Covid: EU unveils 'digital green certificate' to allow citizens to travel. *The Guardian*, March 17. https://www.theguardian.com/world/2021/mar/17/covid-eu-unveils-digital-green-certificate-to-allow-citizens-to-travel.

Herstel-NL. 2021. *Het plan van herstel-NL*. https://www.herstel-nl.nl/het-plan.

Herszenhorn, D.M., and J.H. Vela. 2021. European Commission to propose tougher vaccine export rules. *Politico*, March 24. https://www.politico.eu/article/commission-to-table-tougher-vaccine-export-rules/.

Hertelendy, A.J., G.R. Ciottone, C.L. Mitchell, et al. 2020. Crisis standards of care in a pandemic: Navigating the ethical, clinical, psychological and policy-making maelstrom. *International Journal for Quality in Health Care* 1–4. https://doi.org/10.1093/intqhc/mzaa094.

Herzog, L.M., O.F. Norheim, E.J. Emanuel, and M.S. McCoy. 2021. Covax must go beyond proportional allocation of covid vaccines to ensure fair and equitable access. *British Medical Journal* 372: m4853.

Heyland, D.K., P. Dodek, G. Rocker, et al. 2006. What matters most in end-of-life care; perceptions of seriously ill patients and their family members. *Canadian Medical Association Journal* 174: 627–633.

Heymann, D.L., and A. Wilder-Smith. 2020. Successful smallpox eradication: What can we learn to control Covid-19? *Journal of Travel Medicine* 27 (4): 1–3. https://doi.org/10.1093/jtm/taaa090.

Hick, J.L., L. Rubinson, D.T. O'Laughlin, and J.C. Farmer. 2007. Clinical review: Allocating ventilators during large-scale disasters – Problems, planning, and process. *Critical Care* 11: 217. https://doi.org/10.1186/cc5929.

Hick, J. L., Hanfling, D., M. K. Wynia, and A.T. Pavia. 2020. Duty to plan: Health care, crisis standards of care, and novel coronavirus SARS-CoV-2. *NAM Perspectives*. Discussion paper. Washington, DC: National Academy of Medicine; https://doi.org/10.31478/202003b.

Higgins, C. 2020. Why we shouldn't be calling our healthcare workers 'heroes'. *The Guardian*, May 27: https://www.theguardian.com/commentisfree/2020/may/27/healthcare-workers-heros-language-heroism.

Hinchliffe, S. 2015. More than one world, more than one health: Re-configuring interspecies health. *Social Science & Medicine* 129: 28–35.

Ho, A., and I. Dascalu. 2020. Global disparity and solidarity in a pandemic. *Hastings Center Report* 50 (3): 65–67.

Hodgson, S.H., K. Mansotta, G. Mallett, et al. 2021. What defines an efficacious Covid-19 vaccine? A review of the changes assessing the clinical efficacy of vaccines against SARS-CoV-2. *Lancet Infectious Diseases* 21 (2): E26–E35.

Hoffman, D. N. 2021. *Vaccine mandates for health care workers raise several ethical dilemmas*. The Hastings Center, August 10. https://www.thehastingscenter.org/vaccine-mandates-for-health-care-workers-raise-several-ethical-dilemmas/.

Hogan, A.B., P. Winskill, Q. Watson, et al. 2020. Report 33. In *Modelling the allocation and impact of a covid-19 vaccine*. London: Imperial College, September 25. https://doi.org/10.25561/82822.

Hollinghurst, J., J. Lyons, R. Fry, et al. 2021. The impact of Covid-19 on adjusted mortality risk in care homes for older adults in Wales, UK: A retrospective population-based cohort study for mortality in 2016–2020. *Age and Ageing* 50: 25–31.

Hollingsworth, H. 2021. More people choosing to die at home as hospitals limit visitations amid pandemic. *HuffPost*, February 7. https://www.huffpost.com/entry/die-at-home-covid-19-pandemic_n_601fffbfc5b6f38d06e485b7.

Holm, S. 2020. Controlled human infection with SARS-CoV-2 to study Covid-19 vaccines and treatments: Bioethics in Utopia. *Journal of Medical Ethics* 46: 569–573.

Holst, J. 2020. Global health – Emergence, hegemonic trends and biomedical reductionism. *Globalization and Health* 16: 42. https://doi.org/10.1186/s12992-020-00573-4.

Holt, E. 2020. Slovakia to test all adults for SARS-CoV-2. *Lancet* 396: 1386–1387.

Homs, O. 2020. How has Israel launched the world's fastest Covid vaccination drive? *The Guardian*, December 30. https://www.theguardian.com/world/2020/dec/30/how-has-israel-launched-the-worlds-fastest-covid-vaccination-drive.

Honigsbaum, M. 2020. *The pandemic century. One hundred years of panic, hysteria, and hubris*. New York: W. W. Norton & Company.

Hopkins, D.R. 2002. *The greatest killer. Smallpox in history. With a new introduction*. Chicago/London: The University of Chicago Press.

Hopman, J., and S. Mehtar. 2020. Country level analysis of Covid-19 policies. *EClinicalMedicine* 25: 100500.

Horii, M. 2014. Why do the Japanese wear masks? A short historical review. *Journal of Contemporary Japanese Studies* 14 (2). https://www.japanesestudies.org.uk/ejcjs/vol14/iss2/horii.html.

Horowitz, J., and E. Bubola. 2020. Italy's coronavirus victims face death alone, with funerals postponed. *New York Times*, March 19. https://www.nytimes.com/2020/03/16/world/europe/italy-coronavirus-funerals.html.

Horton, R. 2004. Rediscovering human dignity. *The Lancet* 364: 1081–1085.

Horton, R., R. Beaglehole, R. Bonita, et al. 2014. From public to planetary health: A manifesto. *Lancet* 383: 847.

Horton, R. 2020. *The Covid-19 catastrophe. What's gone wrong and how to stop it happening again*. Cambridge: Polity Press.

Hossain, M., A. Sultana, and N. Purohit. 2020. Mental health outcomes of quarantine and isolation for infection prevention: A systematic umbrella review of the global evidence. *Epidemiology and Health* 42: e2020038.

Hotez, P.J. 2021. *Preventing the next pandemic. Vaccine diplomacy in a time of anti-science*. Baltimore: Johns Hopkins University Press.
Hotez, P. 2021. Covid vaccines: Time to confront antivax aggression. *Nature* 592: 661.
Howard-Jones, N. 1975. *The scientific background of the International Sanitary Conferences, 1851–1938*. Geneva: World Health Organization.
Howard, J., A. Huang, Z. Li, et al. 2020. Face masks against Covid-19: An evidence review. *Preprints*, 2020040203; https://doi.org/10.20944/preprints202004.0203.v1.
Hristova, B. 2020. Recovering COVID-19 patient describes what it was like to have the virus. *CBC News*, 20 March. https://www.cbc.ca/news/canada/toronto/coronavirus-patient-1.5502501.
Huang, C., Y. Wang, X. Li, et al. 2020. Clinical features of patients infected with 2019 novel coronavirus in Wuhan, China. *Lancet* 395: 497–506.
Huang, I.Y. 2020. Fighting Covid-19 through government initiatives and collaborative governance: The Taiwan experience. *Public Administration Review* 80 (4): 665–670.
Huang, C., L. Huang, Y. Wang, et al. 2021. 6-month consequences of Covid-19 in patients discharged from hospital: A cohort study. *Lancet* 397: 220–232.
Huber, V. 2006. The unification of the globe by disease? The International Sanitary Conferences on cholera, 1851–1894. *The Historical Journal* 49 (2): 453–476.
Hűbner, M., T. Zingg, D. Martin, et al. 2020. Surgery for non-Covid patients during the pandemic. *PLoS ONE* 15 (10): e0241331. https://doi.org/10.1371/journal.pone.0241331.
Hughes, M.T., J. Kahn, and A. Kachalia. 2021. Who goes first? Government leaders and prioritization of SARS-CoV-2 vaccines. *New England Journal of Medicine* 384 (5): e15. (1–2).
Huizinga, J. 1924. *The waning of the Middle Ages*. New York: St. Martin's Press.
Humphreys, G. 2021. Opening up with Covid-19 passes. *Bulletin of the World Health Organization* 99: 546–547.
Hunt, R.W. 1993. A critique of using age to ration health care. *Journal of Medical Ethics* 19: 19–23.
Huxtable, R. 2020. Covid-19: Where is the national ethical guidance? *BMC Medical Ethics* 21: 32. https://doi.org/10.1186/s12910-020-00478-2.
Ibarrondo, F.J., J.A. Fulcher, D. Goodman-Meza, et al. 2020. Rapid decay of anti-SARS-CoV-2 antibodies in persons with mild Covid-19. *New England Journal of Medicine* 383 (11): 1085–1087.
IHME COVID-19 Forecasting Team, R. C. Reiner, R.M. Barber, et al. 2021. Modeling COVID-19 scenarios for the United States. *Nature Medicine* 27: 94–105.
IMF. 2020. *World economic outlook. A long and difficult ascent*. Washington, DC: International Monetary Fund, October. https://www.imf.org/en/Publications/WEO/Issues/2020/09/30/world-economic-outlook-october-2020.
Independent Panel for Pandemic Preparedness & Response. 2021. *Covid-19: Make it the last pandemic*. https://theindependentpanel.org/wp-content/uploads/2021/05/COVID-19-Make-it-the-Last-Pandemic_final.pdf.
———. 2021. *Covid-19: Make it the last pandemic. A summary*, 1. https://theindependentpanel.org/wp-content/uploads/2021/05/Summary_COVID-19-Make-it-the-Last-Pandemic_final.pdf.
Inglesby, T. 2020. Make pandemics lose their power. *In Covid-19 and world order. The future of conflict, competition, and cooperation*, ed. H. Brands and F.J. Gavin, 75–92. Baltimore: Johns Hopkins University Press.
Ingold, T. 2000. Globes and spheres. The topology of environmentalism. In *The perception of the environment. Essays on livelihood, dwelling and skill*, ed. T. Ingold, 209–218. London/New York: Routledge.
Inman, P. 2021. Drop Covid vaccine patent rules to save lives in poorest countries, UK and Germany told. *The Guardian*, June 12. https://www.theguardian.com/world/2021/jun/12/drop-covid-vaccine-patent-rules-to-save-lives-in-worlds-poorest-countries-britain-and-germany-told.
Institute of Medicine. 1992. *Emerging infections. Microbial threats to health in the United States*. Washington, DC: National Academy Press.
———. 2009. *Guidance for establishing crisis standards of care for use in disaster situations: A letter report*. Washington, DC: The National Academies Press.

Intergovernmental Panel on Climate Change. 2018. *Global warming of 1.5°C*. Switzerland: IPCC. https://report.ipcc.ch/sr15/pdf/sr15_spm_final.pdf.

Ioannidis, J.P.A. 2020. A fiasco in the making? As the coronavirus pandemic takes hold, we are making decisions without reliable data. *STAT*. https://www.statnews.com/2020/03/17/a-fiasco-in-the-making-as-the-coronavirus-pandemic-takes-hold-we-are-making-decisions-without-reliable-data/.

Isaacs, D., and A. Priesz. 2021. Covid-19 and the metaphor of war. *Journal of Paediatrics and Child Health* 57: 6–8.

Iserson, K.V. 2020. Healthcare ethics during a pandemic. *Western Journal of Emergency Medicine* 21 (3): 477–483.

Iserson, K.V., and J.C. Moskop. 2007. Triage in medicine, part I: Concept, history, and types. *Annals of Emergency Medicine* 49 (3): 275–281.

Iserson, K.V. 2021. SARS-CoV-2 (Covid-19) vaccine development and production: An ethical way forward. *Cambridge Quarterly of Healthcare Ethics* 30: 59–68.

Islam, T., A.H. Pitafi, V. Arya, et al. 2020. Panic buying in the Covid-19 pandemic: A multi-country examination. *Journal of Retailing and Consumer Services*, October 23; https://doi.org/10.1016/j.jretconser.2020.102357.

Iyengar, K., S. Bahl, R. Vaishya, and A. Vaish. 2020. Challenges and solutions in meeting up the urgent requirement of ventilators for Covid-19 patients. *Diabetes & Metabolic Syndrome: Clinical Research & Reviews* 14: 499–501.

Jabbari, P., and N. Rezaei. 2020. With the risk of reinfection, is Covid-19 here to stay? *Disaster Medicine and Public Health Preparedness* 14 (4): e33. https://doi.org/10.1017/dmp.2020.274.

Jacobson, K. 2009. A developed nature: A phenomenological account of the experience of home. *Continental Philosophy Review* 42: 355–373.

Jambroes, M., T. Nederland, M. Kaljouw, et al. 2016. Implications of health as 'the ability to adapt and self-manage' for public health policy: A qualitative study. *European Journal of Public Health* 26 (3): 412–416.

Jansen, M.O., P. Angelos, S.J. Schrantz, et al. 2020. Fair and equitable subject selection in concurrent Covid-19 clinical trials. *Journal of Medical Ethics* 47: 7–11.

Janwadkar, A.S., and T.M. Bibler. 2020. Ethical challenges in advance care planning during the Covid-19 pandemic. *American Journal of Bioethics* 20 (7): 202–204.

Jefferson, T., C.B. Del Mar, L. Dooley, et al. 2011. Physical interventions to interrupt or reduce the spread of respiratory viruses. *Cochrane Database of Systematic Reviews* 2011, Issue 7. Art. No.: CD006207. https://doi.org/10.1002/14651858.CD006207.pub4.

Jeffrey, D.I. 2020. Relational ethical approaches to the Covid-19 pandemic. *Journal of Medical Ethics* 46: 495–498.

Jennings, B., and A. Dawson. 2015. Solidarity in the moral imagination of bioethics. *Hastings Center Report* 45 (5): 31–38.

Jennings, B. 2020. *Beyond the Covid crisis – A new social contract with public health*. The Hastings Center. https://www.thehastingscenter.org/beyond-the-covid-crisis-a-new-social-contract-with-public-health/.

Jerolmack, C. 2013. Who's worried about turkeys? How 'organisational silos' impede zoonotic disease surveillance. In *Pandemics and emerging infectious diseases. The sociological agenda*, ed. R. Dingwal, L.M. Hoffman, and K. Staniland, 33–45. Chichester: Wiley-Blackwell.

Ji, D., X. Li, and S. Ramakrishna. 2020. Addressing the worldwide shortage of face masks. *BMC Materials* 2: 9. https://doi.org/10.1186/s42833-020-00015-w.

Johnson, S. 2006. *The ghost map. The story of London's most terrifying epidemic – And how it changed science, cities, and the modern world*. New York: Riverhead Books.

———. 2021. Big pharma fuelling human rights crisis over Covid vaccine inequity – Amnesty. *The Guardian*, September 22. https://www.theguardian.com/global-development/2021/sep/22/big-pharma-fuelling-human-rights-crisis-over-covid-vaccine-inequity-amnesty.

Johnson, W. 2020. The photograph that shocked a nation. *National Geographic*, July 24. https://www.nationalgeographic.com/newsletters/photography/2020/07/photograph-shocked-indonesia-july-24/.

Jones, K.E., N. G. Patel, M.A. Levy, A. Storeygard, D. Balk, J. L. Gittleman, and P. Daszak. 2008. Global trends in emerging infectious diseases. *Nature* 451 (no.7181): 990-993.

Jones, N.R., Z.U. Qureshi, R.J. Temple, et al. 2020. Two metres or one: What is the evidence for physical distancing in Covid-19? *British Medical Journal* 370: m3223. https://doi.org/10.1136/bmj.m3223.

Jonsen, A. 1998. *The birth of bioethics*. New York/Oxford: Oxford University Press.

Jung, J., H. Jang, H.K. Kim, et al. 2020. The importance of mandatory Covid-19 diagnostic testing prior to release from quarantine. *Journal of Korean Medical Science* 35 (34): e314.

Kahn, J.P., L.M. Henry, A.C. Mastroianni, et al. 2020. For now, it's unethical to use human challenge studies for SARS-CoV-2 vaccine development. *Proceedings of the National Academy of Sciences of the United States of America* 117 (46): 28538–28542.

Kalil, A.C. 2020. Treating Covid-19 – Off-label drug use, compassionate use, and randomized clinical trials during pandemics. *JAMA* 323 (19): 1897–1898.

Kaminsky, C. 2020. Normality "ex post": Social conditions of moral responsibility. In *Medicine and ethics in times of corona*, ed. M. Woesler and H.-M. Sass, 63–74. Zürich: LIT Verlag.

Kandel, N., S. Chungong, A. Omaar, and J. Xing. 2020. Health security capacities in the context of Covid-19 outbreak: An analysis of International Health Regulations annual report data from 182 countries. *Lancet* 395: 1047–1053.

Karagiannidis, C., C. Mostert, C. Hentschker, et al. 2020. Case characteristics, resource use, and outcomes of 10.021 patients with Covid-19 admitted to 920 German hospitals: An observational study. *Lancet Respiratory Medicine* 8 (9): 853–862.

Karp, P. 2020. Former WHO board member warns world against coronavirus 'vaccine nationalism'. *The Guardian*, May 18. https://www.theguardian.com/world/2020/may/18/former-who-board-member-warns-world-against-coronavirus-vaccine-nationalism.

Kashyap, A., and M. Wurth. 2021. Rich countries must stop 'vaccine apartheid'. View. *Euronews*, March 11. https://www.euronews.com/2021/03/11/rich-countries-must-stop-vaccine-apartheid-view.

Katib, A.A. 2020. Research ethics challenges during the Covid-19 pandemic: What should and what should not be done. *Journal of Ideas in Health* 3: 183–187.

Keulemans, M. 2020. Steeds meer besmettingen, maar waar zit het lek? *De Volkskrant*, 9 September: 4.

Keulemans, M., E. de Visser, and T.N. Jansen. 2020. Het virus tot zover. *De Volkskrant*, 17 October, 4–11.

Khamsi, R. 2020. If a coronavirus vaccine arrives, can the world make enough? *Nature* 580: 578–580.

Khan, A.S. 2020. *The next pandemic. On the front lines against humankind's gravest dangers*. New York: PublicAffairs.

Kiaghadi, A., H.S. Rifai, and W. Liaw. 2020. Assessing COVID-19 risk, vulnerability and infection prevalence in communities. *PLoS ONE* 15 (10): e0241166. https://doi.org/10.1371/journal.pone.0241166.

Kim, S.Y.H., and C. Grady. 2020. Ethics in the time of Covid. What remains the same and what is different. *Neurology* 94 (23): 1007–1008.

King, C.S., D. Sahjwani, A.W. Brown, et al. 2020. Outcomes of mechanically ventilated patients with COVID-19 associated respiratory failure. *PLoS One* 15 (11): e0242651. https://doi.org/10.1371/journal.pone.0242651.

Kirchhoffer, D.G. 2020. Dignity, autonomy, and allocation of scarce medical resources during Covid-19. *Journal of Bioethical Inquiry* 17: 691–696.

Kiros, M., H. Anudalem, R. Kiros, et al. 2020. Covid-19 pandemic: Current knowledge about the role of pets and other animals in disease transmission. *Virology Journal* 17: 143. https://doi.org/10.1186/s12985-020-01416-9.

Kissler, S.M., C. Tedijante, E. Goldstein, Y.H. Grad, and M. Lipsitch. 2020. Projecting the transmission dynamics of SARS-CoV-2 through the postpandemic period. *Science* 368: 860–868.

Koff, W.C., T. Schenkelberg, T. Williams, and at al. 2021. Development and deployment of Covid-19 vaccines for those most vulnerable. *Science Translational Medicine* 13: eabd1525.

Kofler, N., and F. Baylis. 2021. *Covid-19 vaccination certificates: Prospects and problems*. The Hastings Center, March 10. https://www.thehastingscenter.org/covid-19-vaccination-certificates-prospects-and-problems/.

Kofman, A., R. Kantor, and E.Y. Adashi. 2021. Potential Covid-19 endgame scenarios. Eradication, elimination, cohabitation, or conflagration? *JAMA* 326 (4): 303–304.

Kolata, G. 2005. *Flu. The story of the great influenza pandemic of 1918 and the search for the virus that caused it*. New York: Simon & Schuster.

Kollewe, J. 2021. From Pfizer to Moderna: Who's making billions from Covid-19 vaccines? *The Guardian*, 6 March. https://www.theguardian.com/business/2021/mar/06/from-pfizer-to-moderna-whos-making-billions-from-covid-vaccines.

Komesaroff, P.A. 2020. Not all bad: Sparks of hope in a global disaster. *Journal of Bioethical Inquiry* 17: 515–518.

Komrad, M.S. 2020. Medical ethics in the time of Covid-19. *Current Psychiatry* 19 (7): 29-32, 46.

Koplan, J.P., T.C. Bond, M.H. Merson, et al. 2009. Towards a common definition of global health. *Lancet* 373: 1993–1995.

Korteweg, N. 2020. Ik maak me het meeste zorgen dat mensen de maatregelen niet vasthouden op vakantie. *NRC* 5 July. https://www.nrc.nl/nieuws/2020/07/05/een-lockdown-zat-nooit-in-ons-arsenaal-a4005060.

Kossakovski, F. 2020. What we've learned about how our immune system fights Covid-19. *National Geographic*, December 29. https://www.nationalgeographic.com/science/article/what-coronavirus-has-taught-us-about-human-immune-system.

Kotalik, J. 2005. Preparing for an influenza pandemic: Ethical issues. *Bioethics* 19 (4): 422–431.

KPMG. 2021. *Dit zijn de lessen van 1,5 jaar coronacrisis*. KPMG Health, October 14. https://home.kpmg/content/dam/kpmg/nl/pdf/2021/sectoren/dit-zijn-de-lessen-van-15-jaar-coronacrisis.pdf.

Kramer, J. 2020. They spend 12 years solving a puzzle. It yielded the first Covid-19 vaccines. *National Geographic*, December 31. https://www.nationalgeographic.com/science/article/these-scientists-spent-twelve-years-solving-puzzle-yielded-coronavirus-vaccines.

———. 2021. Experts torn over changing vaccine doses to speed up lagging rollout. *National Geographic*, January 12. https://www.nationalgeographic.com/science/article/experts-debate-half-doses-and-delayed-boosters-for-covid-vaccines.

———. 2021. Why people latch on to conspiracy theories, according to science. *National Geographic*, January 8. https://www.nationalgeographic.com/science/article/why-people-latch-on-to-conspiracy-theories-according-to-science.

Kretzschmar, M.E., G. Roshnova, M.C.J. Bootsma, et al. 2020. Impact of delays on effectiveness of contact tracing strategies for COVID-19: A modelling study. *Lancet Public Health* 5: e452–e459. https://doi.org/10.1016/S2468-2667(20)30157-2.

Krütli, P., T. Rosemann, K.Y. Törnblom, and T. Smieszek. 2016. How to fairly allocate scarce medical resources: Ethical argumentation under scrutiny by health professionals and lay people. *PLoS ONE* 11 (7): e0159086. https://doi.org/10.1371/journal.pone.0159086.

Kucharski, A. 2020. *The rules of contagion. Why things spread – And why they stop*. London: Profile Books.

Kumar, O.R.V., B.S. Ramkumar, B.S. Pruthvishree, et al. 2020. SARS-CoV-2 (Covid-19): Zoonotic origin and susceptibility of domestic and wild animals. *Journal of Pure and Applied Microbiology* 14 (suppl 1): 741–747.

Kupferschmidt, K., and J. Cohen. 2020. Can China's Covid-19 strategy work elsewhere? *Science* 367 (6482): 1061–1062.

Kuylen, M.N.I., S.Y. Kim, A.R. Keene, and G.S. Owen. 2021. Should age matter in Covid-19 triage? A deliberative study. *Journal of Medical Ethics* 47: 291–295.

Lam, T.T., M. H. Shum, H-C. Zhu et al. 2020. Identifying SARS-CoV-2 related coronaviruses in Malayan Pangolins. *Nature* 583 (no.7815): https://doi.org/10.1038/s41586-020-2169-0.

Landau, M. D. 2021. How Merck's antiviral pill could change the game for Covid-19. *National Geographic*, October 2. https://www.nationalgeographic.com/science/article/how-mercks-antiviral-pill-could-change-the-game-for-covid-19.

Langreth, R. 2021. Five steps to prevent the next pandemic. *NDTV*, February 4. https://www.ndtv.com/world-news/five-steps-to-prevent-the-next-pandemic-2362755.

Lapolla, P., A. Mingoli, and R. Lee. 2020. Deaths from Covid-19 in healthcare workers in Italy – What can we learn? *Infection Control & Hospital Epidemiology* 1–2. https://doi.org/10.1017/ice.2020.241.

Larson, H.J. 2018. The biggest pandemic risk? Viral misinformation. *Nature* 562: 309.

Lavine, J.S., O.N. Björnstad, and R. Antia. 2021. Immunological characteristics govern the transition of Covid-19 to endemicity. *Science* 371: 741–745.

Lazarus, J.V., S.C. Ratzan, A. Palayew, et al. 2021. A global survey of potential acceptance of a Covid-19 vaccine. *Nature Medicine* 27 (2): 225–228.

Lebrecht, N. 2020. *Concertgebouw chorus is devastated after pre-Covid Bach Passion*. https://slippedisc.com/2020/05/concertgebouw-chorus-is-devastated-after-pre-covid-bach-passion/.

Lederberg, J., R.E. Shope, and S.C. Oaks. 1992. *Emerging infections: Microbial threats to health in the United States*. Washington, DC: National Academies Press.

Ledford, H. 2020. Coronavirus shuts down trials of drugs for multiple other diseases. *Nature* 580 (7801): 15–16.

———. 2020. Why do Covid death rates appear to be falling? *Nature* 587: 190–192.

———. 2021. Why Covid vaccines are so difficult to compare. *Nature* 591: 16–17.

———. 2021. Covid vaccines and blood clots: Five key questions. *Nature* 592 (7855): 495–496.

Lednicky, J.A., M. Lauzardo, Z.H. Fan, et al. 2020. Viable SARS-CoV-2 in the air of a hospital room with Covid-19 patients. *International Journal of Infectious Diseases* 100: 476–482.

Lee, K.D., S.B. Lee, J.K. Lim, et al. 2020. Providing essential clinical care for non-COVID-19 patients in a Seoul metropolitan acute care hospital amidst ongoing treatment of COVID-19. *Journal of Hospital Infection* 106 (4): 673–677.

Lee, T.H., and A.H. Chen. 2021. Last-mile logistics of Covid vaccination – The role of health care organizations. *New England Journal of Medicine* 384 (8): 685–687.

Lee, V.J., C.J. Chiew, and W.X. Khong. 2020. Interrupting transmission of Covid-19: Lessons from containment efforts in Singapore. *International Journal of Travel Medicine* 27: 1–5.

Lei, R., and R. Qiu. 2020. *Report from China: Ethical questions on the response to the coronavirus*. The Hastings Center, January 31. https://www.thehastingscenter.org/report-from-china-ethical-questions-on-the-response-to-the-coronavirus/.

Lei, R., R. Qiu, and P. Jia. 2021. *WHO-China report on Covid: Important step forward, more to be done*. The Hastings Center, April 9. https://www.thehastingscenter.org/who-china-report-on-covid-important-step-forward-more-to-be-done/.

Lemus-Delgado, D. 2020. China and the battle to win the scientific narrative about the origin of Covid-19. *Journal of Science Communication* 19 (05). https://doi.org/10.22323/2.19050206.

Le Page, M. 2021. Threats from new variants. *New Scientist* 249: 8–9.

Levinas, E. 1991. *Totality and Infinity: An essay on exteriority*. Dordrecht: Kluwer Academic Publishers.

Lewandowsky, S., U.K.H. Ecker, C.M. Seifert, et al. 2012. Misinformation and its correction: Continued influence and successful debiasing. *Psychological Science in the Public Interest* 13 (3): 106–131.

Lewis, D. 2021. Covid-19 rarely infects through surfaces. So why are we still deep cleaning? *Nature* 590: 26–28.

Lewis, T. 2021. Slovakia offers a lesson in how rapid testing van fight Covid. *Scientific American*, April 8. https://www.scientificamerican.com/article/slovakia-offers-a-lesson-in-how-rapid-testing-can-fight-covid/.

———. 2021. How the U.S. pandemic response went wrong – And what went right – During a year of Covid. *Scientific American*, March 11. https://www.scientificamerican.com/article/how-the-u-s-pandemic-response-went-wrong-and-what-went-right-during-a-year-of-covid/.

Lewis, J., and U. Schuklenk. 2021. Bioethics met its Covid-19 Waterloo: The doctor knows best again. *Bioethics* 35: 3–5.

Lie, R.K., and F.G. Miller. 2020. *Global allocation of coronavirus vaccines*. The Hastings Center, December 12. http://www.bioethics.net/2020/12/global-allocation-of-coronavirus-vaccines/.

Lipsitch, M. 2020. We know enough now to act decisively against Covid-19. Social distancing is a good place to start. *STAT*, March 18. https://www.statnews.com/2020/03/18/we-know-enough-now-to-act-decisively-against-covid-19/.

Lipworth, W., M. Gentgall, I. Kerridge, and C. Stewart. 2020. Science at warp speed: Medical research, publication, and translation during the Covid-19 pandemic. *Journal of Bioethical Inquiry* 17: 555–561.

Lipworth, W. 2020. Beyond duty: Medical 'heroes' and the Covid-19 pandemic. *Journal of Bioethical Inquiry* 17: 723–730.

Litewska, S.G., and J.D. Moreno. 2021. *Covid-19 in Argentina and the abuse of bioethics*. The Hastings Center, May 20. https://www.thehastingscenter.org/covid-19-in-argentina-and-the-abuse-of-bioethics/https://www.thehastingscenter.org/covid-19-in-argentina-and-the-abuse-of-bioethics/.

Liu, J., and R. Chung. 2021. *Capitalist philanthropy and vaccine imperialism*. The Hastings Center, September 10. https://www.thehastingscenter.org/capitalist-philanthropy-and-vaccine-imperialism/.

Liu, L. 2020. Sustainable Covid-19 mitigation: Wuhan lockdowns, health inequities, and patient evacuation. *International Journal of Health Policy and Management* 9 (10): 415–418.

Liu, Y., S. Salwi, and B. Drolet. 2020. Multivalue ethical framework for fair global allocation of a Covid-19 vaccine. *Journal of Medical Ethics* 46: 499–501.

Lo, B., and M.H. Katz. 2005. Clinical decision making during public health emergencies: Ethical considerations. *Annals of Internal Medicine* 143 (7): 493–498.

Logunov, D.Y., I.V. Dolzhikova, D.V. Shcheblyakov, et al. 2021. Safety and efficacy of an rAd26 and rAd5 vector-based heterologous prime-boost COVID-19 vaccine: An interim analysis of a randomised controlled phase 3 trial in Russia. *Lancet* 397: 671–681.

Long, N.J. 2020. From social distancing to social containment. Reimagining sociality for the coronavirus pandemic. *Medicine Anthropology Theory* 7 (2): 247–260.

London, A.J., and J. Kimmelman. 2020. Against pandemic research exceptionalism. Crises are no excuse for lowering scientific standards. *Science* 368 (6490): 476–477.

Longrigg, J. 1980. The great plague of Athens. *History of Science* 18: 209–225.

Luscombe, R. 2021. Michigan must tell Johnson & Johnson vaccine recipients that it was developed using stem cells. *The Guardian*, March 29. https://www.theguardian.com/us-news/2021/mar/29/johnson-and-johnson-covid-vaccine-stem-cells-michigan-law.

Looi, M.-K. 2020. Covid-19: Is a second wave hitting Europe? *British Medical Journal* 371: m4113. https://doi.org/10.1136/bmj.m4113.

Luscombe, R. 2021. Biden says up to 90% of adults will be eligible for Covid vaccine by 19 April. *The Guardian*, March 29. https://www.theguardian.com/us-news/2021/mar/29/joe-biden-covid-vaccine-program-expansion.

Lynteris, C. 2018. Plague masks: The visual emergence of anti-epidemic personal protection equipment. *Medical Anthropology* 37 (6): 442–457.

Ma, X., Y. Wang, T. Gao, et al. 2020. Challenges and strategies to research ethics in conducting Covid-19 research. *Journal of Evidence Based Medicine* 13 (2): 173–177.

Maboloc, C.R. 2021. Global ethics and the right to universal access to Covid-19 vaccines. *Eubios Journal of Asian and International Bioethics* 31 (3): 169–171.

MacDonald, N. E., and SAGE Working Group on Vaccine Hesitancy. 2015. Vaccine hesitancy: Definition, scope and determinants. *Vaccine* 33: 4161–4164.

Macer, D. 2020. Wearing masks in Covid-19 pandemic, the precautionary principle, and the relationships between individual responsibility and group solidarity. *Eubios Journal of Asian and International Bioethics* 30 (4): 129–132.

———. 2020. The foundation and functioning of the world emergency Covid-19 pandemic ethics committee. In *Medicine and ethics in times of corona*, ed. M. Woesler and H.-M. Sass, 115–125. Zürich: LIT Verlag.

Macip, S. 2020. *Modern epidemics. From the Spanish flu to Covid-19*. Cambridge: Polity Press.

Mackenzie, C., and N. Stoljar. 2000. *Relational autonomy: Feminist perspectives on autonomy, agency, and the social self*. New York: Oxford University Press.

Mackenzie, D. 2020. *Covid-19. The pandemic that never should have happened, and how to stop it*. London: The Bridge Street Press.
Mahase, E. 2020. Covid-19: Remdesivir probably reduces recovery time, but evidence is uncertain, panel finds. *British Medical Journal* 380: m3049. https://doi.org/10.1136/bmj.m3049.
Makoni, M. 2020. Covid-19 in Africa: Half a year later. *The Lancet Infectious Diseases* 20 (10): 1127.
Mallapaty, S. 2020. The coronavirus is most deadly if you are old and male. *Nature* 585: 16–17.
Mallapaty, S. 2020. Where did COVID come from? WHO investigation begins but faces challenges. *Nature* 587: 341–342. https://doi.org/10.1038/d41586-020-03165-9.
———. 2020. Meet the scientists investigating the origins of the Covid pandemic. *Nature* 588: 208. https://doi.org/10.1038/d41586-020-03402-1.
———. 2021. After the WHO report. What's next in the search for Covid's origins. *Nature* 592: 337–338.
———. 2021. Are Covid vaccination programmes working? Scientists seek first clues. *Nature* 589: 504–505.
Mallapaty, S. 2021. Can Covid vaccines stop transmission? Scientists race to find answers. *Nature*, February 19. https://www.nature.com/articles/d41586-021-00450-z.
Mallapaty, S. 2021. Laos bats host closest known relatives of virus behind Covid. *Nature* 597: 603.
Mallet, S. 2004. Understanding home. A critical review of the literature. *Sociological Review* 52 (1): 62–89.
Mansnerus, E. 2013. Using model-based evidence in the governance of pandemics. In *Pandemics and emerging infectious diseases: The sociological agenda*, ed. R. Dingwall, L.M. Hoffman, and K. Staniland, 110–121. Chichester: Wiley.
Marcel, G. 1962. *Homo viator. Introduction to a metaphysic of hope*. New York: Harper & Row.
———. 1984. Reply to John E. Smith. In *The philosophy of Gabriel Marcel*, ed. P.A. Schilpp and L.E. Hahn, 350–353. La Salle: Open Court.
Marinthe, G., G. Brown, S. Delouvée, and F. Jolley. 2020. Looking out for myself: Exploring the relationship between conspiracy mentality, perceived personal risk, and Covid-19 prevention measures. *British Journal of Health Psychology* 25: 957–980.
Marsa, L. 2021. Can AstraZeneca dispel doubts about its shots? *National Geographic*, April 2. https://www.nationalgeographic.com/science/article/can-astrazeneca-dispel-doubts-about-its-shots.
Martin-Delgado, J., E. Viteri, A. Mula, et al. 2020. Availability of personal protective equipment and diagnostic and treatment facilities for healthcare workers in volved in Covid-19 care: A cross-sectional study in Brazil, Colombia, and Ecuador. *PLoS ONE* 15 (11): e0242185. https://doi.org/10.1371/journal.pone.0242185.
Martins, A.A. 2021. *Global bioethics in a pandemic: A dialogical approach*. Health Care Ethics USA. https://www.chausa.org/publications/health-care-ethics-usa/article/winter-2021/global-bioethics-in-a-pandemic-a-dialogical-approach.
Maschke, K.J., and M.K. Gusmano. 2020. *Ethics and evidence in the search for vaccine and treatments for Covid-19*. The Hastings Center, April 15. https://www.thehastingscenter.org/ethics-and-evidence-in-the-search-for-a-vaccine-and-treatments-for-covid-19/.
Mason, B. 2020. *Covid-19 vaccine review – Pre-orders, approvals, and prioritizations*, December 17. https://www.fxempire.com/forecasts/article/covid-19-vaccine-review-pre-orders-approvals-and-prioritizations-689038.
Maxmen, A. 2021. Why did the world's pandemic warning system fail when Covid hit? *Nature* 589: 499–500.
———. 2021. US Covid origins report: Researchers pleased with scientific approach. *Nature* 597: 159–160.
Maxmen, A., and S. Mallapaty. 2021. The Covid lab-leak hypothesis: What scientists do and don't know. *Nature* 594: 313–315.
McBride Folkers, K., and A. Caplan. 2020. *False hope about coronavirus treatments*. The Hastings Center, March 20. https://www.thehastingscenter.org/false-hope-about-coronavirus-treatments/.
McCarthy, M., and A. Caplan. 2020. *Coronavirus mutation panic*. The Hastings Center, December 22. https://www.thehastingscenter.org/coronavirus-mutation-panic/.

McKeever, A. 2021. We still don't know the origins of the coronavirus. Here are 4 scenarios. *National Geographic*, April 2. https://www.nationalgeographic.com/science/article/we-still-dont-know-the-origins-of-the-coronavirus-here-are-four-scenarios.
McKibben, B. 2020. The end of the world as we know it. *Times Literary Supplement* July 31: 4–5.
McKie, R. 2020. What is the new Covid strain – And will vaccines work against it? *The Guardian*, December 19. https://www.theguardian.com/world/2020/dec/19/what-is-the-new-covid-strain-and-will-vaccines-work-against-it.
McLaws, M.-L. 2020. What is the Covid 'bubble' concept, and could it work in Australia? *The Conversation*, August 31. https://theconversation.com/what-is-the-covid-bubble-concept-and-could-it-work-in-australia-144938.
McMahon, D.E., G.A. Peters, L.C. Ivers, and E.E. Freeman. 2020. Global resource shortages during Covid-19: Bad news for low-income countries. *PLoS Neglected Tropical Diseases* 14 (7): e0008412. https://doi.org/10.1371/journal.pntd.0008412.
McNairy, M., B. Bullington, and K. Bloom-Feshbach. 2020. Searching for human connectedness during Covid-19. *Journal of General Internal Medicine* 35 (10): 3043–3044.
McNeill, W.H. 1998. *Plagues and people*. New York: Anchor Books.
McVeigh, K. 2020. Rich countries leaving the rest of the world behind on Covid vaccines, warns Gates Foundation. *The Guardian*, December 10. https://www.theguardian.com/global-development/2020/dec/10/rich-countries-leaving-rest-of-the-world-behind-on-covid-vaccines-warns-gates-foundation.
Meagher, K.M., N.W. Cummins, A.E. Bharuha, et al. 2020. Covid-19 ethics and research. *Mayo Clinic Proceedings* 95 (6): 1119–1123.
Meek, J. 2020. Red pill, blue pill. *London Review of Books*, 19–23.
Mello, M.M., R.D. Silverman, and S.B. Omer. 2020. Ensuring uptake of vaccines against SARS-CoV-2. *New England Journal of Medicine* 383 (14): 1296–1298.
Menachery, V.D., B.L. Yount Jr., K. Debbink, et al. 2015. A Sars-like cluster of circulating bat coronaviruses shows potential for human emergence. *Nature Medicine* 21: 1508–1513.
Mercatelli, D., and F.M. Giorgi. 2020. Geographic and Genomic Distribution of SARS-CoV-2 Mutations. *Frontiers of Microbiology* 11: 1800. https://doi.org/10.3389/fmicb.2020.01800.
Midgley, M. 2014. *Are you an illusion?* London/New York: Routledge.
Migone, A.R. 2020. The influence of national policy characteristics on Covid-19 containment policies: A comparative analysis. *Policy Design and Practice* 3 (3): 259–276.
Miller, F.G. 2020. *Pandemic language*. Hastings Center, July; www.thehastingscenter.org/pandemic-language/.
Millstein, J.H., and S. Kindt. 2020. Reimagining the patient experience during the Covid-19 pandemic. *NEJM Catalyst/Innovations in Care Delivery*; https://doi.org/10.1056/CAT.20.0349.
Milne, G., T. Hames, C. Scotton, N. Gent, A. Johnsen, R.M. Anderson, and T. Ward. 2021. Does infection with or vaccination against SARS-CoV-2 lead to lasting infection? *The Lancet Respiratory Medicine*, October 21; https://doi.org/10.1016/S2213(21)00407-0.
Moatti, J.-P. 2020. The French reponse to Covid-19: Intrinsic difficulties at the interface of science, public health, and policy. *Lancet Public Health* 5 (5): e255.
Moghadas, S.M., M.C. Fitzpatrick, P. Sah, et al. 2020. The implications of silent transmission for the control of Covid-19 outbreaks. *Proceedings of the National Academy of Sciences of the United States of America* 117 (30): 17513–17515.
Mohamed, K., E. Rodriguez-Roman, F. Rahmani, et al. 2020. Borderless collaboration is needed for Covid-19 – A disease that knows no borders. *Infection Control & Hospital Epidemiology* 10: 1–2.
Möhlenkamp, S., and H. Thiele. 2020. Ventilation of COVID-19 patients in intensive care units. *Herz* 45 (4): 329–331.
Momtazmanesh, H., D. Ochs, L.Q. Uddin, et al. 2020. All together to fight Covid-19. *American Journal of Tropical Medicine and Hygiene* 102 (6): 1181–1183.
Mondelli, M.U., M. Colaneri, E.M. Seminari, et al. 2021. Low risks of SARS-Co-V-2 transmission by fomites in real-life conditions. *Lancet Infectious Diseases* 21: e112.

Moon, S., D. Sridhar, M.A. Pate, et al. 2015. Will Ebola change the game? Ten essential reforms before the next pandemic. The report of the Harvard-LSHTM Independent Panel on the Global Response to Ebola. *Lancet* 386: 2204–2221.

Moore, J. 2000. Placing *home* in context. *Journal of Environmental Psychology* 20: 207–217.

Moore, E.H., and A. Burgess. 2020. Risk rituals? *Journal of Risk Research* 14 (1): 111–124.

Morawska, L., and D.K. Milton. 2020. It is time to address airborne transmission of Coronavirus Disease 2019 (Covid-19). *Clinical Infectious Diseases*, ciaa939; https://doi.org/10.1093/cid/ciaa939.

Morens, D.M., and A.S. Fauci. 2007. The 1918 influenza pandemic: Insights for the 21st century. *Journal of Infectious Diseases* 195 (7): 1018–1028.

Morens, D.M., G.K. Folkers, and A.S. Fauci. 2009. What is a pandemic? *The Journal of Infectious Diseases* 200: 1018–1021.

Morens, D.M., J.G. Breman, C.H. Calisher, et al. 2020. The origin of Covid-19 and why it matters. *American Journal of Tropical Medicine and Hygiene* 103 (3): 955–959.

Moroni, F., M. Gramegna, S. Ajello, et al. 2020. Collateral damage: Medical care avoidance behavior among patients with myocardial infarction during the Covid-19 pandemic. *JACC: Care Reports* 2 (10): 1620–1624.

Morse, S.S. 1991. Emerging viruses: Defining the rules for viral traffic. *Perspectives in Biology and Medicine* 34 (3): 387–409.

———. 1995. Factors in the emergence of infectious disease. *Emerging Infectious Diseases* 1: 7–15.

Mullard, A. 2020. Covid-19 vaccine development pipeline gears up. *Lancet* 395 (10239): 1751–1752.

———. 2020. How Covid vaccines are being divvied up around the world. *Nature*, 30 November. https://www.nature.com/articles/d41586-020-03370-6.

Műller, M., P.M. Derlet, C. Mudry, and A. Aeppli. 2020. Testing of asymptomatic individuals for fast feedback-control of Covid-19 pandemic. *Physical Biology* 17 (6): 065007. https://doi.org/10.1088/1478-3975/aba6d0.

Műller, O., G. Lu, A. Jahn, and O. Razum. 2020. Covid-19 control: Can Germany learn from China? *International Journal of Health Policy and Management* 9 (10): 432–435.

Murphy, K., H. Williamson, E. Sargeant, and M. McCarthy. 2020. Why people comply with Covid-19 social distancing restrictions: Self-interest or duty? *Australian & New Zealand Journal of Criminology* 53 (4): 477–496.

Murray, J. 2020. Care workers move into Sheffield dementia home to shield residents. *The Guardian*, March 25. https://www.theguardian.com/world/2020/mar/25/care-workers-move-into-sheffield-dementia-home-to-shield-residents.

Nacoti, M., A. Ciocca, A. Giupponi, et al. 2020. At the epicenter of the Covid-19 pandemic and humanitarian crises in Italy: Changing perspectives on preparation and mitigation. *NEJM Catalyst Innovations in Care Delivery* 1–5; https://doi.org/10.1056/CAT.20.0080.

National Academies of Sciences, Engineering, and Medicine. 2020. *Rapid expert consultation on crisis standards of care for the Covid-19 pandemic (March 28, 2020)*. Washington, DC: The National Academies Press. https://doi.org/10.17226/25765.

———. 2020. *Framework for Equitable Allocation of COVID-19 Vaccine*. Washington, DC: The National Academies Press. https://doi.org/10.17226/25917.

National Academy of Medicine. 2020. *National organizations call for action to implement crisis standards of care during Covid-19 surge*. December 18, 2020. https://nam.edu/national-organizations-call-for-action-to-implement-crisis-standards-of-care-during-covid-19-surge/.

Nehme, M., S. Stringhini, and I. Guessous. 2020. Perceptions of immunity and vaccination certificates among the general population: A nested study within a serosurvey of anti-SARS-CVoV-2 antibodies (SEROCoV-POP). *Swiss Medical Weekly* 150: w20398.

Neushul, P. 1993. Science, government, and the mass production of penicillin. *Journal of the History of Medicine and Allied Sciences* 48: 371–395.

Neuteboom, N., P. Golec, and S. Phlippen. 2020. *De Nederlandse economie tijdens Covid-19*. October 7. https://insights.abnamro.nl/2020/10/de-nederlandse-economie-tijdens-covid-19-economische-gevolgen-van-de-tweede-golf/.

Newton, P.N., K.C. Bond, et al. 2020. Covid-19 and risks to the supply and quality of tests, drugs, and vaccines. *The Lancet Global Health* 8 (6): e754–e755.

Nie, J.-B., A. Gilbertson, M. de Roubaux, et al. 2016. Healing without waging war: Beyond military metaphors in medicine and HIV cure research. *American Journal of Bioethics* 16 (10): 3–11.

Nielsen, D.A. 1996. Pericles and the plague: Civil religion, anomie, and injustice in Thucydides. *Sociology of Religion* 57 (4): 397–407.

NIH. 2020. *Coronavirus Disease 2019 (COVID-19) Treatment Guidelines*. https://www.covid19treatmentguidelines.nih.gov/.

Nogrady, B. 2020. What the data say about asymptomatic Covid infections. *Nature* 587: 534–535.

Nolen, S. 2021. Merck will share formula of its Covid pill with poor countries. *New York Times*, October 27. https://www.nytimes.com/2021/10/27/health/covid-pill-access-molnupiravir.html.

Nuffield Council on Bioethics. 2020. *Research in global health emergencies – Ethical issues*. January 28. https://www.nuffieldbioethics.org/publications/research-in-global-health-emergencies.

OECD. 2020. *Flattening the Covid-19 peak: Containment and mitigation policies*. https://read.oecd-ilibrary.org/view/?ref=124_124999-yt5ggxirhc&title=Flattening_the_COVID-19_peak-Containment_and_mitigation_policies.

———. 2020. *Hospital beds (indicator)*; https://doi.org/10.1787/0191328e-en.

———. 2020. *Intensive care bed capacity*, April 20. https://www.oecd.org/coronavirus/en/data-insights/intensive-care-beds-capacity.

Olival, K.J., P. R. Hosseini, C. Zambrana-Torrelio, N. Ross, T. L. Bogich, and P. Daszak. 2017. Host and viral traits predict zoonotic spillover from mammals. *Nature* 546 (no. 7660): 646–650.

Oltermann, P., and A. Giuffrida. 2021. Russia's Sputnik V Covid vaccine gaining acceptance in Europe. *The Guardian*, March 10. https://www.theguardian.com/world/2021/mar/10/russias-sputnik-v-covid-vaccine-gaining-acceptance-in-europe.

Oltermann, P. 2021. Scepticism over Oxford vaccine threatens Europe's immunization push. *The Guardian*, February 19. https://www.theguardian.com/world/2021/feb/19/german-politicians-counter-astrazeneca-covid-vaccine-scepticism-with-show-of-support.

Onder, G., I. Carpenter, H. Finne-Soveri, et al. 2012. Assessment of nursing home residents in Europe: The Services and Health for Elderly in Long TERm care (SHELTER) study. *BMC Health Service Research* 12 (5) 1-10; https://doi.org/10.1186/1472-6963-12-5.

Opillard, F., A. Palle, and L. Michelis. 2020. Discourse and strategic use of the military in France and Europe in the Covid-19 crisis. *Tijdschrift voor Economische en Sociale Geografie* 111 (3): 239–259.

Oran, D.P., and E.J. Topol. 2020. Prevalence of asymptomatic SARS-CVoV-2 infection. A narrative review. *Annals of Internal Medicine* 173 (5): 362–367.

Orange, R. 2020. As Covid death toll soars ever higher, Sweden wonders who to blame. *The Guardian*, December 20. https://www.theguardian.com/world/2020/dec/20/as-covid-death-toll-soars-ever-higher-sweden-wonders-who-to-blame?CMP=Share_iOSApp_Other.

Orfali, K. 2020. What triage issues reveal: Ethics in the Covid-19 pandemic in Italy and France. *Journal of Bioethical Inquiry* 17: 675–679.

Osama, T., M.S. Razai, and A. Majeed. 2021. Covid-19 vaccine passports: Access, equity, and ethics. *British Medical Journal* 373: n861.

Oswick, C., D. Grant, and R. Oswick. 2020. Categories, crossroads, control, connectedness, and change: A metaphorical exploration of Covid-19. *The Journal of Applied Behavioral Science* 56 (3): 284–288.

Oude Munnink, B.B., R.S. Sikkema, D.F. Nieuwenhuijse, et al. 2020. Transmission of SARS-CoV-2 on mink farms between humans and mink and back to humans. *Science*, 10 November; https://doi.org/10.1126/science.abe5901.

Our World In Data. 2021. *Statistics and research. Coronavirus (Covid-19) vaccinations*. https://ourworldindata.org/covid-vaccinations.

Pagliarini, A. 2021. Latin America's lack of a united front on Covid has had disastrous consequences. *The Guardian*, April 23. https://www.theguardian.com/commentisfree/2021/apr/23/latin-america-united-front-covid-disastrous-pink-tide.

Park, S.W., B.M. Bolker, D. Champredon, et al. 2020. Reconciling early outbreak estimates of the basic reproductive number and its uncertainty: Framework and applications to the novel coronavirus (SARS-CoV2) outbreak. *Journal of the Royal Society Interface* 17: 20200144. https://doi.org/10.1098/rsif.2020.0144.

Park, A. 2020. Covid-19 vaccines are coming: Here's what to expect. *Time*, November 19. https://time.com/5913372/covid-19-vaccines-guide/.

Parker, L. 2020. To find a vaccine for Covid-19, will we have to deliberately infect people? *National Geographic*, September 16. https://www.nationalgeographic.com/science/article/to-make-a-coronavirus-vaccine-we-may-need-to-deliberately-infect-people.

Parkinson, J., C. Deng, and L. Lin. 2021. China deploys Covid-19 vaccine to build influence, with U.S. on sidelines. *The Wall Street Journal*, February 21. https://www.wsj.com/articles/china-covid-vaccine-africa-developing-nations-11613598170.

Parshley, L. 2020. The magnitude of America's contact tracing crisis is hard to overstate. *National Geographic*, September 1. https://www.nationalgeographic.com/science/2020/09/contact-tracing-crisis-magnitude-hot-mess-america-fixes-coronavirus-cvd/.

———. 2020. 'Super antigens' tied to mysterious Covid-19 syndrome in children. *National Geographic*, October 16. https://www.nationalgeographic.com/science/2020/10/super-antigen-tied-multisystem-inflammatory-syndrome-children-long-covid/.

Pasquier, P., A. Luft, J. Gillard, et al. 2020. How do we fight Covid-19? Military medical actions in the war against the Covid-19 pandemic in France. *BMJ Military Health*; https://doi.org/10.1136/bmjmilitary-2020-001569.

Pavelka, M., K. Van-Zandvoort, S. Abbott, et al. 2021. The impact of population-wide rapid antigen testing on SARS-CoV-2 prevalence in Slovakia. *Science*: eabf9648; https://doi.org/10.1126/science.abf9648.

Peeples, L. 2020. What the data say about wearing face masks. *Nature* 586: 186–189.

Pegg, D. 2020. What was Exercise Cygnus and what did it find? *The Guardian*, May 7. https://www.theguardian.com/world/2020/may/07/what-was-exercise-cygnus-and-what-did-it-find.

Pei, S., S. Kandula, and J. Shaman. 2020. Differential effects of intervention timing on COVID-19 spread in the United States. *medRxiv preprint*. https://www.medrxiv.org/content/10.1101/2020.05.15.20103655v1.full.pdf. https://doi.org/10.1101/2020.05.15.20103655.

Peisah, C., A. Byrnes, I. Doron, et al. 2020. Advocacy for the human rights of older people in the Covid pandemic and beyond: A call to mental health professionals. *International Psychogeriatrics* 32 (10): 1199–1204.

Peiser, J., and M. Boorstein. 2021. U.S. Bishops splinter on the morality of taking coronavirus vaccines. *Washington Post*, March 4. https://www.washingtonpost.com/nation/2021/03/02/archdiocese-new-orleans-johnson-vaccine/.

Pence, G.E. 2021. *Pandemic bioethics*. Peterborough: Broadview Press.

Pennella, A.R., and A. Ragonese. 2020. Health professionals and Covid-19 pandemic: Heroes in a new war? *Journal of Health and Social Sciences* 5 (2): 169–168.

Pennings, S., and X. Symons. 2021. Persuasion, not coercion or incentivization, is the best means of promoting Covid-19 vaccination. *Journal of Medical Ethics*; https://doi.org/10.1136/medethics-2020-107076.

Perez, G.I.P., and A.T.B. Abadi. 2020. Ongoing challenges faced in the global control of Covid-19 pandemic. *Archives of Medical Research* 51: 574–576.

Persad, G., M.E. Peek, and E.J. Emanuel. 2020. Fairly prioritizing groups for access to Covid-19 vaccines. *JAMA* 324 (16): 1601–1602.

Peto, J. 2020. Covid-19 mass testing facilities could end the epidemic rapidly. *British Medical Journal*: 368, m2263. https://doi.org/10.1136/bmj.m1163.

———., N.A. Alwan, K.M. Godfrey, et al. 2020. Universal testing as the UK Covid-19 lockdown exit strategy. *Lancet* 395 (10234): 1420–1421.

Pfeiffer, J., and R.R. Chapman. 2019. NGOs, austerity, and universal health coverage in Mozambique. *Global Health* 15: 0. https://doi.org/10.1186/s12992-019-0520-8.
Phelan, A.I. 2020. Covid-19 immunity passports and vaccination certificates: Scientific, equitable, and legal challenges. *Lancet* 395: 1595–1598.
Philips, N. 2021. The coronavirus will become endemic. *Nature* 390: 382–384.
Phillips, T. 2021. Political chaos and poverty leave a continent at virus's mercy. *The Guardian*, May 2. https://www.theguardian.com/world/2021/may/02/political-chaos-and-poverty-leave-south-america-at-viruss-mercy.
Pisano, G.P., R. Sadun, and M. Zanini. 2020. Lessons from Italy's response to coronavirus. *Harvard Business Review*. March 27. https://hbr.org/2020/03/lessons-from-italys-response-to-coronavirus.
Piscitello, G.M., E.M. Kapania, W.D. Miller, et al. 2020. Variation in ventilator allocation guidelines by US state during the coronavirus disease 2019 pandemic. *JAMA Network Open* 3 (6): e2012606. https://doi.org/10.1001/jamanetworkopen.2020.12606.
Platform Containment Nu. 2020. *Covid-19 in the Netherlands. A time line*. https://www.containmentnu.nl/en/articles/timeline.
Plohl, N., and B. Musil. 2021. Modeling compliance with Covid-19 prevention guidelines: The critical role of trust in science. *Psychology, Health & Medicine* 26 (1): 1–12.
Plotkin, S.A., A.A.F. Mahmoud, and J. Farrar. 2015. Establishing a global vaccine-development fund. *New England Journal of Medicine* 373 (4): 297–300.
Plotkin, S.A., and A. Caplan. 2020. Extraordinary diseases require extraordinary solutions. *Vaccine* 38: 3987–3988.
Pollock, A.M., P. Roderick, and B. Pankhania. 2020. Covid-19: Why is the UK government ignoring WHO's advice? *British Medical Journal*, 30 March; https://doi.org/10.1136/bmj.m1284.
Pontifical Academy for Life. 2005. Moral reflections on vaccines prepared from cells derived from aborted human foetuses. *The Linacre Quarterly* 86 (2–3): 182–187.
———. 2017. *Note on Italian vaccine issue*. http://www.academyforlife.va/content/pav/en/the-academy/activity-academy/note-vaccini.pdf.
———. 2020. *Humana communitas in the age of pandemic: Untimely meditations on life's rebirth*. Vatican City, July 22. http://www.academyforlife.va/content/dam/pav/documenti%20pdf/2020/Nota%20Covid19%2022%20luglio/testo%20pdf/HUMANA%20COMMUNITAS%20IN%20THE%20AGE%20OF%20PANDEMIC%20def%20ENG.pdf,
Pope Francis. 2005. *Encyclical letter Laudato Si' – On care for our common home*. Rome: Vatican.
Potter, V.R. 1971. *Bioethics: Bridge to the future*. Englewood Cliffs: Prentice Hall.
———. 1988. *Global bioethics: Building on the Leopold legacy*. East Lansing: Michigan State University Press.
Powell, V.D., and M.J. Silveira. 2020. What should palliative care's response be to the Covid-19 pandemic? *Journal of Pain and Symptom Management* 60 (1): e1–e3.
Powell, T., and E. Chuang. 2020. Covid and NYC: What we could do better. *American Journal of Bioethics* 20 (7): 62–66.
Price, W.N., A.K. Rai, and T. Minssen. 2020. Knowledge transfer for large-scale vaccine manufacturing. *Science* 369 (6506): 912–914.
Public Health England. 2021. *Covid-19 vaccination first phase priority groups*, February 23. https://www.gov.uk/government/publications/covid-19-vaccination-care-home-and-healthcare-settings-posters/covid-19-vaccination-first-phase-priority-groups.
Pursell, E., D. Gould, and J. Chudleigh. 2020. Impact of isolation on hospitalized patients who are infectious: Systematic review with meta-analysis. *BMJ Open* 10: e030371. https://doi.org/10.1136/bmjopen-2019-030371.
Radbruch, L., F.M. Knaul, L. de Lima, et al. 2020. The key role of palliative care in response to the Covid-19 tsunami of suffering. *Lancet* 395 (10235): 1467–1469.
Raffle, A.E., A.M. Pollock, and L. Harding-Edgar. 2020. Covid-19 mass testing programmes. *British Medical Journal*: 370, m3263. https://doi.org/10.1136/bmj.m3262.

Rajan, S., J. Cylus, and M. McKee. 2020. Successful find-test-trace-isolate-support systems. How to win at snakes and ladders. *Eurohealth* 26 (2): 34–39.

Rajandran, K. 2020. 'A long battle ahead': Malaysian and Singaporean prime ministers employ metaphors for Covid-19. *GEMA Online Journal of Language Studies* 17 (2): 163–176.

Ramney, M.J., V. Griffeth, and A.K. Jha. 2020. Critical supply shortages – The need for ventilators and protective equipment during the Covid-19 pandemic. *New England Journal of Medicine* 382 (18): e41.

Rankin, J. 2020. Belgium experiments with 'corona bubbles' to ease social restrictions. *The Guardian*, May 10. https://www.theguardian.com/world/2020/may/10/belgium-experiments-with-corona-bubbles-to-ease-social-restrictions.

Ravitsky, V. 2020. *Post-Covid bioethics*. The Hastings Center, May 20. https://www.thehastingscenter.org/post-covid-bioethics/.

Reardon, S. 2021. The most worrying mutations in five emerging coronavirus variants. *Scientific American*, January 29. https://www.scientificamerican.com/article/the-most-worrying-mutations-in-five-emerging-coronavirus-variants/.

Rechel, B. 2019. Funding for public health in Europe in decline? *Health Policy* 123 (1): 21–26.

Remuzzi, A., and G. Remuzzi. 2020. Covid-19 and Italy: What next? *Lancet* 395: 1225–1228.

Ren, X. 2020. Pandemic and lockdown: A territorial approach to Covid-19 in China, Italy and the United States. *Eurasian Geography and Economics* 51 (2): 162–183.

Retmann, A. 2021. EU blasts UK and Russia in 'vaccine propaganda' war. *EUObserver*, March 10. https://euobserver.com/science/151183.

Reuters. 2021. New coronavirus variant, described as 'double mutant', reported in India. *The Guardian*, March 25. https://www.theguardian.com/world/2021/mar/25/new-coronavirus-variant-described-as-double-mutant-reported-in-india.

Rhodes, J. 2021. *How to make a vaccine. An essential guide for Covid-19 & beyond*. Chicago/London: The University of Chicago Press.

Ribes, M. 2020. Covid-19 retrospective, a disaster that should have been averted. *Bioethics Observatory*, September 18. https://bioethicsobservatory.org/2020/09/coronavirus-crisis-responsibilities-exoneration/36839/.

Richards, T., and H. Scowcroft. 2020. Patient and public involvement in covid-19 policy making. *British Medical Journal* 370: m2575. https://doi.org/10.1136/bmj.m2575.

Rifkin, J. 2009. *The empathic civilization. The race to global consciousness in a world in crisis*. Cambridge: Polity Press.

Rio Declaration on Environment and Development. 1992. http://www.unesco.org/education/pdf/RIO_E.PDF.

Rivlin, M.M. 2000. Why the fair innings argument is not persuasive. *BMC Medical Ethics* 1: 1. https://doi.org/10.1186/1472-6939-1-1.

RIVM Corona Gedragsunit. 2020. *Analyse thuisblijven, testen en quarantaine*. July. https://www.rivm.nl/documenten/onderzoek-thuisblijven-testen-en-quarantaine.

———. 2020. *Naleven van quarantaine en isolatie advies*, November. https://www.rivm.nl/documenten/notitie-naleven-quarantaine-en-isolatie-advies.

Rodrigues, R.A.L., A.C. dos S.P. Andrade, P.V. de M. Boratto, et al. 2017. An anthropocentric view of the virosphere-host relationship. *Frontiers in Microbiology* 8: 1673. https://doi.org/10.3389/fmicb.2017.01673.

Rohela, P., A. Bhan, D. Ravindranath, et al. 2020. Must there be a 'war' against coronavirus? *Indian Journal of Medical Ethics* 5 (3): 10.20529/IJME.2020.070.

Rohwer, F., and K. Barott. 2013. Viral information. *Biology and Philosophy* 28 (2): 283–297.

Romer, D., and K.H. Jamieson. 2020. Conspiracy theories as barriers to controlling the spread of Covid-19 in the U.S. *Social Science & Medicine* 263: 113356. https://doi.org/10.1016/j.socscimed.2020.113356.

Rosa, H. 2020. *The uncontrollability of the world*. Cambridge: Polity Press.

Rose, J. 2020. Pointing the finger. *London Review of Books* 42 (9): 3, 6–8, 10.

Rosenbaum, L. 2020. Facing Covid-19 in Italy – Ethics, logistics, and therapeutics on the epidemic's front line. *New England Journal of Medicine* 382 (20): 1873–1875.
———. 2020. The untold toll – The pandemic's effects on patients without Covid-19. *New England Journal of Medicine* 382 (24): 2368–2371.
Rosenberg, C.E. 1987. *The cholera years. The United States in 1832, 1849, and 1866*. Chicago/London: The University of Chicago Press.
———. 1992. *Explaining epidemics and other studies in the history of medicine*. Cambridge: Cambridge University Press.
Rosenthal, M.S., and A. Caplan. 2021. Why we need a Covid-19 commission. *The Hastings Center*, March 3. https://www.thehastingscenter.org/why-we-need-a-covid-19-commission/.
———. 2021. *How to make it right: Covid reparations*. The Hastings Center, March 15. https://www.thehastingscenter.org/how-to-make-it-right-covid-reparations/.
Rubin, R. 2020. Investigating whether blood type is linked to Covid-19 risk. *JAMA* 324 (13): 1273.
Rubin, E.J., and D.L. Longo. 2020. SARS-CoV-2 vaccination – An ounce (actually, much less) of prevention. *New England Journal of Medicine* 383 (27): 2677–2678.
Ruderman, C., C. Shawn Tracy, C.M. Bensimon, et al. 2006. On pandemics and the duty to care: Whose duty? Who cares? *BMC Medical Ethics* 7: 5. https://doi.org/10.1186/1472-6939-7-5.
Russell, F.M., and B. Greenwood. 2021. Who should be prioritized for Covid-19 vaccination? *Human Vaccines & Immunotherapeutics*; https://doi.org/10.1080/21645515.2020.1827882.
Rutkow, L. 2020. Origins of the Covid-19 pandemic and the path forward. A global public health policy perspective. In *Covid-19 and world order. The future of conflict, competition, and cooperation*, ed. H. Brand and F.J. Gavin, 93–113. Baltimore: Johns Hopkins University Press.
Rutter, H., M. Wolpert, and T. Greenhalgh. 2020. Managing uncertainty in the Covid-19 era. *British Medical Journal* 320: m3349.
Ryan, F. 2020. *Virusphere. Ebola, AIDS, influenza and the hidden world of the virus*. London: William Collins.
Saag, M.S. 2020. Misguided use of hydroxychloroquine for Covid-19. The infusion of politics into science. *JAMA* 324 (21): 2161–2162.
Sabino, E.C., L.F. Buss, M.P.S. Carvalho, et al. 2021. Resurgence of Covid-19 in Manaus, Brazil, despite high seroprevalence. *Lancet* 397: 452–455.
Sabucedo, J.-M., M. Alzate, and D. Hur. 2020. Covid-19 and the metaphor of war. *Revista de Psicologia Social/International Journal of Social Psychology* 35 (3): 6–8. https://doi.org/10.1080/02134748.2020.1783840.
Safi, M. 2021. WHO platform for pharmaceutical firms unused since pandemic began. *The Guardian*, January 22. https://www.theguardian.com/world/2021/jan/22/who-platform-for-pharmaceutical-firms-unused-since-pandemic-began.
Safi, M., and M. Pantovic. 2021. Vaccine diplomacy: West falling behind in race for influence. *The Guardian*, February 19. https://www.theguardian.com/world/2021/feb/19/coronavirus-vaccine-diplomacy-west-falling-behind-russia-china-race-influence.
Sahm, S. 2021. "Vaccine for all" – Prüfstein globaler Bioethik. *Zeitschrift für Medizinische Ethik* 67: 201–206.
Sahoo, S., A. Mehra, V. Suri, et al. 2020. Lived experiences of the corona survivors (patients admitted in Covid wards): A narrative real-life documented summaries of internalized guilt, shame, stigma, anger. *Asian Journal of Psychiatry* 53: 102187.
Sample, I. 2021. Delay in giving second jab of Pfizer vaccine improves immunity. *The Guardian*, May 14. https://www.theguardian.com/science/2021/may/14/delay-in-giving-second-jabs-of-pfizer-vaccine-improves-immunity.
Sample, I., and P. Walker. 2021. Covid response 'one of the UK's worst ever public health failures.' *The Guardian*, October 12. https://www.theguardian.com/politics/2021/oct/12/covid-response-one-of-uks-worst-ever-public-health-failures.
Sand, J. 2020. We share what we exhale. *Times Literary Supplement*, May 1:22–23.
Santarpia, J.L., D.N. Rivera, V.L. Herrera, et al. 2020. Aerosol and surface contamination of SARS-CoV-2 observed in quarantine and isolation care. *Scientific Reports* 10: 12732. https://doi.org/10.1038/s41598-020-69286-3.

Savulescu, J. 2021. Good reasons to vaccinate: Mandatory or payment for risk? *Journal of Medical Ethics* 47: 78–85.

Sayer, A. 2011. *Why things matter to people. Social science, values and ethical life*. Cambridge: Cambridge University Press.

Schaefer, G.O., R.J. Leland, and E.J. Emanuel. 2021. Making vaccines available to other countries before offering domestic booster vaccinations. *JAMA* 326 (10): 903–904.

Scheres, J., and K. Kuszewski. 2019. The ten threats to global health in 2018 and 2019. A welcome and informative communication of WHO to everybody. *Zdrowie Publiczne i Zarządzanie* 17 (1): 2–8.

Schinkel, W. 2021. Het Nederlandse coronabeleid is een vorm van necropolitiek. *NRC*, July 26. https://www.nrc.nl/nieuws/2021/07/25/het-nederlandse-coronabeleid-is-een-vorm-van-necropolitiek-a4052299.

Schlagenhauf, P., D. Patel, A.J. Rodriguez-Morales, et al. 2021. Variants, vaccines and vaccination passports: Challenges and chances for travel medicine. *Travel Medicine and Infectious Disease* 40: 101996.

Schlegel, T. 2020. *Didier Sicard: "Il est urgent d'enquêter sur l'origine animale de l'épidémie de Covid-19."* https://www.franceculture.fr/sciences/didier-sicard-il-est-urgent-denqueter-sur-lorigine-animale-de-lepidemie-de-covid-19.

Schmidt, H., P. Pathak, T. Sönmez, and M.U. Űnver. 2020. Covid-19: How to prioritize worse-off populations in allocating safe and effective vaccines. *British Medical Journal* 371: m3795.

Schuetz, A.N., P. Hemarajata, N. Mehta, et al. 2020. When should asymptomatic persons be tested for Covid-19? *Journal of Clinical Microbiology*, 6 October; https://doi.org/10.1128/JCM.02563-20.

Schwab, K., and T. Malleret. 2020. *Covid-19: The great reset*. Cologny/Geneva: World Economic Forum.

Scully, J.L. 2020. Disability, disablism, and Covid-19 pandemic triage. *Journal of Bioethical Inquiry* 17: 601–605.

Seale, H., A.E. Heywood, J. Leask, et al. 2020. Covid-19 is rapidly changing: Examining public perceptions and behaviors in response to this pandemic. *PLoS One* 15 (6): e0235112. https://doi.org/10.1371/journal.pone.0235112.

Selâl Şengör, A.M. 1991. Our home, the planet earth. *Diogenes* 39 (155): 25–51.

Shaban, R.Z., S. Nahidi, C. Sotomayor-Castillo, et al. 2020. SARS-CoV-2 infection and Covid-19: The lives experience and perceptions of patients in isolation and care in an Australian Healthcare setting. *American Journal of Infection Control* 48: 1445–1450.

Shaw, D. 2020. The many meanings of 'stay safe' in a pandemic: Sympathy, duty, and threat. *Journal of Medical Ethics Blog*, May 13. https://blogs.bmj.com/medical-ethics/2020/05/13/the-many-meanings-of-stay-safe-in-a-pandemic-sympathy-duty-and-threat/.

Sheahan, L., and F. Brennan. 2020. What matters? Palliative care, ethics, and the Covid-19 pandemic. *Journal of Bioethical Inquiry* 17: 793–796.

Sheahan, L., and S. Lamont. 2020. Understanding ethical and legal obligations in a pandemic: A taxonomy of 'duty' for health practitioners. *Journal of Bioethical Inquiry* 17: 697–701.

Shevzov-Zebrun, N., A. Caplan, and B. Parent. 2021. Should Covid vaccination schedules deviate from the status que – As a last resort? *The Hastings Center*, February 1. https://www.thehastingscenter.org/should-covid-vaccination-schedules-deviate-from-the-status-quo-as-a-last-resort/.

Shi, Z., H.J. de Vries, A.P.J. Vlaar, et al. 2021. Diaphragm pathology in critically ill patients with Covid-19 and postmortem findings from 3 medical centers. *JAMA Internal Medicine* 181 (1): 122–124.

Shipman, P.L. 2014. The bright side of the Black Death. *American Scientist* 102 (6): 410–413.

Shokoohi, M., M. Osooli, and S. Stranges. 2020. Covid-19 pandemic: What can the West learn from the East? *International Journal of Health Policy and Management* 9 (10): 436–438.

Shortt, S. 2001. Venerable or vulnerable? Ageism in health care. *Journal of Health Services Research & Policy* 6 (1): 1–2.

Siebenhaar, K.U., A.K. Köther, and G.W. Alpers. 2020. Dealing with the Covid-19 infodemic: Distress by information, information avoidance, and compliance with preventive measures. *Frontiers in Psychology* 11: 567905. https://doi.org/10.3389/psyg.2020.567905.

Silva, D.S., M.J. Smith, and R.E.G. Upshur. 2013. Disadvantaging the disadvantaged: When health policies and practices negatively affect marginalized populations. *Canadian Journal of Public Health* 104 (5): 410–412.

Silva, D.S. 2020. Ventilators by lottery. The least unjust form of allocation in the coronavirus disease 2019 pandemic. *Chest* 158 (3): 890–891.

Singanayagam, A., S. Hakki, J. Dunning, et al. 2021. Community transmission and viral load kinetics of the SARS-CoV-2 delta (B.1.617.2) variant in vaccinated and unvaccinated individuals in the UK: A prospective, longitudinal, cohort study. *Lancet Infectious Diseases*, October 28; https://doi.org/10.1016/S1473-3099(21)00648-4.

Sisk, B.A., and J. DuBois. 2020. Research ethics during a pandemic: A call for normative and empirical analysis. *American Journal of Bioethics* 20 (7): 82–84.

Siva, N. 2020. Experts call to include prisons in Covid-19 vaccine plans. *Lancet* 396: 1870.

Slingeland, A. 2020. Dr. Samuel Sarphati. *Hektoen International. A Journal of Medical Humanities* 12(3). https://hekint.org/2020/05/05/dr-samuel-sarphati/.

Smith, J., M. Lipsitch, and J.W. Almond. 2011. Vaccine production, distribution, access, and uptake. *Lancet* 378 (9789): 428–438.

Smith, M.J., and R.E.G. Upshur. 2015. Ebola and learning lessons from moral failures: Who cares about ethics? *Public Health Ethics* 8 (3): 305–318.

———. 2020. Learning lessons from Covid-19 requires recognizing moral failures. *Journal of Bioethical Inquiry* 17: 563–566.

Smith, S.G. 1994. The essential qualities of a home. *Journal of Environmental Psychology* 14: 31–46.

Sokol, D., and B. Gray. 2020. Should we give priority care to healthcare workers in the Covid-19 pandemic? *BMJOpinion*, April 1. https://blogs.bmj.com/bmj/2020/04/01/should-we-give-priority-care-to-healthcare-workers-in-the-covid-19-pandemic/.

Solbakk, J.H., H.-B. Bentzen, S. Holm, et al. 2021. Back to WHAT? The role of research ethics in pandemic times. *Medicine, Health Care and Philosophy* 24: 3–18.

Solis, J., C. Franco-Paredes, A.F. Henao-Martinez, et al. 2020. Structural vulnerability in the U.S. revealed in three waves of Covid-19. *American Journal of Tropical Medicine and Hygiene* 103 (1): 25–27.

Solomon, M.Z., M.K. Wynia, and L.O. Gostin. 2020. Covid-19 crisis triage – Optimizing health outcomes and disability rights. *New England Journal of Medicine* 383 (5): e27. https://doi.org/10.1056/NEJMp2008300.

Soupios, M.A. 2004. Impact of the plague in Ancient Greece. *Infectious Disease Clinics of North America* 18: 45–51.

Soy, A. 2020. Coronavirus in Africa; Five reasons why Covid-19 has been less deadly than elsewhere. *BBC News*, 7 October. https://www.bbc.com/news/world-africa-54418613.

Spencer, J., and C. Jewett. 2021. Twelve months of trauma: More than 3,600 US health workers died in Covid's first year. *The Guardian*, April 8. https://www.theguardian.com/us-news/2021/apr/08/us-health-workers-deaths-covid-lost-on-the-frontline.

Spicer, N., I. Agyepong, I. Ottersen, A. Jahn, and G. Ooms. 2020. "It's far too complicated": Why fragmentation persists in global health. *Globalization and Health* 16 (1): 60. https://doi.org/10.1186/s12992-020-00592-1.

Spinney, L. 2020. Why is Europe yet again at the centre of the coronavirus pandemic? *The Guardian*, 2 November. https://www.theguardian.com/commentisfree/2020/nov/02/europe-coronavirus-pandemic-lockdowns-infection-rates.

Spooner, J.L. 1967. History of surgical face masks. *AORN Journal* 5 (1): 76–80.

Stafford, N. 2020. Covid-19: Why Germany's case fatality rate seems so low. *British Medical Journal* 369: m1395. https://doi.org/10.1136/bmj.m1395.

Steinbrook, R. 2020. Contact tracing, testing, and control of Covid-19; learning from Taiwan. *JAMA Internal Medicine* 180 (9): 1163–1164.

Stillwell, C. 2021. 'Vaccine interlopers' exploit chaotic policies to skip line in rural Tennessee. *The Guardian*, March 29. https://www.theguardian.com/us-news/2021/mar/29/tennessee-vaccine-interlopers-skip-line.

Stokmans, D., and M.L. Adriaanse. 2020. Hoe Nederland de controle verloor. De corona uitbraak van dag tot dag. *NRC* 20-21 June. https://www.nrc.nl/nieuws/2020/06/19/hoe-nederland-reageerde-op-het-nieuwe-virus-uit-china-van-niks-aan-de-hand-tot-blinde-paniek-a4003075.

Subbaraman, N. 2020. Who gets a Covid vaccine first? Access plans are taking shape. *Nature* 585: 492–493.

Sudre, C.H., B. Murray, T. Varsavsky, et al. 2021. Attributes and predictors of long Covid. *Nature Medicine* 27: 626–631.

Sulaman, M., X. Long, and M. Salman. 2020. COVID-19 pandemic and environmental pollution: A blessing in disguise? *Science of the Total Environment* 728: 138820. https://doi.org/10.1016/j.scitotenv.2020.138820.

Sullivan, D.R., and J.R. Curtis. 2020. A view from the frontline: Palliative and ethical considerations of the Covid-19 pandemic. *Journal of Palliative Medicine* 24 (2): 293–295.

Sullivan, H. 2021. South Africa paying more than double EU price for Oxford vaccine. *The Guardian*, January 22. https://www.theguardian.com/world/2021/jan/22/south-africa-paying-more-than-double-eu-price-for-oxford-astrazeneca-vaccine.

Sun, N., L. Wei, H. Wang, et al. 2021. Qualitative study of the psychological experience of Covid-19 patients during hospitalization. *Journal of Affective Disorders* 278: 15–22.

Sunstein, C.R. 2020. The Meaning of masks. *Journal of Behavioral Economics for Policy* 4 (S): 5–8.

Surendra, H., I.R.F. Elyazar, B.A. Djaafra, et al. 2021. Clinical characteristics and mortality associated with Covid-19 in Jakarta, Indonesia: A hospital-based retrospective cohort study. *The Lancet Regional Health – Western Pacific,* March 2; https://doi.org/10.1016/j.;anwpc.2021.100108.

Suttle, C. 2005. The viriosphere: The greatest biological diversity on Earth and driver of global processes. *Environmental Microbiology* 7 (4): 481–482.

Swazo, N.K., M.H. Talukder, and M.K. Ahsan. 2020. A *duty* to treat? A *right* to refrain? Bangladeshi physicians in moral dilemma during Covid-19. *Philosophy, Ethics, and Humanities in Medicine* 15: 7. https://doi.org/10.1186/s13010-020-00091-6.

Syed, Q., W. Sopwith, M. Regan, and M.A. Bellis. 2003. Behind the mask. Journey through an epidemic: Some observations of contrasting health responses to SARS. *Journal of Epidemiology and Community Health* 57: 855–856.

Symons, X. 2020. Is seeking 'herd immunity' ethical? *BioEdge*, May 9. https://www.bioedge.org/indepth/view/is-seeking-herd-immunity-ethical/13428.

———. 2020. Should we sacrifice older people to save the economy? *BioEdge*, March 28. https://www.bioedge.org/bioethics/should-we-sacrifice-older-people-to-save-the-economy/13377.

Szalinski, C. 2021. Fringe doctors' group promote ivermectin for Covid despite a lack of evidence. *Scientific American*, September 29. https://www.scientificamerican.com/article/fringe-doctors-groups-promote-ivermectin-for-covid-despite-a-lack-of-evidence/.

Tallès, O., et al. 2020. Ils se sont sacrifiés soigner les maladies du Covid. *La Croix*, September 28. https://www.la-croix.com/Monde/covid-19-million-morts-soignants-sacrifies-soigner-maladies-2020-09-28-1201116386.

Tampa, V. 2021. It's time for Africa to rein in Tanzania's anti-vaxxer president. *The Guardian*, February 8. https://www.theguardian.com/global-development/2021/feb/08/its-time-for-africa-to-rein-in-tanzanias-anti-vaxxer-president.

Tangwa, G.B., and N.S. Munung. 2020. Covid-19: Africa's relation with epidemics and some imperative ethics considerations of the moment. *Research Ethics* 16: 1–11. https://doi.org/10.1177/1747016120937391.

Tashiro, A., and R. Shaw. 2020. Covid-19 pandemic response in Japan: What is behind the initial flattening of the curve? *Sustainability* 12 (13): 5250. https://doi.org/10.3390/su12135250.

Taylor, L. 2021. Covid 19: Is Manaus the final nail in the coffin for natural herd immunity? *British Medical Journal* 372: m394.

't Hoen, E. 2020. Protect against market exclusivity in the fight against Covid-19. *Nature Medicine* 26 (6): 813.

Ten Have, H. 1983. *Geneeskunde en filosofie. De invloed van Jeremy Bentham op het medisch denken en handelen*. Lochem-Poperinge: Uitgeversmaatschappij De Tijdstroom.

Ten Have, H.A.M.J. 1990. Knowledge and practice in European medicine: The case of infectious diseases. In *The growth of medical knowledge*, ed. H.A.M.J. ten Have, G.K. Kimsma, and S. Spicker, 15–40. Dordrecht/Boston/London: Kluwer Academic Publishers.

Ten Have, H. 1993. Physicians' priorities – Patients' expectations. In *Solidarity, justice and health care priorities*, ed. Z. Szawarski and D. Evans, 42–52. Linköping: Linköping University.

———. 2014. Macro-triage in disaster planning. In *Disaster bioethics: Normative issues when nothing is normal*, ed. D.P. O'Mathuna, M. Clarke, and B. Gordijn, 13–32. Dordrecht: Springer.

———. 2014. Vulnerability as the antidote to neoliberalism in bioethics. *Revista Redbioetica/UNESCO* 5 (1; no.9): 87–92.

———. 2015. Respect for human vulnerability: The emergence of a new principle in bioethics. *Journal of Bioethical Inquiry* 12 (3): 395–408.

———. 2016. *Global bioethics. An introduction*. London/New York: Routledge.

———. 2019. *Wounded planet. How declining biodiversity endangers health and how bioethics can help*. Baltimore: Johns Hopkins University Press.

———. 2020. La pandemia de COVID-19 vista por los expertos de bioética. *Bioetica Complutense* 39, June: 18–21.

———. 2020. Sheltering at our common home. *Journal of Bioethical Inquiry* 17: 525–529.

Ten Have, H., and R. Pegoraro. 2021. *Bioethics, healthcare and the soul*. London: Routledge.

Tham, J., L. Melahn, and M. Baggot. 2021. Withdrawing critical care from patients in a triage situation. *Medicine, Health Care and Philosophy* 24: 205–211.

The Earth Charter. 2000. https://earthcharter.org/invent/images/uploads/echarter_english.pdf.

The Royal Society. 2021. *Twelve criteria for the development and use of COVID-19 vaccine passports*. https://royalsociety.org/-/media/policy/projects/set-c/set-c-vaccine-passports.pdf?la=en-GB&hash=A3319C914245F73795AB163AD15E9021.

Thibodeau, P.H., C. McPherson Frantz, and M. Berretta. 2017. The earth is our home: Systemic metaphors to redefine our relationship with nature. *Climatic Change* 142 (1): 287–300.

Thomas, A.J. 2020. *Cholera. The Victorian plague*. Barnsley: Pen and Sword Books.

Thompson, A.K., K. Faith, J.L. Gibson, and R.E.G. Upshur. 2006. Pandemic influenza preparedness. An ethical framework to guide decision-making. *BMC Medical Ethics* 7: 12. https://doi.org/10.1186/1472-6939-7-12.

Thompson, D.-C., M.-G. Barbu, C. Beiu, et al. 2020. The impact of Covid-19 pandemic on long-term care facilities worldwide: An overview of international issues. *BioMed Research International*, Article ID 8870249; https://doi.org/10.1155/2020/8870249.

Thoradeniya, T., and S. Jayasinghe. 2021. Covid-19 and future pandemics: A global systems approach and relevance to SDGs. *Globalization and Health* 17: 59. https://doi.org/10.1186/s12992-021-00711-6.

Thorlby, R., C. Fraser, and T. Gardner. 2020. Non-Covid-19 NHS care during the pandemic. Activity trends for key NHS services in England. *The Health Foundation*, December 12. https://www.health.org.uk/news-and-comment/charts-and-infographics/non-covid-19-nhs-care-during-the-pandemic.

Thornhill, C., and R. Miron. 2020. Karl Jaspers. In *The Stanford Encyclopedia of Philosophy*, ed. E.N. Zalta. https://plato.stanford.edu/archives/spr2020/entries/jaspers/.

Thucydides, *The history of the Peloponnesian War*. Second Book, Chapter VI. http://classics.mit.edu/Thucydides/pelopwar.2.second.html.

Torreele, E. 2020. The rush to create a Covid-19 vaccine may do more harm than good. *British Medical Journal* 370: m3209.

Travica, B. 2020. *Containment strategies for Covid-19 pandemic*. May 18. Available at SSRN. https://ssrn.com/abstract=3604519 or https://doi.org/10.2139/ssrn.3604519.

TRTWorld. 2020. *Why Turkey is better-equipped to tackle coronavirus*, March 16. https://www.trtworld.com/turkey/why-turkey-is-better-equipped-to-tackle-coronavirus-34610.

Truog, R.D., C. Mitchell, and G.Q. Daley. 2020. The toughest triage – Allocating ventilators in a pandemic. *New England Journal of Medicine* 382 (21): 1973–1975.

Trust for America's Health and Robert Wood Johnson Foundation. 2015. *Outbreaks: Protecting Americans from infectious diseases*. Washington. https://www.tfah.org/report-details/outbreaks/.

Tsao, S.-F., H. Chen, T. Tisseverasinghe, Y. Yang, L. Li, and Z.A. Butt. 2021. What social media told us in the time of Covid-19: A scoping review. *Lancet Digit Health* 3: e175–e194.

Tuchman, B. 1978. *A distant mirror. The calamitous 14th century*. London: Penguin Books.

Turale, S., C. Meechamnan, and W. Kunaviktikul. 2020. Challenging time: Ethics, nursing and the Covid-19 pandemic. *International Nursing Review* 67 (2): 164–167.

UNESCO. 2005. *Universal declaration on bioethics and Human Rights*. http://portal.unesco.org/en/ev.php-URL_ID=31058&URL_DO=DO_TOPIC&URL_SECTION=201.html.

———. 2020. *Statement on Covid-19: Ethical considerations from a global perspective*. Paris: UNESCO. https://unesdoc.unesco.org/ark:/48223/pf0000373115.

United Nations. 2020. *COVID-19 Response. This war needs a war-time plan to fight it*. https://www.un.org/en/coronavirus/war-needs-war-time-plan-fight-it.

———. 2020. *Everyone, everywhere must have access to eventual Covid-19 immunization*, Secretary-General says in video message to global vaccine summit. Press release, June 4. https://www.un.org/press/en/2020/sgsm20108.doc.htm.

Van den Brink, G. 2020. *Ruw ontwaken uit een neoliberale droom en de eigenheid van het Europese continent*. Amsterdam: Prometheus.

Van der Vliet, N., K. van der Swaluw, M. Zonneveld, et al. 2020. *Gedragswetenschappelijke literatuur rond mondkapjesgebruik. Een rapid review van de literatuur*. Bilthoven: Gedragsexpertiseteam RIVM. https://www.rivm.nl/sites/default/files/2020-05/Gedragswetenschappelijke%20literatuur%20mondkapjes%20-%20Corona%20Gedragsunit%20beveiligd.pdf.

Van Doremalen, N., T. Bushmaker, D.H. Morris, et al. 2020. Aerosol and surface stability of SARS-CoV-2 as compared to SARS-CoV-1. *New England Journal of Medicine* 382 (16): 1564–1567.

Van Kempen, J., and B. Soetenhorst. 2021. Hoe ga je verder als land als niet iedereen de inenting wil? *Het Parool*, September 3. https://www.parool.nl/nederland/hoe-ga-je-verder-als-land-als-niet-iedereen-de-inenting-wil~b0a14ada/.

Van Westen-Lagerweij, N.A., E. Meijer, E.G. Meeuwsen, et al. 2021. Are smokers protected against SARS-CoV-2 infection (Covid-19)? The origins of the myth. *NPJ Primary Care Respiratory Medicine* 31: 10. https://doi.org/10.1038/s41533-021-00223-1.

Vardavas, C., and K. Nikitara. 2020. COVID-19 and smoking: A systematic review of the evidence. *Tobacco Induced Diseases*, March 20; https://doi.org/10.18332/tid/119324.

Vatican Covid-19 Commission/Pontifical Academy for Life. 2020. *Vaccine for all. 20 points for a fairer and healthier world*. December 29. https://press.vatican.va/content/salastampa/it/bollettino/pubblico/2020/12/29/0697/01628.html#notaing.

Venkatapuram, S. 2020. *Covid-19 and the global ethics freefall*. The Hastings Center, March 19. https://www.thehastingscenter.org/covid-19-and-the-global-ethics-freefall/.

Venkatapuram, S., and A.C. Zielinska. 2021. *Covid vaccine patent waivers are for health sovereignty*. The Hastings Center, June 1. https://www.thehastingscenter.org/covid-vaccine-patent-waivers-are-for-health-sovereignty/.

Vervaeke, L. 2020. Waarom luisterde het Westen niet naar China? *De Volkskrant* 7 September: 6–7.

Vidal, J. 2020. 'Tip of the iceberg': Is our destruction of nature responsible for Covid-19? *The Guardian*, March 18. https://www.theguardian.com/environment/2020/mar/18/tip-of-the-iceberg-is-our-destruction-of-nature-responsible-for-covid-19-aoe.

Vincent, J.-L., and J. Creteur. 2020. Ethical aspects of the Covid-19 crisis: How to deal with an overwhelming shortage of acute beds. *European Heart Journal: Acute Cardiovascular Care* 0 (0): 1–5.

Visser, M. 2020. Eén of de tien Nederlanders gelooft dat er rond corona vieze spelletjes worden gespeeld. *Trouw*, August 15. https://www.trouw.nl/binnenland/een-op-de-tien-nederlanders-gelooft-dat-er-rond-corona-vieze-spelletjes-worden-gespeeld~bd98ce41/?referrer=https%3A%2F%2Fwww.google.com%2F.

Vlaskamp, M. 2021. Chinees vaccin verovert de wereld. *De Volkskrant*, February 23.

Voo, T.C., H. Clapham, and C.C. Tam. 2020. Ethical implication of immunity passports during the Covid-19 pandemic. *The Journal of Infectious Diseases* 222: 715–718.

Voo, T.C., A.A. Reis, B. Thomé, et al. 2021. Immunity certification for Covid-19: Ethical considerations. *Bulletin of the World Health Organization* 199: 155–161.

Voormolen, S. 2020. De eerste noodkreten kwamen rond Kerst. Die gingen in de doofpot. *NRC*, 19–20 December.

Waldman, R.J., E.D. Mintz, and H.E. Papowitz. 2013. The cure for cholera – Improving access to safe water and sanitation. *New England Journal Medicine* 368 (7): 592–594.

Walker, I.F. 2020. Beyond the military metaphor. *Medicine Anthropology Theory* 7 (2): 261–271.

Wallis, C. 2021. 7 ways to reduce reluctance to take Covid vaccines. *Scientific American*, March 1. https://www.scientificamerican.com/article/7-ways-to-reduce-reluctance-to-take-covid-vaccines/.

———. 2021. Few would fear Covid vaccines if policy makers explained their risks better. *Scientific American*, April 30. https://www.scientificamerican.com/article/few-would-fear-covid-vaccines-if-policy-makers-explained-their-risks-better/.

Wang, C.J., C.Y. Ng, and R.H. Brook. 2020. Response to Covid-19 in Taiwan. Big data analytics, new technology, and proactive testing. *JAMA* 323 (14): 1341–1342.

Ward, H., G. Cooke, C. Atchison, et al. 2020. Declining prevalence of antibody positivity to SARS-CoV-2: A community study of 365,000 adults. *Preprint*; https://doi.org/10.1101/2020.10.26.20219725.

Wareham, C.S. 2015. Youngest first? Why it is wrong to discriminate against the elderly in healthcare. *South African Journal of Bioethics and Law* 8 (1): 37–39.

Warnock, G.J. 1971. *The object of morality*. London: Methuen & Co.

Webb, K. 2016. *Ethical life. Its natural and social histories*. Princeton/Oxford: Princeton University Press.

Weber, M. 1958. Science as a vocation. *Daedalus* 87 (1): 111–134.

Wei-Haas, M. 2021. Why some coronavirus variants are more contagious – And how we can stop them. *National Geographic*, January 27. https://www.nationalgeographic.com/science/article/why-some-coronavirus-variants-are-more-contagious.

Weintraub, K. 2020. Continuing Covid-19 vaccine trials may put some volunteers at unnecessary risk. Is that ethical? *USA Today*, December 4. https://eu.usatoday.com/story/news/health/2020/12/04/vaccine-ethics-does-continuing-covid-19-trials-put-volunteers-risk/6473436002/.

Wenham, C. 2021. What went wrong in the global governance of covid-19? *British Medical Journal* 372: n303.

Werner, P., and R. Landau. 2020. Laypersons' priority-setting preferences for allocating Covid-19 patients to a ventilator: Does a diagnosis of Alzheimer's disease matter? *Clinical Interventions in Aging* 15: 2407–2414.

Werner, R.M., A.K. Hoffman, and N.B. Coe. 2020. Long-term care policy after Covid-19 – Solving the nursing home crisis. *New England Journal of Medicine* 383 (10): 903–905.

West-Oram, P.G.N., and A. Buyx. 2017. Global health solidarity. *Public Health Ethics* 10 (2): 212–224.

Whang, O., and K. Elliott. 2020. Poll finds more Americans than ever think we should wear masks. *National Geographic* (October 5) https://www.nationalgeographic.com/history/2020/10/poll-increasing-bipartisan-majority-americans-support-mask-wearing.

Whatley, Z., and T. Shodiya. 2020. Why so many Americans are skeptical of a coronavirus vaccine. *Scientific American*, October 12. https://www.scientificamerican.com/article/why-so-many-americans-are-skeptical-of-a-coronavirus-vaccine/.

White, D.B., M.H. Katz, J.M. Luce, and B. Lo. 2009. Who should receive life support during a public health emergency? Using ethical principles to improve allocation decisions. *Annals of Internal Medicine* 150 (2): 132–138.

White, D.B., and B. Lo. 2020. A framework for rationing ventilators and critical care beds during the Covid-19 pandemic. *JAMA* 323 (18): 1773–1774.

White House. 2020. *Executive Order on ensuring access to United States Government Covid-19 vaccines*. December 8. https://trumpwhitehouse.archives.gov/presidential-actions/executive-order-ensuring-access-united-states-government-covid-19-vaccines/.

Whitmee, S., A. Haines, C. Beyrer, et al. 2015. Safeguarding human health in the Anthropocene epoch. Report of the Rockefeller Foundation-*Lancet* Commission on planetary health. *Lancet* 386: 1978.

Wicke, P., and N.M. Bolognesi. 2020. Framing COVID-19: How we conceptualize and discuss the pandemic on Twitter. *PLoS ONE* 15 (9): e0240010. https://doi.org/10.1371/journal.pone.0240010.

Wilder-Smith, A., and D.O. Freedman. 2020. Isolation, quarantine, social distancing and community containment: Pivotal role for old-style public health measures in the novel coronavirus (2019-nCoV) outbreak. *International Journal of Travel Medicine* 27: 1–4.

Wilkinson, D., H. Zohny, A. Kappes, et al. 2020. Which factors should be included in triage? An online survey of the attitudes of the UK general public to pandemic triage dilemmas. *BMJ Open* 10: e045593. https://doi.org/10.1136/bmjopen-2020-045593.

Williams, A. 1997. Intergenerational equity: An exploration of the 'fair innings' argument. *Health Economics* 6: 117–132.

Willsher, K. 2021. Vaccine scepticism in France reflects 'dissatisfaction with political class.' *The Guardian*, January 11. https://www.theguardian.com/global/2021/jan/11/vaccine-scepticism-in-france-reflects-dissatisfaction-with-political-class.

Wilson, D., and W. Dixon. 2012. *A history of Homo Economicus. The nature of the moral in economic theory*. London/New York: Routledge.

Wilson, S. 2020. Pandemic leadership: Lessons from New Zealand's approach to Covid-19. *Leadership* 16 (3): 279–293.

Wilson, K., S. Halabi, and L.O. Gosting. 2020. The International Health Regulations (2005), the threat of populism and the Covid-19 pandemic. *Globalization and Health* 16: 70. https://doi.org/10.1186/s12992-020-00600-4.

Windsor, L.C., G.Y. Reinhardt, A.J. Windsor, et al. 2020. Gender in the time of Covid-19: Evaluating national leadership and Covid-19 fatalities. *PLoS ONE* 15 (12): e0244531.

Wirth, M., L. Rauschenback, B. Hurwitz, et al. 2020. The meaning of care and ethics to mitigate the harshness of triage in second-wave scenario planning during the Covid-19 pandemic. *American Journal of Bioethics* 20 (7): W17–W19.

Wise, J. 2021. Covid-19: The E484K mutation and the risks it poses. *British Medical Journal* 372: n359.

Witt, J.F. 2020. *American contagions. Epidemics and the law from smallpox to Covid-19*. New Haven/London: Yale University Press.

Woesler, M., and H.-M. Sass, eds. 2020. *Medicine and ethics in times of corona*. Zürich: LIT Verlag.

Wolfe, N. 2011. *The viral storm. The dawn of a new pandemic age*. New York: St. Martin's Griffin.

Wolfe, N.D., C.P. Dunavan, and J. Diamond. 2007. Origins of major infectious diseases. *Nature* 447: 279–283.

Wong, A., S. Ho, O. Olunsanya, M.V. Antonini, and D. Lyness. 2021. The use of social media and online communications in times of pandemic Covid-19. *Journal of the Intensive Care Society* 22 (3): 255–260.

Wong, C.M.L., and O. Jensen. 2020. The paradox of trust: Perceived risk and public compliance during the Covid-19 pandemic in Singapore. *Journal of Risk Research* 23 (7–8): 1021–1030.

Woo, J.J. 2020. Policy capacity and Singapore's response to the Covid-19 pandemic. *Policy and Society* 39 (3): 345–362.

World Emergency Covid-19 Pandemic Ethics (WeCope) Committee. 2020. Statement on individual autonomy and social responsibility within a public health emergency. In *Medicine and ethics in times of corona*, ed. M. Woesler and H.-M. Sass, 419–425. Zürich, LIT Verlag.

World Health Organization. 2004. *SARS risk assessment and preparedness framework*. Geneva: WHO. https://www.who.int/csr/resources/publications/CDS_CSR_ARO_2004_2.pdf.

———. 2005. *WHO global influenza preparedness plan: The role of WHO and recommendations for national measures before and during pandemics*. Geneva: World Health Organization. https://apps.who.int/iris/handle/10665/68998.

World Health Organization. Regional Office for the Western Pacific. 2006. *SARS: How a global epidemic was stopped*. Manila: WHO Regional Office for the Western Pacific. https://apps.who.int/iris/handle/10665/207501.

World Health Organization. 2007. *World health report 2007 – A safer future: Global public health security in the 21st century*. Geneva: WHO.

———. 2009. *Global health risks. Mortality and burden of disease attributable to selected major risks*. Geneva: WHO Press.

———. 2015. *Report of the Ebola Interim Assessment Panel*. https://www.who.int/csr/resources/publications/ebola/report-by-panel.pdf?ua=1.

———. 2016. *An R&D Blueprint for action to prevent epidemics. Funding & coordination models for preparedness and response*. https://www.who.int/blueprint/what/improving-coordination/workstream_5_document_on_financing.pdf?ua=1.

———. 2018. *Prioritizing diseases for research and development in emergency contexts*. http://www.emro.who.int/pandemic-epidemic-diseases/news/list-of-blueprint-priority-diseases.html.

———. 2019. *Cholera fact sheet*, January. https://www.who.int/news-room/fact-sheets/detail/cholera.

———. 2020. *Coronavirus disease (Covid-19): Similarities and difference with influenza*. March 17. https://www.who.int/emergencies/diseases/novel-coronavirus-2019/question-and-answers-hub/q-a-detail/coronavirus-disease-covid-19-similarities-and-differences-with-influenza.

———. 2020. *Ethical standards for research during public health emergencies: Distilling existing guidance to support Covid-19 R&D*. March 20. https://www.who.int/publications/i/item/WHO-RFH-20.1.

———. 2020. *Coronavirus Outbreak (COVID-19): WHO Update* (13 May 2020). United Nations. 13 May. https://www.youtube.com/watch?v=geVCLfdItHc.

———. 2020. *Coronavirus disease (Covid-19): Tobacco*. May 27. https://www.who.int/emergencies/diseases/novel-coronavirus-2019/question-and-answers-hub/q-a-detail/coronavirus-disease-covid-19-tobacco#.

———. 2020. *Maintaining essential health services: Operational guidance for the Covid-19 context. Interim Guidance*, June 1. https://www.who.int/publications/i/item/WHO-2019-nCoV-essential-health-services-2020.1.

———. 2020. *Advice on the use of masks in the context of Covid-19. Interim Guidance*, June 5. https://apps.who.int/iris/bitstream/handle/10665/332293/WHO-2019-nCov-IPC_Masks-2020.4-eng.pdf?sequence=1&isAllowed=y.

———. 2020. *Critical preparedness, readiness and response actions for Covid-19: Interim guidance*. https://apps.who.int/iris/bitstream/handle/10665/331511/CriticalpreparednessreadinessandresponseactionsCOVID-102020-03-22_FINAL-eng.pdf?sequence=1&isAllowed=y.

———. 2020. *Shortage of personal protective equipment endangering health workers worldwide*. https://www.who.int/news/item/03-03-2020-shortage-of-personal-protective-equipment-endangering-health-workers-worldwide.

———. 2020. *Urgent health challenges for the next decade*. Geneva: WHO. https://www.who.int/news-room/photo-story/photo-story-detail/urgent-health-challenges-for-the-next-decade.

———. 2020. *Origin of SARS-CoV-2*; https://www.who.int/publications/i/item/origin-of-sars-cov-2.

———. 2020. *Pandemic fatigue – Reinvigorating the public to prevent COVID-19. Policy framework for supporting pandemic prevention and management*. Copenhagen: WHO Regional Office for Europe; 2020. Licence: CC BY-NC-SA 3.0 IGO. https://apps.who.int/iris/handle/10665/335820.

———. 2020. *Tuberculosis. Key facts*. https://www.who.int/news-room/fact-sheets/detail/tuberculosis.

———. 2021. *Malaria. Key facts*. https://www.who.int/news-room/fact-sheets/detail/malaria.

———. 2020. *World malaria report 2020. 20 years of global progress & challenges.* Geneva: WHO. https://cdn.who.int/media/docs/default-source/malaria/world-malaria-reports/9789240015791-eng.pdf?sfvrsn=d7a8ec53_3&download=true.

———. 2020. *Dengue and severe dengue.* Geneva: WHO. https://www.who.int/news-room/fact-sheets/detail/dengue-and-severe-dengue.

———. 2020. *Coronavirus disease (Covid-19) advice for the public: Mythbusters.* November 23. https://www.who.int/emergencies/diseases/novel-coronavirus-2019/advice-for-public/myth-busters.

———. 2020. *WHO SAGE values framework for the allocation and prioritization of COVID-19 vaccination.* Geneva: WHO, September 14. https://apps.who.int/iris/bitstream/handle/10665/334299/WHO-2019-nCoV-SAGE_Framework-Allocation_and_prioritization-2020.1-eng.pdf?sequence=1&isAllowed=y.

———. 2020. *Therapeutics and Covid-19: Living guideline*, 17 December 2020; https://www.who.int/publications/i/item/WHO-2019-nCoV-therapeutics-2022.2.

———. 2021. *WHO-convened Global Study of Origins of SARS-CoV-2: China part.* Geneva: WHO.

———. 2021. *Call to action: Vaccine equity declaration.* Geneva: WHO. https://www.who.int/campaigns/annual-theme/year-of-health-and-care-workers-2021/vaccine-equity-declaration.

———. 2021. *WHO Director-General's statement on Tanzania and Covid-19.* https://www.who.int/news/item/20-02-2021-who-director-general-s-statement-on-tanzania-and-covid-19.

———. 2021. *Coronovirus (Covid-19) Dashboard.* https://covid19.who.int/.

———. 2021. *Global report on ageism.* Geneva: WHO.

———. 2021. *WHO Director-General's opening remarks at the media briefing on Covid-19 – 8 January 2021.* https://www.who.int/director-general/speeches/detail/who-director-general-s-opening-remarks-at-the-media-briefing-on-covid-19-8-january-2021.

———. 2021. *Health and care worker deaths during Covid-19.* October 21. https://www.who.int/news/item/20-10-2021-health-and-care-worker-deaths-during-covid-19.

WHO Rapid Evidence Appraisal for Covid-19 Therapies (REACT) Working Group. 2020. Association between administration of systemic Corticosteroids and mortality among critically Ill patients with COVID-19 a meta-analysis. *JAMA* 324 (13): 1330–1341.

WHO Ad hoc expert group on the next steps for Covid-19 vaccine evaluation. 2021. Placebo-controlled trials of Covid-19 vaccines – Why we still need them. *New England Journal of Medicine* 384 (2): e2. (1–3).

World Wide Fund for Nature. 2020. *Living planet report 2020 – Bending the curve of biodiversity loss.* Gland: WWF. https://f.hubspotusercontent20.net/hubfs/4783129/LPR/PDFs/ENGLISH-FULL.pdf.

Wouters, O.J., K.C. Shadlen, M. Salcher-Konrad, et al. 2021. Challenges to ensuring global access to Covid-19 vaccines: Production, affordability, allocation, and deployment. *Lancet* 397: 1024–1034.

Wright, L. 2021. *The plague year. America in the time of Covid.* London: Allen Lane.

Wu, J., F. Xu, W. Zhou, et al. 2004. Risk factors for SARS among persons without known contact with SARS patients, Beijing, China. *Emerging Infectious Diseases* 10 (2): 210–216.

Wu, H., J. Huang, C.J.P. Zhang, Z. He, and E. Ming. 2020. Facemask shortage and the novel coronavirus disease (Covid-19) outbreak: Reflections on public health measures. *EClinicalMedicine* 21: 100329.

Wynne, K.J., M. Petrova, and R. Coghlan. 2020. Dying individuals and suffering populations: Applying a population-level bioethics lens to palliative care in humanitarian contexts: Before, during and after the Covid-19 pandemic. *Journal of Medical Ethics* 46: 514–525.

Yeoh, K.-W., and K. Shah. 2020. Research ethics during a pandemic (Covid-19). *International Health* 13: 1–2.

Zarocostas, J. 2020. How to fight an infodemic. *Lancet* 395: 676.

Zeberg, H., and S. Pääbo. 2020. The major genetic risk factor for severe COVID-19 is inherited from Neanderthals. *Nature* 587: 610–612.

Zhang, L. 2021. *The origins of Covid-19. China and global capitalism.* Stanford: Stanford Briefs.

Zhang, S. 2020. A vaccine reality check. *The Atlantic*, July 24. https://www.theatlantic.com/health/archive/2020/07/covid-19-vaccine-reality-check/614566/.
Ziegler, M. 2014. The Black Death and the future of the plague. *The Medieval Globe* 1: 259–283.
Ziegler, P. 2009. *The Black Death*. New York: HarperCollins Publishers.
Zimmer, C. 2020. Welcome to the virosphere. *New York Times*, March 24. https://www.nytimes.com/2020/03/24/science/viruses-coranavirus-biology.html.
———. 2021. *A planet of viruses*. 3rd ed. Chicago/London: The University of Chicago Press.
Žižek, S. 2020. *Pandemic! COVID-19 shakes the world*. New York: Polity Press.
Zwart, H. 2020. Emerging viral threats and the simultaneity of the non-simultaneous: Zooming out in times of Corona. *Medicine, Health Care and Philosophy* 23: 589–602.

Index

H. ten Have, *The Covid-19 Pandemic and Global Bioethics*, Advancing Global Bioethics 18, https://doi.org/10.1007/978-3-030-91491-2

T

U

V

The manufacturer's authorised representative in the EU is Springer Nature Customer Service Centre GmbH, Europaplatz 3, 69115 Heidelberg, Germany. If you have any concerns regarding our products, please contact ProductSafety@springernature.com

Printed and bound by CPI Group (UK) Ltd, Croydon, CR0 4YY
01/10/2026
02225861-0016